Selected Titles in This Series

291 Bruce C. Berndt and Ken Ono, Editors, q-Series with applications to combinatorics, number theory, and physics, 2001

290 Michel L. Lapidus and Machiel van Frankenhuysen, Editors, Dynamical, spectral, and arithmetic zeta functions, 2001

289 Salvador Pérez-Esteva and Carlos Villegas-Blas, Editors, Second summer school in analysis and mathematical physics: Topics in analysis: Harmonic, complex, nonlinear and quantization, 2001

288 Marisa Fernández and Joseph A. Wolf, Editors, Global differential geometry: The mathematical legacy of Alfred Gray, 2001

287 Marlos A. G. Viana and Donald St. P. Richards, Editors, Algebraic methods in statistics and probability, 2001

286 Edward L. Green, Serkan Hoşten, Reinhard C. Laubenbacher, and Victoria Ann Powers, Editors, Symbolic computation: Solving equations in algebra, geometry, and engineering, 2001

285 Joshua A. Leslie and Thierry P. Robart, Editors, The geometrical study of differential equations, 2001

284 Gaston M. N'Guérékata and Asamoah Nkwanta, Editors, Council for African American researchers in the mathematical sciences: Volume IV, 2001

283 Paul A. Milewski, Leslie M. Smith, Fabian Waleffe, and Esteban G. Tabak, Editors, Advances in wave interaction and turbulence, 2001

282 Arlan Ramsay and Jean Renault, Editors, Groupoids in analysis, geometry, and physics, 2001

281 Vadim Olshevsky, Editor, Structured matrices in mathematics, computer science, and engineering II, 2001

280 Vadim Olshevsky, Editor, Structured matrices in mathematics, computer science, and engineering I, 2001

279 Alejandro Adem, Gunnar Carlsson, and Ralph Cohen, Editors, Topology, geometry, and algebra: Interactions and new directions, 2001

278 Eric Todd Quinto, Leon Ehrenpreis, Adel Faridani, Fulton Gonzalez, and Eric Grinberg, Editors, Radon transforms and tomography, 2001

277 Luca Capogna and Loredana Lanzani, Editors, Harmonic analysis and boundary value problems, 2001

276 Emma Previato, Editor, Advances in algebraic geometry motivated by physics, 2001

275 Alfred G. Noël, Earl Barnes, and Sonya A. F. Stephens, Editors, Council for African American researchers in the mathematical sciences: Volume III, 2001

274 Ken-ichi Maruyama and John W. Rutter, Editors, Groups of homotopy self-equivalences and related topics, 2001

273 A. V. Kelarev, R. Göbel, K. M. Rangaswamy, P. Schultz, and C. Vinsonhaler, Editors, Abelian groups, rings and modules, 2001

272 Eva Bayer-Fluckiger, David Lewis, and Andrew Ranicki, Editors, Quadratic forms and their applications, 2000

271 J. P. C. Greenlees, Robert R. Bruner, and Nicholas Kuhn, Editors, Homotopy methods in algebraic topology, 2001

270 Jan Denef, Leonard Lipschitz, Thanases Pheidas, and Jan Van Geel, Editors, Hilbert's tenth problem: Relations with arithmetic and algebraic geometry, 2000

269 Mikhail Lyubich, John W. Milnor, and Yair N. Minsky, Editors, Laminations and foliations in dynamics, geometry and topology, 2001

For a complete list of titles in this series, visit the AMS Bookstore at **www.ams.org/bookstore/**.

CONTEMPORARY MATHEMATICS

291

q-Series with Applications to Combinatorics, Number Theory, and Physics

A Conference on
q-Series with Applications to Combinatorics,
Number Theory, and Physics
October 26–28, 2000
University of Illinois

Bruce C. Berndt
Ken Ono
Editors

American Mathematical Society
Providence, Rhode Island

Editorial Board
Dennis DeTurck, managing editor

Andreas Blass   Andy R. Magid   Michael Vogelius

This volume contains the proceedings of a conference on q-Series with Applications to Combinatorics, Number Theory, and Physics, which was held at the University of Illinois on October 26–28, 2000.

2000 *Mathematics Subject Classification.* Primary 05Axx, 05Exx, 11Exx, 11Fxx, 11Mxx, 11Pxx, 33Dxx, 33Exx, 81Rxx, 82Bxx.

Library of Congress Cataloging-in-Publication Data

q-series with applications to combinatorics, number theory, and physics : a conference on q-series with applications to combinatorics, number theory, and physics, October 26–28, 2000, University of Illinois / Bruce C. Berndt, Ken Ono, editors.
   p. cm. — (Contemporary mathematics, ISSN 0271-4132 ; 291)
   Includes bibliographical references.
   ISBN 0-8218-2746-4 (alk. paper)
   1. q-series—Congresses.   I. Berndt, Bruce C., 1939–   II. Ono, Ken.   III. Contemporary mathematics (American Mathematical Society) ; v. 291.

QA295.Q29 2001
515′.243—dc21
                                  2001053662

Copying and reprinting. Material in this book may be reproduced by any means for educational and scientific purposes without fee or permission with the exception of reproduction by services that collect fees for delivery of documents and provided that the customary acknowledgment of the source is given. This consent does not extend to other kinds of copying for general distribution, for advertising or promotional purposes, or for resale. Requests for permission for commercial use of material should be addressed to the Assistant to the Publisher, American Mathematical Society, P. O. Box 6248, Providence, Rhode Island 02940-6248. Requests can also be made by e-mail to reprint-permission@ams.org.

    Excluded from these provisions is material in articles for which the author holds copyright. In such cases, requests for permission to use or reprint should be addressed directly to the author(s). (Copyright ownership is indicated in the notice in the lower right-hand corner of the first page of each article.)

© 2001 by the American Mathematical Society. All rights reserved.
The American Mathematical Society retains all rights
except those granted to the United States Government.
Printed in the United States of America.

∞ The paper used in this book is acid-free and falls within the guidelines
established to ensure permanence and durability.
Visit the AMS home page at URL: http://www.ams.org/

10 9 8 7 6 5 4 3 2 1   06 05 04 03 02 01

Contents

Preface	vii
Program for q-series piano recital	ix
Congruences and conjectures for the partition function S. AHLGREN AND K. ONO	1
MacMahon's partition analysis VII: Constrained compositions G. E. ANDREWS, P. PAULE, AND A. RIESE	11
Crystal bases and q-identities M. OKADO, A. SCHILLING AND M. SHIMOZONO	29
The Bailey-Rogers-Ramanujan group D. STANTON	55
Multiple polylogarithms: A brief survey D. BOWMAN AND D. M. BRADLEY	71
Swinnerton-Dyer type congruences for certain Eisenstein series M. BOYLAN	93
More generating functions for L-function values G. G. H. COOGAN	109
On sums of an even number of squares, and an even number of triangular numbers: an elementary approach based on Ramanujan's $_1\psi_1$ summation formula S. COOPER	115
Some remarks on multiple Sears transformations Y. KAJIHARA	139
Another way to count colored Frobenius partitions L. W. KOLITSCH	147
Proof of a summation formula for an $\tilde{A}_n$ basic hypergeometric series conjectured by Warnaar C. KRATTENTHALER	153
On the representation of integers as sums of squares Z.-G. LIU	163

3-regular partitions and a modular $K3$ surface
    J. LOVEJOY AND D. PENNISTON   177

A new look at Hecke's indefinite theta series
    A. POLISHCHUK   183

A proof of a multivariable elliptic summation formula conjectured by Warnaar
    H. ROSENGREN   193

Multilateral transformations of q-series with quotients of parameters that are nonnegative integral powers of q
    M. SCHLOSSER   203

Completeness of basic trigonometric system in $\mathcal{L}^p$
    S. K. SUSLOV   229

The generalized Borwein conjecture. I. The Burge transform
    S. O. WARNAAR   243

Mock ϑ-functions and real analytic modular forms
    S. P. ZWEGERS   269

Preface

Those of us who use q-series in our mathematical research are often asked the question, "What is a q-series?" The quickest and simplest (but not so accurate or informative) answer is: "It is a series with q's in the summands." More informatively, we might say q-series contain products $(a;q)_n$, where

$$(0.1) \qquad (a;q)_0 := 1, \quad (a;q)_n := (1-a)(1-aq)\cdots(1-aq^{n-1}), \quad \text{if } n \geq 1.$$

This is not entirely accurate, because in such series one often lets parameters tend to 0 or ∞, and so products of the type (0.1) may no longer appear. Theta functions frequently arise and so are also thought of as q-series, even though they contain no products of the form (0.1). Lambert series, or generalized Lambert series, often make appearances, especially in applications to number theory, and are also regarded as part of the subject of q-series. In arithmetic applications of modular forms, which include theta functions, one often needs their q-expansions. Thus, a component of the vast theory of modular forms also has a home in the theory of q-series. In conclusion, to paraphrase a senator who once claimed that he could not define pornography, but he knew it when he saw it, most of us working with q-series cannot give a good definition of a q-series, but we know a q-series when we see it.

The subject of q-series can be said to begin with Euler and his pentagonal number theorem. In fact, q-series are sometimes called Eulerian series. Contributions were made by Gauss, Jacobi, and Cauchy, but the first attempt at a systematic development, especially from the point of view of studying series with the products (0.1) in the summands, was made by E. Heine in 1847. In the latter part of the nineteenth and in the early portions of the twentieth centuries, two English mathematicians, L. J. Rogers and F. H. Jackson, made fundamental contributions. Their work was largely ignored by the mathematical community, and so for many years the subject of q-series was considered to be an unimportant, obscure topic on the fringes of respectable mathematics. To illustrate the humble position occupied by the subject for several years, we offer two testimonies. In 1940, G. H. Hardy, on page 222 of his famous book, *Ramanujan*, described what we now call Ramanujan's famous $_1\psi_1$ summation theorem as "a remarkable formula with many parameters." This is now one of the fundamental theorems of the subject, but Hardy, as well as other mathematicians during his time, could not foresee its importance. T. W. Chaundy, in his obituary of F. H. Jackson (J. London Math. Soc. 37 (1942) 126–128), uncivilly claimed, "The problem [q-series] is very much that of unscrambling an egg."

Despite such humble beginnings, the subject of q-series has flourished in the past three decades. There are several reasons for this. It took mathematicians several years before most of Rogers and Ramanujan's contributions to q-series were appropriately appreciated, and now the subject is a fundamental and active branch of analysis. Many theorems in the theory of hypergeometric series have analogues in basic or q-hypergeometric series, and undoubtedly Chaundy was referring to what he considered to be the esoteric pastime of finding such q-analogues. But irrespective of whether q-analogues exist or not, the subject is replete with beautiful, elegant, and sometimes surprising theorems. Secondly, beginning with the work of Jacobi, q-series have found applications to number theory, and the closeness of these two subjects continues to be ever stronger. Thirdly, increasingly numerous applications to combinatorics are being made, especially in the theory of partitions. Fourthly, through the pioneering work of Rodney Baxter, Barry McCoy, and other theoretical physicists in the past three decades, q-series are now a necessary tool in their subject.

During the year 2000, the Mathematics Department at the University of Illinois embraced *The Millennial Year in Number Theory*. In view of the increasing importance and visibility of q-series, it seemed appropriate that as one of the events in this auspicious *Year*, a conference, *q-series with Applications to Combinatorics, Number Theory, and Physics*, should be held. On October 26–28, sixty-two mathematicians representing twelve countries gathered at the University of Illinois to lecture about and discuss the latest findings. It also seemed appropriate to emphasize survey lectures to help us better chart a course for the future. A total of thirty-nine lectures, highlighted by five plenary survey talks, were given. The plenary lecturers were Scott Ahlgren, George Andrews, Richard Askey, Anne Schilling, and Dennis Stanton. All four aspects (analysis, combinatorics, number theory, and physics) of the subject were featured in these five lectures, as well as in the shorter talks. All the participants helped to make the conference a very successful one, and we thank all of them for their participation. These proceedings contain nineteen of the papers presented at the conference. We hope that they will convey to readers the richness, beauty, and efficacy of the subject.

Special thanks are given to Christian Krattenthaler for a superb evening piano recital. His program can be found before the first paper in this volume.

Many organizations graciously helped to finance the conference. In particular, we are grateful to the National Science Foundation, the National Security Agency, the Number Theory Foundation, The University of Illinois, the Alfred P. Sloan Foundation, and the David and Lucile Packard Foundation for their generous support.

Bruce C. Berndt
Ken Ono
July, 2001

q-Series Piano Recital: Levis Faculty Center

Christian Krattenthaler
8 p.m., Friday, October 27, 2000

PROGRAM

Joseph Haydn (1732–1809)   Sonata in D major, Hob. XVI/37
         Allegro con brio
         Largo e sostenuto
         Finale. Presto ma non troppo

Franz Schubert (1797–1828)   4 Impromptus, D 899
         c minor
         E flat major
         G flat major
         A flat major

Intermission

Sergey Rachmaninov (1873–1943)   Morceaux de Fantaisie (Phantasy Pieces), op. 3
         Élégie
         Prélude
         Mélodie
         Polichinelle
         Sérénade

Congruences and conjectures for the partition function

Scott Ahlgren and Ken Ono

1. Introduction

The topic of congruences for the partition function $p(n)$ has been widely studied. The purpose of this paper is threefold. In the first part we give an account of some of the contributions which the two authors have made to the area in the past several years. In the second part we present a new construction of certain modular forms related to the partition function; this gives a new (and particularly simple) framework in which to consider congruences for the partition function modulo primes $\ell \geq 5$. Finally, we will pose some conjectures which we hope will clarify some of the interesting remaining questions on the congruential distribution of values of $p(n)$.

2. Recent results

A *partition* of a positive integer n is a non-increasing sequence of positive integers whose sum is n. Let $p(n)$ denote the number of partitions of n (we define $p(0) = 1$ and $p(\alpha) = 0$ if $\alpha \notin \mathbb{Z}_{\geq 0}$). We are concerned with the topic of linear congruences for the partition function; i.e. relations of the form

$$p(An + B) \equiv 0 \pmod{M} \text{ for all } n,$$

where A, B, and M are integers. Such congruences were, of course, first discovered by Ramanujan. Throughout the paper, $\ell \geq 5$ will denote a prime number, and δ_ℓ will denote the integer

(2.1) $$\delta_\ell := \frac{\ell^2 - 1}{24}.$$

With this notation, Ramanujan's famous congruences take the form

(2.2) $$p(\ell n - \delta_\ell) \equiv 0 \pmod{\ell} \text{ if } \ell = 5, 7, \text{ or } 11.$$

1991 *Mathematics Subject Classification*. Primary 11P83; Secondary 05A17.

Key words and phrases. Ramanujan-type congruences for the partition function.

Both authors thank the National Science Foundation for its generous support. The second author also thanks the Alfred P. Sloan Foundation and the David and Lucile Packard Foundation for their generous support.

© 2001 American Mathematical Society

Ramanujan conjectured (and in some cases proved) extensions of these congruences to powers of 5, 7, and 11. In fact, after his work [**Be-O, R1, R2, R3**] and subsequent work of Watson [**Wa**] and Atkin [**At1**], it is now known that if $24m \equiv 1 \pmod{5^a 7^b 11^c}$, then we have

$$(2.3) \qquad p(m) \equiv 0 \pmod{5^a 7^{\lfloor (b+2)/2 \rfloor} 11^c}.$$

Since the work of Ramanujan, further examples of congruences involving primes $\ell \leq 31$ have been found (see the works of [**At2, At-O, At-SwD1, At-SwD2, N4, L-O, W**]). The first systematic treatment of such congruences came only recently, when the second author [**O2**] proved that for any prime $\ell \geq 5$, there exist infinitely many (non-nested) congruences of the form

$$(2.4) \qquad p(An + B) \equiv 0 \pmod{\ell} \text{ for all } n.$$

The first author extended the method to prove that if $\ell \geq 5$ is prime and m is a positive integer, then there exist infinitely many congruences of the form

$$(2.5) \qquad p(An + B) \equiv 0 \pmod{\ell^m} \text{ for all } n.$$

In fact, it is shown that such congruences exist for any modulus M which is coprime to 6.

The residue class $-\delta_\ell \pmod{\ell}$ has always played a distinguished role in the theory. Indeed, all of Ramanujan's congruences and their extensions (2.3) lie within the progressions $\ell n - \delta_\ell$. Further, all of the congruences (2.4) and (2.5) whose existence is proven in [**A2, O2**] necessarily lie within this progression. To further highlight the importance of this class, we mention work of Kiming and Olsson [**K-O**], who proved that if $\ell \geq 5$ is prime and

$$p(\ell n + \beta) \equiv 0 \pmod{\ell} \text{ for all } n,$$

then $\beta \equiv -\delta_\ell \pmod{\ell}$.

However, some of the examples alluded to above–in particular those given in [**At2, N4**]–illustrate that congruences may indeed lie outside of this class. Atkin [**At2**], for example, proved that

$$p(17303n + 237) \equiv 0 \pmod{13}.$$

These examples call into question the true importance of the class $-\delta_\ell \pmod{\ell}$.

In a recent paper [**A-O**], the two authors have shown that congruences for $p(n)$ are much more widespread than was previously known. In fact, they show that the class $-\delta_\ell \pmod{\ell}$ is just one of $(\ell + 1)/2$ residue classes modulo ℓ in which the partition function has similar congruence properties.

To state the main result in [**A-O**] requires some notation. For each prime $\ell \geq 5$, define the integer $\epsilon_\ell \in \{\pm 1\}$ by

$$(2.6) \qquad \epsilon_\ell := \left(\frac{-6}{\ell} \right),$$

and let S_ℓ denote the set of $(\ell + 1)/2$ integers

$$(2.7) \qquad S_\ell := \left\{ \beta \in \{0, 1, \ldots, \ell - 1\} \ : \ \left(\frac{\beta + \delta_\ell}{\ell} \right) = 0 \text{ or } -\epsilon_\ell \right\}.$$

Then we have

THEOREM 1. *If $\ell \geq 5$ is prime, m is a positive integer, and $\beta \in S_\ell$, then a positive proportion of the primes $Q \equiv -1 \pmod{24\ell}$ have the property that*

$$p\left(\frac{Q^3 n + 1}{24}\right) \equiv 0 \pmod{\ell^m}$$

for all $n \equiv 1 - 24\beta \pmod{24\ell}$ with $\gcd(Q, n) = 1$.

We remark that the case when $\beta \equiv -\delta_\ell \pmod{\ell}$ contains the main results in [**O2**] and [**A2**]. Further, we note that given $\beta \in S_\ell$ and a prime Q as in Theorem 1, fixing n in an appropriate residue class modulo $24\ell Q$ gives a Ramanujan-type congruence within the progression $\ell n + \beta$. This yields the following

THEOREM 2. *If $\ell \geq 5$ is prime, m is a positive integer, and $\beta \in S_\ell$, then there are infinitely many non-nested arithmetic progressions $\{An + B\} \subseteq \{\ell n + \beta\}$ such that*

$$p(An + B) \equiv 0 \pmod{\ell^m}$$

for every integer n.

Finally, we note that if M is an integer coprime to 6, then Theorem 2 and the Chinese Remainder Theorem guarantee the existence of infinitely many congruences modulo M. The results in [**A-O**] provide a theoretical framework which (to our knowledge) explains every known partition function congruence.

3. A new construction

All of the results in [**A-O**] rely on the construction of half-integral weight modular forms whose coefficients capture values of $p(n)$ modulo ℓ^m. In this section we present an alternate construction in the case $m = 1$ using the theory of modular forms modulo ℓ as developed by Serre and Swinnerton-Dyer [**SwD**]. This approach yields an elegant proof of Theorem 1 in the case when $m = 1$, and is particularly convenient for constructing examples. The main advantage of this approach is that it allows us to work with modular forms on $\mathrm{SL}_2(\mathbb{Z})$; although the construction in [**A-O**] is more general, it requires the use of modular forms of much higher level. Throughout we use the notation $q := e^{2\pi i z}$, and we adopt standard notation from the theory of modular forms. Define the character χ_{12} by $\chi_{12}(d) := \left(\frac{12}{d}\right)$. Our aim in this section is to prove the following:

THEOREM 3.1. *Suppose that $\ell \geq 5$ is prime. Then there exists a cusp form $P_\ell(z)$ in $S_{(\ell^2 - 2)/2}(\Gamma_0(576), \chi_{12}) \cap \mathbb{Z}[[q]]$ such that*

$$P_\ell(z) \equiv \sum_{n \equiv 0 \pmod{\ell}} p(n - \delta_\ell) q^{24n - \ell^2} + 2 \sum_{\left(\frac{n}{\ell}\right) = -\epsilon_\ell} p(n - \delta_\ell) q^{24n - \ell^2} \pmod{\ell}.$$

We recall (see, for example [**S-St**]) that if $f(z) = \sum_{n=1}^{\infty} a(n) q^n \in S_{\lambda + \frac{1}{2}}(\Gamma_1(N))$, and r and t are positive integers, then

$$\sum_{n \equiv r \pmod{t}} a(n) q^n \in S_{\lambda + \frac{1}{2}}(\Gamma_1(Nt^2)).$$

Let $P_\ell(z)$ be as in Theorem 3.1, and suppose that $\beta \in S_\ell$, where S_ℓ is defined in (2.7). Extracting those terms from $P_\ell(z)$ whose exponents are congruent to $24\beta - 1 \pmod{\ell}$, we obtain the following corollary, which implies Theorem 2.1 of [**A-O**] in the case $m = 1$.

COROLLARY 3.2. *Suppose that $\ell \geq 5$ is prime and that $\beta \in S_\ell$. Then there exists a cusp form $F_{\ell,\beta}(z) \in S_{(\ell^2-2)/2}(\Gamma_1(576\ell^2)) \cap \mathbb{Z}[[q]]$ for which*

$$F_{\ell,\beta}(z) \equiv \sum_{n=0}^{\infty} p(\ell n + \beta) q^{24\ell n + 24\beta - 1} \pmod{\ell}.$$

Applying the arguments in Section 3 of [**A-O**] (which rely on certain facts arising from the theory of Galois representations associated to modular forms and Shimura's theory of half integral weight modular forms), the forms given in Corollary 3.2 yield a proof of Theorem 1 in the case when $m = 1$.

Before beginning the proof of Theorem 3.1, we briefly recall certain facts about the theory of modular forms modulo ℓ (see [**Sw-D**] for details). If k is an even integer, then let M_k (resp. S_k) denote the $\mathbb{C}$-vector space of weight k modular (resp. cusp) forms with respect to $\mathrm{SL}_2(\mathbb{Z})$. Let $M_{k,\ell}$ and $S_{k,\ell}$ denote the $\mathbb{F}_\ell$-vector spaces given by

$$M_{k,\ell} := \left\{ f(z) = \sum_{n=0}^{\infty} a(n)q^n \pmod{\ell} \; : \; f(z) \in M_k \cap \mathbb{Z}[[q]] \right\},$$

$$S_{k,\ell} := \left\{ f(z) = \sum_{n=1}^{\infty} a(n)q^n \pmod{\ell} \; : \; f(z) \in S_k \cap \mathbb{Z}[[q]] \right\}.$$

As usual, let $E_k(z)$ denote the normalized weight k Eisenstein series on $\mathrm{SL}_2(\mathbb{Z})$. Using the fact that

$$E_{\ell-1}(z) \equiv 1 \pmod{\ell},$$

one sees that the set of modular forms modulo ℓ forms a graded algebra.

We recall that if $f(z) = \sum_{n=0}^{\infty} a(n)q^n$ is a modular form with integral coefficients, then $\omega_\ell(f)$, the filtration of f modulo ℓ, is defined by

$$\omega_\ell(f) := \min\{k \; : \; f \pmod{\ell} \in M_{k,\ell}\}.$$

Also, if $f(z) = \sum_{n=0}^{\infty} a(n)q^n$ has integral coefficients, then the Ramanujan theta operator is defined by

$$\Theta_\ell(f) :\equiv \sum_{n=0}^{\infty} n a(n) q^n \pmod{\ell}.$$

Finally, we record the following

PROPOSITION 3.3. [**Sw-D**, §3, Lemma 5] *Suppose that $f(z) = \sum_{n=0}^{\infty} a(n)q^n$ is a modular form with integral coefficients and that $\omega_\ell(f) \not\equiv 0 \pmod{\ell}$. Then*

$$\omega_\ell(\Theta_\ell(f)) = \omega_\ell(f) + \ell + 1.$$

PROOF OF THEOREM 3.1. We begin by recalling Dedekind's eta function

(3.1) $$\eta(z) := q^{1/24} \prod_{n=1}^{\infty} (1 - q^n).$$

If $\ell \geq 5$ is prime, then define $f_\ell(z) = \sum_{n=1}^{\infty} a_\ell(n) q^n$ by

(3.2) $$f_\ell(z) = \sum_{n=1}^{\infty} a_\ell(n) q^n := \frac{\eta^\ell(\ell z)}{\eta(z)}.$$

Using classical facts (see, for example [**G-H, N1, N2**], we see that $f_\ell(z)$ is a modular form in $M_{(\ell-1)/2}\left(\Gamma_0(\ell), \left(\frac{\bullet}{\ell}\right)\right)$ with integral coefficients. We have (see, for example, [**An, Th. 1.1**]), the generating function

$$\sum_{n=0}^\infty p(n)q^n = \prod_{n=1}^\infty \frac{1}{1-q^n}.$$

Using this with (3.1) and (3.2), we find that

(3.3) $$\sum_{n=1}^\infty a_\ell(n)q^n = \left(\sum_{n=0}^\infty p(n)q^{n+\delta_\ell}\right) \cdot \prod_{n=1}^\infty (1-q^{\ell n})^\ell.$$

Recall that $\delta_\ell = (\ell^2 - 1)/24$. We have $f_\ell(z) \equiv \Delta^{\delta_\ell}(z) \pmod{\ell}$, where

$$\Delta(z) = \eta^{24}(z)$$

is the unique normalized cusp form of weight 12 on $\mathrm{SL}_2(\mathbb{Z})$. By [**SwD, §3, Lemma 6**], we see that $\omega_\ell(\Delta^{\delta_\ell}) = (\ell^2-1)/2$; therefore Proposition 3.3 implies that

$$\omega_\ell\left(\Theta_\ell^{(\ell-1)/2}\left(\Delta^{\delta_\ell}(z)\right)\right) = \ell^2 - 1.$$

Further, notice that

$$\Theta_\ell^{(\ell-1)/2}\left(\Delta^{\delta_\ell}(z)\right) \equiv \sum_{n=0}^\infty \left(\frac{n}{\ell}\right) a_\ell(n) q^n \pmod{\ell}.$$

Therefore, there is a cusp form $P_{0,\ell}(z)$ in $S_{\ell^2-1} \cap \mathbb{Z}[[q]]$ for which

$$P_{0,\ell}(z) \equiv \sum_{n=0}^\infty \left(\frac{n}{\ell}\right) a_\ell(n) q^n \pmod{\ell}.$$

Let $P_{1,\ell}(z) \in S_{\ell^2-1} \cap \mathbb{Z}[[q]]$ be the cusp form defined by

$$P_{1,\ell}(z) := \Delta^{\delta_\ell}(z) \cdot E_{\ell-1}^{(\ell+1)/2}(z).$$

Define the cusp form $P_{2,\ell}(z)$ in $S_{\ell^2-1} \cap \mathbb{Z}[[q]]$ by

$$P_{2,\ell}(z) := P_{1,\ell}(z) - \epsilon_\ell P_{0,\ell}(z).$$

Using (3.3), we obtain
(3.4)
$$P_{2,\ell} \equiv \left\{\sum_{n \equiv 0 \pmod{\ell}} p(n-\delta_\ell)q^n + 2\sum_{\left(\frac{n}{\ell}\right)=-\epsilon_\ell} p(n-\delta_\ell)q^n\right\} \prod_{n=1}^\infty (1-q^{\ell n})^\ell \pmod{\ell}.$$

By construction, the first exponent in $P_{2,\ell}(z) \in S_{\ell^2-1}$ whose coefficient could be non-zero is $\delta_\ell + 1$. Since S_{ℓ^2-1} has a basis of the form

$$\left\{\Delta(z)^j E_4(z)^{\frac{\ell^2-1}{4}-3j} \ : \ 1 \leq j \leq \frac{\ell^2-1}{12}\right\},$$

we see that there exists $C(z) \in M_{(\ell^2-25)/2} \cap \mathbb{Z}[[q]]$ such that

$$P_{2,\ell}(z) = \Delta^{\delta_\ell+1}(z) \cdot C(z).$$

It follows that
$$P_{2,\ell}(z)/\eta^{\ell^2}(z) = \eta^{23}(z)C(z).$$

Recall that $\eta(24z) \in S_{1/2}(\Gamma_0(576), \chi_{12})$, and that $\eta^\ell(\ell z) \equiv \eta^{\ell^2}(z) \pmod{\ell}$. Using these facts together with (3.4), we conclude that there exists a cusp form $P_\ell(z)$ in $S_{(\ell^2-2)/2}(\Gamma_0(576), \chi_{12}) \cap \mathbb{Z}[[q]]$ for which

$$P_\ell(z) \equiv \sum_{n \equiv 0 \pmod{\ell}} p(n-\delta_\ell) q^{24n-\ell^2} + 2 \sum_{\left(\frac{n}{\ell}\right)=-\epsilon_\ell} p(n-\delta_\ell) q^{24n-\ell^2} \pmod{\ell}.$$

This gives Theorem 3.1. $\square$

4. Examples

Here we present examples of Theorem 3.1 for the primes $\ell = 5, 7,$ and 11. Using [**SwD, §3, Lemma 6**], one can verify that

$$P_{2,5} \equiv 2\Delta^2 \pmod 5,$$

$$P_{2,7} \equiv 2\Delta^3 E_4^3 + 6\Delta^4 \pmod 7,$$

$$P_{2,11} \equiv 2\Delta^6 E_4^{12} + 10\Delta^7 E_4^9 + 8\Delta^8 E_4^6 + 2\Delta^9 E_4^3 + 5\Delta^{10} \pmod{11}.$$

Consequently, we have

(4.1) $$\sum_{n=0}^\infty p(5n+1) q^{120n+23} + \sum_{n=0}^\infty p(5n+2) q^{120n+47} \equiv \eta^{23}(24z) \pmod 5,$$

(4.2) $$\sum_{n=0}^\infty p(7n+1) q^{168n+23} + \sum_{n=0}^\infty p(7n+3) q^{168n+71} + \sum_{n=0}^\infty p(7n+4) q^{168n+95}$$
$$\equiv \eta^{23}(24z) E_4^3(24z) + 3\eta^{47}(24z) \pmod 7,$$

(4.3) $$\sum_{n=0}^\infty p(11n+1) q^{264n+23} + \sum_{n=0}^\infty p(11n+2) q^{264n+47} + \sum_{n=0}^\infty p(11n+3) q^{264n+71}$$
$$+ \sum_{n=0}^\infty p(11n+5) q^{264n+119} + \sum_{n=0}^\infty p(11n+8) q^{264n+191}$$
$$\equiv \eta^{23}(24z) E_4^{12}(24z) + 5\eta^{47}(24z) E_4^9(24z) + 4\eta^{71}(24z) E_4^6(24z)$$
$$+ \eta^{95}(24z) E_4^3(24z) + 8\eta^{119}(24z) \pmod{11}.$$

5. Conjectures

We conclude with a variety of conjectures and open problems. We begin with questions related to the existence of further Ramanujan-type congruences.

CONJECTURE 5.1. (Subbarao [**Su**]) *If A and B are integers with $0 \leq B < A$, then there are infinitely many integers n for which*

$$p(An + B) \not\equiv 0 \pmod{2}.$$

This conjecture is known for every arithmetic progression $B \pmod{A}$ for which there is at least one n with $p(An + B) \equiv 1 \pmod{2}$ [**Th. 2, O1**]. The conjecture is also known for every arithmetic progression $B \pmod{A}$ where A is a power of 2 [**Th. 1, B-O**].

Unfortunately, very little is known about the partition function modulo 3. For example, it is not even known that there are infinitely many n for which $3 \mid p(n)$. As an analogue to Conjecture 5.1, it seems reasonable to make the following conjecture.

CONJECTURE 5.2. *If A and B are integers with $0 \leq B < A$, then there are infinitely many integers n for which*

$$p(An + B) \not\equiv 0 \pmod{3}.$$

Apart from Ramanujan's original congruences (2.2), (2.3), no others are known where the modulus of the congruence equals the modulus of the arithmetic progression. In view of this and the work of Kiming and Olsson mentioned in the introduction, we pose the following.

CONJECTURE 5.3. *If $\ell \geq 13$ is prime, and β is an integer, then there are infinitely many integers n for which*

$$p(\ell n + \beta) \not\equiv 0 \pmod{\ell}.$$

Based on the results in [**A-O**], it seems reasonable to make the following conjecture.

CONJECTURE 5.4. *Suppose that $\ell \geq 5$ is prime and that*

$$p(An + B) \equiv 0 \pmod{\ell}$$

for every integer n. Then there exists $\beta \in S_\ell$ such that $\{An + B\} \subseteq \{\ell n + \beta\}$.

We recall the following important conjecture of Newman [**N3**].

CONJECTURE 5.5. *If M is a positive integer, then for every residue class $r \pmod{M}$ there are infinitely many integers n for which $p(n) \equiv r \pmod{M}$.*

Although the results in [**A2, O1**] provide a simple criterion for deducing Conjecture 5.5 for any M coprime to 6, it remains open. In fact, Conjecture 5.5 has not been proven for infinitely many M.

The remaining conjectures and problems are devoted to questions involving the distribution of $p(n)$ modulo integers M.

CONJECTURE 5.6. *If $0 \leq r < M$, then define $\delta_r(M, X)$ by*

$$\delta_r(M, X) := \frac{\#\{0 \leq n < X \ : \ p(n) \equiv r \pmod{M}\}}{X}.$$

1. *If $0 \leq r < M$, then there is a real number $0 < d_r(M) < 1$ for which*

$$\lim_{X \to \infty} \delta_r(M, X) = d_r(M).$$

2. If $s \geq 1$ and $M = 2^s$, then for every $0 \leq i < 2^s$ we have
$$d_i(2^s) = \frac{1}{2^s}.$$

3. If $s \geq 1$ and $M = 3^s$, then for every $0 \leq i < 3^s$ we have
$$d_i(3^s) = \frac{1}{3^s}.$$

4. If there is a prime $\ell \geq 5$ for which $\ell \mid M$, then for every $0 \leq r < M$ we have
$$d_r(M) \neq \frac{1}{M}.$$

Virtually nothing is known about Conjecture 5.6. Part (1) is not known for any values of r and M. If M is coprime to 6, then Theorem 2 implies that
$$\liminf_{X \to \infty} \delta_0(M, X) > 0.$$
There are no other pairs of integers $0 < r < M$ for which it is known that
$$\liminf_{X \to \infty} \delta_r(M, X) > 0.$$

When $M = 2$, part (2) is the well known "folklore conjecture" studied by Parkin and Shanks in the 1960s [**P-S**]. In this direction, the best results are due to Serre [**N-R-S**] and the first author [**A1**]. It is now known that
$$\#\{n \leq X : p(n) \equiv 0 \pmod{2}\} \gg \sqrt{X}$$
$$\#\{n \leq X : p(n) \equiv 1 \pmod{2}\} \gg \sqrt{X}/\log X.$$

Obviously, this falls far short of Conjecture 5.6. The table below provides data supporting parts (2) and (3) of Conjecture 5.6 when $s = 1$.

X	$\delta_0(2;X)$	$\delta_1(2;X)$	$\delta_0(3;X)$	$\delta_1(3;X)$	$\delta_2(3;X)$
200,000	0.5012	0.4988	0.3332	0.3331	0.3337
400,000	0.5000	0.5000	0.3339	0.3324	0.3336
600,000	0.5000	0.5000	0.3337	0.3326	0.3337
800,000	0.5006	0.4994	0.3331	0.3333	0.3336
1,000,000	0.5004	0.4996	0.3330	0.3336	0.3334

Although we have insufficient data to conjecture a value for $d_r(M)$ for any $0 \leq r < M$ with M coprime to 6, it is natural to consider the following problem.

PROBLEM 5.7. *Find a lower bound for $d_0(M)$ when M is coprime to 6.*

Suppose that $\ell \geq 5$ is prime. In view of Theorem 1, Theorem 2, and Conjecture 5.4, it is natural to consider the distribution of $p(n) \pmod{\ell}$ for those $n \pmod{\ell}$ which do not belong to S_ℓ. Based on preliminary calculations, the following speculation does not appear to be too far-fetched.

SPECULATION 5.8. *If $\ell \geq 5$ is prime and $0 \leq r < \ell$, then define $\delta'_r(\ell, X)$ by*

$$\delta'_r(\ell, X) := \frac{\#\{n < X \ : \ p(n) \equiv r \pmod{\ell} \text{ and } n \pmod{\ell} \notin S_\ell\}}{\#\{n < X \ : \ n \pmod{\ell} \notin S_\ell\}}.$$

For every $0 \leq r < \ell$, is it true that $\lim_{X \to \infty} \delta'_r(\ell, X) = \frac{1}{\ell}$?

To support this speculation, we give data describing these functions when $\ell = 5$.

X	$\delta'_0(5; X)$	$\delta'_1(5; X)$	$\delta'_2(5; X)$	$\delta'_3(5; X)$	$\delta'_4(5; X)$
200,000	0.2006	0.1987	0.2011	0.2007	0.1988
400,000	0.2000	0.1996	0.2001	0.2007	0.1996
600,000	0.1999	0.1992	0.1995	0.2009	0.2006
800,000	0.1995	0.1996	0.1994	0.2010	0.2005
1,000,000	0.1999	0.1996	0.1997	0.2006	0.2002

References

[A1] S. Ahlgren, *Distribution of parity of the partition function in arithmetic progressions*, Indag. Math. **10** (1999), 173-181.

[A2] S. Ahlgren, *Distribution of the partition function modulo composite integers M*, Math. Ann. **318, no. 4** (2000), 795-803.

[A-O] S. Ahlgren and K. Ono, *Congruence properties for the partition function*, Proc. Nat. Acad. Sci., U.S.A., accepted for publication.

[An] G. E. Andrews, *The theory of partitions*, Cambridge University Press, Cambridge, 1984.

[At1] A. O. L. Atkin, *Proof of a conjecture of Ramanujan*, Glasgow Math. J. **8** (1967), 14-32.

[At2] A. O. L. Atkin, *Multiplicative congruence properties and density problems for $p(n)$*, Proc. London Math. Soc. (3) **18** (1968), 563-576.

[At-O] A. O. L. Atkin and J. N. O'Brien, *Some properties of $p(n)$ and $c(n)$ modulo powers of 13*, Trans. Amer. Math. Soc. **126** (1967), 442-459.

[At-SwD1] A. O. L. Atkin and H. P. F. Swinnerton-Dyer, *Some properties of partitions*, Proc. London Math. Soc. (3) **4** (1954), 84-106.

[At-SwD2] A. O. L. Atkin and H. P. F. Swinnerton Dyer, *Modular forms on noncongruence subgroups*, Combinatorics, Proc. Sympos. Pure Math., UCLA, 1968 **19** (1971), 1-25.

[Be-O] B. C. Berndt and K. Ono, *Ramanujan's unpublished manuscript on the partition and tau functions with proofs and commentary*, Sém. Lothar. Combin., The Andrews Festchrift **42** (199), Art. B42c.

[B-O] M. Boylan and K. Ono, *Parity of the partition function in arithmetic progressions, II*, Bull. London Math. Soc., accepted for publication.

[G-H] B. Gordon and K. Hughes, *Multiplicative properties of η-products*, Cont. Math. **143** (1993), 415-430.

[K-O] I. Kiming and J. Olsson, *Congruences like Ramanujan's for powers of the partition function*, Arch. Math. (Basel) **59** (1992), 348-360.

[L-O] J. Lovejoy and K. Ono, *Extension of Ramanujan's congruences for the partition function modulo powers of 5*, J. reine angew. math., accepted for publication.

[N1] M. Newman, *Construction and application of a certain class of modular functions*, Proc. London Math. Soc. (3) **7** (1956), 334-350.

[N2] M. Newman, *Construction and application of a certain class of modular functions, II*, Proc. London Math. Soc. (3) **9** (1959), 373-387.

[N3] M. Newman, *Periodicity modulo m and divisibility properties of the partition function*, Trans. Amer. Math. Soc. **97** (1960), 225-236.

[N4] N. Newman, *Note on partitions modulo 5*, Math. Comp. **21** (1967), 481-482.

[N-R-S] J.-L. Nicolas, I. Z. Ruzsa and A. Sárközy (with an appendix by J.-P. Serre), *On the parity of additive representation function*, J. Number Th. **73**, 292-317.

[O1] K. Ono, *Parity of the partition function in arithmetic progressions*, J. reine angew. math. **472** (1996), 1-15.

[O2] K. Ono, *Distribution of the partition function modulo m*, Ann. of Math. **151** (2000), 293-307.

[P-S] T. R. Parkin and D. Shanks, *On the distribution of parity in the partition function*, Math. Comp. **21** (1967), 466-480.

[R1] S. Ramanujan, *Some properties of p(n), the number of partitions of n*, Proc. Cambridge Philos. Soc. **19** (1919), 207–210.

[R2] S. Ramanujan, *Congruence properties of partitions*, Proc. London Math. Soc. **18** (1920), xix.

[R3] S. Ramanujan, *Congruence properties of partitions*, Math. Z. **9** (1921), 147–153.

[S-St] J.-P. Serre and H. Stark, *Modular forms of weight 1/2*, Springer Lect. Notes in Math. **627** (1971), 27-67.

[Su] M. Subbarao, *Some remarks on the partition function*, Amer. Math. Monthly **73** (1966), 851-854.

[SwD] H. P. F. Swinnerton-Dyer, *On ℓ-adic representations and congruences for coefficients of modular forms, Modular functions of one variable*, Springer Lect. Notes in Math. **350** (1973), 1-55.

[Wa] G.N. Watson, *Ramanujan's Vermutung über Zerfällungsanzahlen*, J. reine angew. Math. **179** (1938), 97-128.

[W] R. L. Weaver, *New congruences for the partition function*, Ramanujan J., accepted for publication.

DEPARTMENT OF MATHEMATICS, UNIVERSITY OF ILLINOIS, CHAMPAIGN-URBANA, ILLINOIS 61801
E-mail address: `ahlgren@math.uiuc.edu`

DEPARTMENT OF MATHEMATICS, UNIVERSITY OF WISCONSIN, MADISON, WISCONSIN 53706
E-mail address: `ono@math.wisc.edu`

MacMahon's Partition Analysis VII: Constrained Compositions

George E. Andrews, Peter Paule, and Axel Riese

ABSTRACT. Our object here is to examine compositions under constraints. From this view, classical partitions arise from one set of constraints. We shall show that other quite different conditions give rise to constrained compositions with elegant generating functions that merit further investigation.

1. Introduction

In classical combinatorics as considered by MacMahon [**6**, Vol. 1, p. 3], the generating function for the homogeneous symmetric functions $h_n(x_1, x_2, \ldots, x_r)$ is given by

$$\sum_{n=0}^{\infty} h_n(x_1, x_2, \ldots, x_r) t^n = \frac{1}{(1-x_1 t)(1-x_2 t)\cdots(1-x_r t)},$$

where r is a fixed positive integer. Or we may say that

$$\sum x_1^{a_1} x_2^{a_2} \cdots x_r^{a_r} = \frac{1}{(1-x_1)(1-x_2)\cdots(1-x_r)}$$

is the fully parameterized generating function for compositions. In other words, every composition (i.e. ordered sum of non-negative integers) has its own unique term in the expansion.

To move from compositions (unordered sums) to partitions (sums with the sequence of parts non-increasing) MacMahon introduced his technique of Partition Analysis. We have provided previous applications of this method [**1, 2, 3, 4**] with an account of our implementation of it in [**2, 3**]. So here we content ourselves with the definition of MacMahon's operator $\Omega_{\geq}$ [**6**, Vol. 2, p. 92]:

$$\underset{\geq}{\Omega} \sum_{n_1,\ldots,n_r=-\infty}^{\infty} A_{n_1 n_2 \ldots n_r} \lambda_1^{n_1} \lambda_2^{n_2} \cdots \lambda_r^{n_r} = \sum_{n_1,\ldots,n_r \geq 0} A_{n_1 n_2 \ldots n_r}.$$

2000 *Mathematics Subject Classification.* Primary 11P82; Secondary 05A17.
The first author was partially supported by NSF Grant DMS-9206993.
The third author was supported by SFB Grant F1305 of the Austrian FWF.

With this tool, we have provided fully parameterized solutions of a number of problems. For example, suppose we want to generate partitions instead of compositions. Then as MacMahon [**6**, Vol. 2, p. 97] has shown us, we may proceed directly:

$$\sum_{a_1 \geq a_2 \geq \cdots \geq a_r \geq 0} x_1^{a_1} x_2^{a_2} \cdots x_r^{a_r}$$

$$= \Omega_{\geq} \sum_{a_1, a_2, \ldots, a_r \geq 0} x_1^{a_1} x_2^{a_2} \cdots x_r^{a_r} \lambda_1^{a_1 - a_2} \lambda_2^{a_2 - a_3} \cdots \lambda_{r-1}^{a_{r-1} - a_r}$$

$$= \Omega_{\geq} \frac{1}{(1 - x_1 \lambda_1)\left(1 - x_2 \frac{\lambda_2}{\lambda_1}\right)\left(1 - x_3 \frac{\lambda_3}{\lambda_2}\right) \cdots \left(1 - x_{r-1} \frac{\lambda_{r-1}}{\lambda_{r-2}}\right)\left(1 - \frac{x_r}{\lambda_{r-1}}\right)}$$

$$= \frac{1}{(1 - x_1)(1 - x_1 x_2)(1 - x_1 x_2 x_3) \cdots (1 - x_1 x_2 \cdots x_r)}$$

$$= \sum_{n_1, n_2, \ldots, n_r \geq 0} x_1^{n_1 + n_2 + \cdots + n_r} x_2^{n_2 + n_3 + \cdots + n_r} \cdots x_{r-1}^{n_{r-1} + n_r} x_r^{n_r}.$$

Or (as we have done in [**1**]) we may choose to generate triples of integers a_1, a_2, a_3 so that they form the sides of a non-degenerate triangle (listed in non-increasing order). Then

$$\sum_{\substack{a_1 \geq a_2 \geq a_3 > 0 \\ a_2 + a_3 > a_1}} x_1^{a_1} x_2^{a_2} x_3^{a_3}$$

$$= \Omega_{\geq} \sum_{a_1, a_2, a_3 \geq 0} x_1^{a_1} x_2^{a_2} x_3^{a_3} \lambda_1^{a_1 - a_2} \lambda_2^{a_2 - a_3} \lambda_3^{a_2 + a_3 - a_1 - 1}$$

$$= \Omega_{\geq} \frac{\lambda_3^{-1}}{\left(1 - x_1 \frac{\lambda_1}{\lambda_3}\right)\left(1 - x_2 \frac{\lambda_2 \lambda_3}{\lambda_1}\right)\left(1 - x_3 \frac{\lambda_3}{\lambda_2}\right)}$$

$$= \frac{x_1 x_2 x_3}{(1 - x_1 x_2)(1 - x_1 x_2 x_3)(1 - x_1^2 x_2 x_3)}$$

(see [**1**] for details)

$$= \sum_{n_1, n_2, n_3 \geq 0} x_1^{n_1 + n_2 + 2n_3 + 1} x_2^{n_1 + n_2 + n_3 + 1} x_3^{n_2 + n_3 + 1}.$$

In many instances we have studied previously such lovely parameterizations; see [**1, 2, 3, 4**]. Indeed this situation in its full generality is discussed by MacMahon [**6**, Vol. 2, pp. 107–114]. To understand fully his view of the general case, we quote extensively [**6**, Vol. 2, pp. 107–108].

"The simple theory of unipartite partitions has been made to depend upon s Diophantine inequalities

$$\alpha_1 \geqslant \alpha_2 \geqslant \alpha_3 \geqslant \cdots \geqslant \alpha_s \geqslant 0,$$

s being an arbitrary integer.

We enlarge the theory by making the integers $\alpha_1, \alpha_2, \alpha_3, \ldots$ depend upon a number of inequalities

$$A_1^{(1)}\alpha_1 + A_2^{(1)}\alpha_2 + \cdots + A_s^{(1)}\alpha_s \geqslant 0,$$
$$A_1^{(2)}\alpha_1 + A_2^{(2)}\alpha_2 + \cdots + A_s^{(2)}\alpha_s \geqslant 0,$$
$$\cdots\cdots\cdots\cdots\cdots\cdots\cdots\cdots\cdots\cdots$$
$$A_1^{(r)}\alpha_1 + A_2^{(r)}\alpha_2 + \cdots + A_s^{(r)}\alpha_s \geqslant 0,$$

which involve at most rs numerical magnitudes A, each of which may be positive, zero, or negative; but in each inequality it is clear that one at least must be positive.

For all sets of numbers $\alpha_1, \alpha_2, \ldots, \alpha_s$ which satisfy the inequalities we seek the sum

$$\sum X_1^{\alpha_1} X_2^{\alpha_2} \ldots X_s^{\alpha_s}.$$

By a theorem of Hilbert it appears that there is in every case a finite number of ground or fundamental solutions of the inequalities, viz.:

$$\begin{array}{cccc} \alpha_1^{(1)} & \alpha_2^{(1)} & \alpha_3^{(1)} & \ldots & \alpha_s^{(1)}, \\ \alpha_1^{(2)} & \alpha_2^{(2)} & \alpha_3^{(2)} & \ldots & \alpha_s^{(2)}, \\ \cdots & \cdots & \cdots & \cdots & \cdots \\ \alpha_1^{(m)} & \alpha_2^{(m)} & \alpha_3^{(m)} & \ldots & \alpha_s^{(m)}, \end{array}$$

such that every solution

$$\alpha_1, \alpha_2, \alpha_3, \ldots, \alpha_s$$

is of the form

$$\alpha_1 = c_1 \alpha_1^{(1)} + c_2 \alpha_1^{(2)} + \cdots + c_m \alpha_1^{(m)},$$
$$\alpha_2 = c_1 \alpha_2^{(1)} + c_2 \alpha_2^{(2)} + \cdots + c_m \alpha_2^{(m)},$$
$$\cdots\cdots\cdots\cdots\cdots\cdots\cdots\cdots\cdots$$
$$\alpha_s = c_1 \alpha_s^{(1)} + c_2 \alpha_s^{(2)} + \cdots + c_m \alpha_s^{(m)},$$

$c_1, c_2, c_3, \ldots, c_m$ being positive integers.

This arises from the circumstance that every term

$$X_1^{\alpha_1} X_2^{\alpha_2} X_3^{\alpha_3} \ldots X_s^{\alpha_s}$$

of the summation is found to be expressible as a product

$$\{X_1^{\alpha_1^{(1)}} X_2^{\alpha_2^{(1)}} X_3^{\alpha_3^{(1)}} \ldots X_s^{\alpha_s^{(1)}}\}^{c_1}$$
$$\times \{X_1^{\alpha_1^{(2)}} X_2^{\alpha_2^{(2)}} X_3^{\alpha_3^{(2)}} \ldots X_s^{\alpha_s^{(2)}}\}^{c_2}$$
$$\times \cdots\cdots\cdots\cdots\cdots\cdots\cdots\cdots$$
$$\times \{X_1^{\alpha_1^{(m)}} X_2^{\alpha_2^{(m)}} X_3^{\alpha_3^{(m)}} \ldots X_s^{\alpha_s^{(m)}}\}^{c_m}.$$

Denoting this product by

$$P_1^{c_1} P_2^{c_2} P_3^{c_3} \ldots P_m^{c_m},$$

the sum (or generating function of solutions) takes the form

$$\left(1 - \{Q_1^{(1)} + Q_1^{(2)} + Q_1^{(3)} + \ldots\} + \{Q_2^{(1)} + Q_2^{(2)} + Q_2^{(3)} + \ldots\} - \{Q_3^{(1)} + Q_3^{(2)} + Q_3^{(3)} + \ldots\} + \ldots\right) / \left((1 - P_1)(1 - P_2)(1 - P_3) \ldots (1 - P_m)\right),$$

and we have what is termed a *syzygetic theory*."

MacMahon calls the $P_1, P_2, \ldots, P_m$ the m ground or fundamental solutions.

Now the examples we have considered so far are effectively syzygy-less. In this paper, we shall restrict our considerations to the instance of MacMahon's diophantine inequalities where $r = s$, i.e. the number of diophantine inequalities equals the number of integer variables. We remark that this restriction applies to our examples of compositions, partitions and integer sided triangles.

In Section 2 we examine the two variable system of diophantine inequalities that exist tacitly in the 2000 Putnam Examination problem B3 [5, p. 728]. We both solve and generalize the problem using Partition Analysis. The resulting solutions naturally suggest a more combinatorial consideration, and we present this exploration in Section 3. The combinatorial analysis leads us to natural T-dimensional generalizations of the original two-dimensional problem in Section 4. Section 5 contains our ruminations on the implications of our Section 4 results for further development of the Omega package, our **Mathematica** implementation of MacMahon's Partition Analysis.

2. The Two-Dimensional Problem and Partition Analysis

Problem B3 on the 2000 Putnam Examination [5, p. 728] reads as follows:

PROBLEM. Let $A = \{(x,y) : 0 \leq x, y < 1\}$. For $(x,y) \in A$, let

$$S(x,y) = \sum_{\frac{1}{2} \leq \frac{m}{n} \leq 2} x^m y^n,$$

where the sum ranges over all pairs (m,n) of positive integers satisfying the indicated inequalities. Evaluate

$$\lim_{\{(x,y) \to (1,1), (x,y) \in A\}} (1 - x^2 y)(1 - xy^2) S(x,y).$$

This problem is easily solved using MacMahon's Partition Analysis; for the corresponding automatic evaluation of the limit see Section 3. Indeed the only formula required is the fourth rule from Section 348 of [6, Vol. 2, p. 102],

$$\underset{\geq}{\Omega} \frac{1}{(1 - \lambda^2 x)\left(1 - \frac{y}{\lambda}\right)} = \frac{1 + xy}{(1 - x)(1 - xy^2)}.$$

Hence
$$1 + S(x,y) = \sum_{\substack{m,n \geq 0 \\ 2m \geq n, 2n \geq m}} x^m y^n$$
$$= \underset{\geq}{\Omega} \sum_{m,n \geq 0} x^m y^n \lambda_1^{2m-n} \lambda_2^{2n-m}$$
$$= \underset{\geq}{\Omega} \frac{1}{\left(1 - x\frac{\lambda_1^2}{\lambda_2}\right)\left(1 - y\frac{\lambda_2^2}{\lambda_1}\right)}$$
$$= \underset{\geq}{\Omega} \frac{1 + xy\lambda_2}{\left(1 - \frac{x}{\lambda_2}\right)(1 - xy^2\lambda_2^3)}$$
$$= \underset{\geq}{\Omega} \left(\frac{1}{\left(1 - \frac{x}{\lambda_2}\right)(1 - y\lambda_2^2)} - \frac{y\lambda_2^2}{(1 - y\lambda_2^2)(1 - xy^2\lambda_2^3)} \right)$$
$$= \frac{1 + xy}{(1-y)(1-x^2y)} - \frac{y}{(1-y)(1-xy^2)}$$
$$= \frac{1 + xy + x^2 y^2}{(1 - xy^2)(1 - x^2 y)}.$$

Immediately we see that
$$\lim_{\{(x,y) \to (1,1), (x,y) \in A\}} (1 - xy^2)(1 - x^2 y) S(x,y) = 3.$$

This suggests immediately that we should consider integers $K, L \geq 2$ and
$$S_{K,L}(x,y) = \sum_{\substack{m,n \geq 0 \\ Km \geq n, Ln \geq m}} x^m y^n$$
$$= \underset{\geq}{\Omega} \sum_{m,n \geq 0} x^m y^n \lambda_1^{Km-n} \lambda_2^{Ln-m}$$
$$= \underset{\geq}{\Omega} \frac{1}{\left(1 - x\frac{\lambda_1^K}{\lambda_2}\right)\left(1 - y\frac{\lambda_2^L}{\lambda_1}\right)}.$$

And now we require the seventh formula in Section 348 of [**6**, Vol. 2, p. 102],
$$\underset{\geq}{\Omega} \frac{1}{(1 - \lambda^s x)\left(1 - \frac{y}{\lambda}\right)} = \frac{1 + xy\frac{1 - y^{s-1}}{1-y}}{(1-x)(1-xy^s)}$$
$$= \frac{1}{(1-x)(1-y)} - \frac{y}{(1-y)(1-xy^s)}, \quad (s \geq 1).$$

Therefore
$$S_{K,L}(x,y) = \underset{\geq}{\Omega} \left(\frac{1}{\left(1 - \frac{x}{\lambda_2}\right)(1 - y\lambda_2^L)} - \frac{y\lambda_2^L}{(1 - y\lambda_2^L)(1 - xy^K \lambda_2^{KL-1})} \right)$$
$$= \frac{1 + xy\frac{1 - x^{L-1}}{1-x}}{(1-y)(1-x^L y)} - \frac{y}{(1-y)(1-xy^K)}$$
$$= \frac{1 + xy\frac{(1-x^L)(1-y^K)}{(1-x)(1-y)} - xy^K - x^L y}{(1 - xy^K)(1 - x^L y)}$$

$$= \left\{ 1 + \sum_{\substack{1 \leq i \leq L,\ 1 \leq j \leq K \\ (L,1) \neq (i,j) \neq (1,K)}} x^i y^j \right\} \bigg/ \left((1 - xy^K)(1 - x^L y)\right).$$

Consequently we now have a generalization of the Putnam Problem

$$\lim_{\{(x,y) \to (1,1), (x,y) \in A\}} (1 - xy^K)(1 - x^L y) S_{K,L}(x, y) = KL - 1.$$

3. Experiments with the Omega Package

A fundamental feature of MacMahon's Partition Analysis consists in the fact that it is ideally suited for being supplemented by computer algebra methods.

In [**2**] we presented a corresponding new algorithmic approach to MacMahon's method together with a **Mathematica** implementation. In [**3**] this approach and the Omega package have been improved significantly. It is this version that will be used below in order to demonstrate how the Omega package can be used for exploration and generalization of the Putnam problem.

Before starting computations one has to load the package by

In[1]:= <<Omega2.m

Out[1]= Axel Riese's Omega implementation version 2.30 loaded

We note that the package is freely available from the Web via *http://www.risc.uni-linz.ac.at/research/combinat/risc/software/Omega*.

As the first application we show how the original Putnam problem discussed in detail in Section 2 can be solved automatically; i.e. without any thought only by applying commands of the package.

First of all, note that $1 + S(x, y) = S_{2,2}(x, y)$. Already the preprocessing step, i.e. the conversion of the $S_{2,2}(x, y)$ sum into an Omega expression is carried out automatically:

In[2]:= OSum[x^m y^n, {2m ≥ n, 2n ≥ m}, λ]

    Assuming m ≥ 0
    Assuming n ≥ 0

Out[2]=
$$\underset{\lambda_1, \lambda_2}{\Omega \atop \geq} \frac{1}{\left(1 - \frac{x \lambda_1^2}{\lambda_2}\right)\left(1 - \frac{y \lambda_2^2}{\lambda_1}\right)}$$

Now the task of eliminating λ_1 and λ_2 is done in one stroke by the command

In[3]:= OR[%]

    Eliminating λ₂...
    Eliminating λ₁...

Out[3]=
$$\frac{1 + x y + x^2 y^2}{(1 - x^2 y)(1 - x y^2)}$$

We have seen in Section 2 that this solves Problem B3 from the 2000 Putnam Exam.

Already from this elementary example it is evident that the (K, L) generalization from Section 2 can be discovered without any effort. We restrict to showing only two cases, $(K, L) = (4, 2)$ and $(K, L) = (3, 4)$:

In[4]:= OR[OSum[x^m y^n, {4m ≥ n, 2n ≥ m}, λ]]

Out[4]=
$$\frac{1 + xy + xy^2 + x^2y^2 + xy^3 + x^2y^3 + x^2y^4}{(1 - x^2y)(1 - xy^4)}$$

In[5]:= OR[OSum[x^m y^n, {3m≥n, 4n≥m}, λ]]

Out[5]=
$$\frac{1 + xy + x^2y + x^3y + xy^2 + x^2y^2 + x^3y^2 + x^4y^2 + x^2y^3 + x^3y^3 + x^4y^3}{(1 - x^4y)(1 - xy^3)}$$

We remark that computations of this type are done by the package within a few seconds on an SGI Octane using Mathematica 4.0.1. However, it turns out that the computations related to Theorem 1 discussed in Section 4 are more involved as we will show later in this section.

At the beginning of our efforts to view the Putnam problem in a more general setting, we tried a number of possible systems of linear diophantine inequalities. For example, one natural way to extend the Putnam problem would be to consider T variables $n_1, n_2, \ldots, n_T$ with $n_i \leq 2n_j$ for each i and j. We can ask Omega to consider the next case, $T = 3$:

In[6]:= OR[OSum[x_1^{n_1} x_2^{n_2} x_3^{n_3}, {n_1≤2n_2, n_2≤2n_1, n_1≤2n_3, n_3≤2n_1, n_2≤2n_3, n_3≤2n_2}, λ]]

Out[6]=
$$(1 + x_1x_2x_3 + x_1^2x_2^2x_3^2 - x_1^3x_2^3x_3^2 - x_1^3x_2^2x_3^3 - x_1^2x_2^3x_3^3 - 2x_1^3x_2^3x_3^3 - x_1^4x_2^3x_3^3 -$$
$$x_1^3x_2^4x_3^3 - x_1^4x_2^4x_3^3 - x_1^3x_2^3x_3^4 - x_1^4x_2^3x_3^4 - x_1^3x_2^4x_3^4 + x_1^5x_2^4x_3^4 + x_1^4x_2^5x_3^4 +$$
$$x_1^5x_2^4x_3^4 + x_1^4x_2^4x_3^5 + x_1^5x_2^4x_3^5 + x_1^4x_2^5x_3^5 + 2x_1^5x_2^5x_3^5 + x_1^6x_2^5x_3^5 + x_1^5x_2^6x_3^5 +$$
$$x_1^5x_2^5x_3^6 - x_1^6x_2^6x_3^6 - x_1^7x_2^7x_3^7 - x_1^8x_2^8x_3^8) /$$
$$((1 - x_1^2x_2x_3)(1 - x_1x_2^2x_3)(1 - x_1^2x_2^2x_3)(1 - x_1x_2x_3^2)(1 - x_1^2x_2x_3^2)(1 - x_1x_2^2x_3^2))$$

Now this does not have the simplicity associated with the original Putnam problem. However, it does suggest that somehow we are confronted with a rather messy amalgam of two nice problems. One having as denominator

$$(1 - x_1^2x_2x_3)(1 - x_1x_2^2x_3)(1 - x_1x_2x_3^2)$$

and the other

$$(1 - x_1^2x_2^2x_3)(1 - x_1^2x_2x_3^2)(1 - x_1x_2^2x_3^2).$$

If we now look back to our solution of the Putnam problem, we see that a fully parameterized solution follows from the Partition Analysis solution

$$\frac{1 + xy + x^2y^2}{(1 - x^2y)(1 - xy^2)}$$
$$= \sum_{r,s \geq 0} x^{2s+r} y^{2r+s} + \sum_{r,s \geq 0} x^{2s+r+1} y^{2r+s+1} + \sum_{r,s \geq 0} x^{2s+r+2} y^{2r+s+2}$$
$$= \sum_{\substack{m,n \geq 0 \\ 2m \geq n,\ 2n \geq m \\ m+n \equiv 0 \,(\mathrm{mod}\ 3)}} x^m y^n + \sum_{\substack{m,n \geq 0 \\ 2m \geq n,\ 2n \geq m \\ m+n \equiv 1 \,(\mathrm{mod}\ 3)}} x^m y^n + \sum_{\substack{m,n \geq 0 \\ 2m \geq n,\ 2n \geq m \\ m+n \equiv 2 \,(\mathrm{mod}\ 3)}} x^m y^n$$
$$= \sum_{\substack{m,n \geq 0 \\ 2m \geq n,\ 2n \geq m}} x^m y^n.$$

Indeed, from
$$m = 2s + r, \quad n = s + 2r$$

we see that equivalently
$$s = \frac{2m-n}{3}, \quad r = \frac{2n-m}{3},$$
and this suggests that the right inequalities equivalent to $r \geq 0$ and $s \geq 0$ are $2m \geq n$ and $2n \geq m$.

So an expansion like
$$\frac{1}{(1-x_1^2x_2x_3)(1-x_1x_2^2x_3)(1-x_1x_2x_3^2)}$$
$$= \sum_{N_1,N_2,N_3 \geq 0} x_1^{2N_1+N_2+N_3} x_2^{N_1+2N_2+N_3} x_3^{N_1+N_2+2N_3}$$
suggests we consider
$$n_1 = 2N_1 + N_2 + N_3, \quad n_2 = N_1 + 2N_2 + N_3, \quad n_3 = N_1 + N_2 + 2N_3$$
which is equivalent to
$$N_1 = \frac{3n_1 - n_2 - n_3}{4}, \quad N_2 = \frac{3n_2 - n_1 - n_3}{4}, \quad N_3 = \frac{3n_3 - n_1 - n_2}{4},$$
and this suggests that the right inequalities equivalent to $N_1, N_2, N_3 \geq 0$ should be
$$3n_1 \geq n_2 + n_3, \quad 3n_2 \geq n_1 + n_3, \quad 3n_3 \geq n_1 + n_2.$$

When we ask Omega to provide the generating function related to these inequalities we find what is shown in (1) below.

In addition, an expansion like
$$\frac{1}{(1-x_1^2x_2^2x_3)(1-x_1^2x_2x_3^2)(1-x_1x_2^2x_3^2)}$$
$$= \sum_{N_1,N_2,N_3 \geq 0} x_1^{2N_1+2N_2+N_3} x_2^{2N_1+N_2+2N_3} x_3^{N_1+2N_2+2N_3}$$
leads from
$$n_1 = 2N_1 + 2N_2 + N_3, \quad n_2 = 2N_1 + N_2 + 2N_3, \quad n_3 = N_1 + 2N_2 + 2N_3$$
to
$$N_1 = \frac{2n_1 + 2n_2 - 3n_3}{5}, \quad N_2 = \frac{2n_1 + 2n_3 - 3n_2}{5}, \quad N_3 = \frac{2n_2 + 2n_3 - 3n_1}{5},$$
and this suggests that the right inequalities equivalent to $N_1, N_2, N_3 \geq 0$ should be
$$2n_1 + 2n_2 \geq 3n_3, \quad 2n_1 + 2n_3 \geq 3n_2, \quad 2n_2 + 2n_3 \geq 3n_1.$$

When we ask Omega to provide the generating function related to these inequalities we find what is shown in (2) below.

In both cases Omega tells us that succinct and appealing generating functions correspond to each of these extensions of the Putnam problem to 3 dimensions.

Once these observations have been provided by Omega, one can try a few more cases which lead directly to conjecturing the results in the next section. This process — and also the run-time behavior mentioned above — is illustrated by the following examples. In all these applications we specify the dimension T and the parameters k and c according to Theorem 1 and the corresponding inequalities (3).

• The case $T = 3$, $k = 1$, $c = 2$ is solved quickly:

```
In[7]:= OR[ OSum[x_1^{n_1} x_2^{n_2} x_3^{n_3}, {n_1+n_2 ≤ 3n_3, n_1+n_3 ≤ 3n_2, n_2+n_3 ≤ 3n_1}, λ] ] //
        Timing
```

Out[7]=

(1) $$\left\{5.24 \text{ Second}, \frac{1 + x_1 x_2 x_3 + x_1^2 x_2^2 x_3^2 + x_1^3 x_2^3 x_3^3}{(1 - x_1^2 x_2 x_3)(1 - x_1 x_2^2 x_3)(1 - x_1 x_2 x_3^2)}\right\}$$

- For $T = 3$, $k = 1$, $c = 3$ the computation takes considerably longer:

  In[8]:= OR[OSum[x₁^n₁ x₂^n₂ x₃^n₃, {n₁+n₂ ≤ 4n₃, n₁+n₃ ≤ 4n₂, n₂+n₃ ≤ 4n₁}, λ]] // Timing

  Out[8]=
  $$\{100.55 \text{ Second},$$
  $$(1 + x_1 x_2 x_3 + x_1^2 x_2 x_3 + x_1 x_2^2 x_3 + x_1^2 x_2^2 x_3 + x_1 x_2 x_3^2 + x_1^2 x_2 x_3^2 + x_1 x_2^2 x_3^2 +$$
  $$x_1^2 x_2^2 x_3^2 + x_1^3 x_2^2 x_3^2 + x_1^2 x_2^3 x_3^2 + x_1^3 x_2^3 x_3^2 + x_1^2 x_2^2 x_3^3 + x_1^3 x_2^2 x_3^3 + x_1^2 x_2^3 x_3^3 +$$
  $$x_1^3 x_2^3 x_3^3 + x_1^4 x_2^4 x_3^3 + x_1^4 x_2^3 x_3^4 + x_1^3 x_2^4 x_3^4 + x_1^4 x_2^4 x_3^4) /$$
  $$((1 - x_1^3 x_2 x_3)(1 - x_1 x_2^3 x_3)(1 - x_1 x_2 x_3^3))\}$$

- For $T = 3$, $k = 1$, $c = 4$ we finally run out of memory.

- Next we turn to the case $T = 3$, $k = 2$, $c = 1$:

  In[9]:= OR[OSum[x₁^n₁ x₂^n₂ x₃^n₃, {2n₁+2n₂ ≥ 3n₃, 2n₁+2n₃ ≥ 3n₂, 2n₂+2n₃ ≥ 3n₁}, λ]] // Timing

  Out[9]=

(2) $$\left\{74.59 \text{ Second}, \frac{1 + x_1 x_2 x_3 + x_1^2 x_2^2 x_3^2 + x_1^3 x_2^3 x_3^3 + x_1^4 x_2^4 x_3^4}{(1 - x_1^2 x_2^2 x_3)(1 - x_1^2 x_2 x_3^2)(1 - x_1 x_2^2 x_3^2)}\right\}$$

- Concerning run-time, things are getting worse for $T = 3$, $k = 2$, $c = 3$:

  In[10]:= OR[OSum[x₁^n₁ x₂^n₂ x₃^n₃, {2n₁+2n₂ ≤ 5n₃, 2n₁+2n₃ ≤ 5n₂, 2n₂+2n₃ ≤ 5n₁}, λ]] // Timing

  Out[10]=
  $$\left\{785.57 \text{ Second}, \frac{1 + x_1 x_2 x_3 + x_1^2 x_2^2 x_3^2 + x_1^3 x_2^3 x_3^3 + x_1^4 x_2^4 x_3^4 + x_1^5 x_2^5 x_3^5 + x_1^6 x_2^6 x_3^6}{(1 - x_1^3 x_2^2 x_3^2)(1 - x_1^2 x_2^3 x_3^2)(1 - x_1^2 x_2^2 x_3^3)}\right\}$$

- For $T = 3$, $k = 3$, $c = 1$ we need more than 30 minutes:

  In[11]:= OR[OSum[x₁^n₁ x₂^n₂ x₃^n₃, {3n₁+3n₂ ≥ 4n₃, 3n₁+3n₃ ≥ 4n₂, 3n₂+3n₃ ≥ 4n₁}, λ]] // Timing

  Out[11]=
  $$\{1959.27 \text{ Second},$$
  $$(1 + x_1 x_2 x_3 + x_1^2 x_2^2 x_3 + x_1^2 x_2 x_3^2 + x_1 x_2^2 x_3^2 + x_1^2 x_2^2 x_3^2 + x_1^2 x_2^3 x_3^2 +$$
  $$x_1^3 x_2^3 x_3^2 + x_1^2 x_2^2 x_3^3 + x_1^3 x_2^2 x_3^3 + x_1^2 x_2^3 x_3^3 + x_1^3 x_2^3 x_3^3 + x_1^4 x_2^3 x_3^3 + x_1^3 x_2^4 x_3^3 +$$
  $$x_1^4 x_2^4 x_3^3 + x_1^3 x_2^3 x_3^4 + x_1^4 x_2^3 x_3^4 + x_1^3 x_2^4 x_3^4 + x_1^4 x_2^4 x_3^4 + x_1^5 x_2^4 x_3^4 + x_1^4 x_2^5 x_3^4 +$$
  $$x_1^4 x_2^4 x_3^5 + x_1^5 x_2^5 x_3^5 + x_1^6 x_2^5 x_3^5 + x_1^5 x_2^6 x_3^5 + x_1^5 x_2^5 x_3^6 + x_1^6 x_2^6 x_3^6) /$$
  $$((1 - x_1^3 x_2^3 x_3)(1 - x_1^3 x_2 x_3^3)(1 - x_1 x_2^3 x_3^3))\}$$

The main reason for this run-time explosion is that in our implementation we eliminate one λ_i after the other. It emerges that even for applications with rather simple solutions, the size of the intermediate results after eliminating some of the λ_i's usually grows enormously. Therefore we are currently working on a new elimination strategy which avoids this problem. For this we split the term which the Ω operator acts on additively into subterms whenever possible. In each of these subterms we then eliminate the λ_i that is optimal in a certain sense and proceed recursively. Clearly elimination works much faster with this approach. However, in the end we are faced with a sum of many multivariate rational functions, each of quite simple form. It turns out that trying to bring the sum over a common denominator is a real bottle-neck that often cannot be accomplished by **Mathematica**.

In the following we demonstrate the speed-up achieved with an unofficial prototype which utilizes this alternative method.

In[1]:= <<Omega3.m

Out[1]= Axel Riese's Omega prototype version 3.3 loaded

- First we take another look at the $T = 3$, $k = 2$, $c = 3$ case:

In[2]:= OR[OSum[x$_1^{n_1}$ x$_2^{n_2}$ x$_3^{n_3}$, {2n$_1$+2n$_2$ ≤ 5n$_3$, 2n$_1$+2n$_3$ ≤ 5n$_2$, 2n$_2$+2n$_3$ ≤ 5n$_1$}, λ]] //
       Timing

Out[2]=

$$\left\{58.95 \text{ Second}, \frac{-1 - x_1 x_2 x_3 - x_1^2 x_2^2 x_3^2 - x_1^3 x_2^3 x_3^3 - x_1^4 x_2^4 x_3^4 - x_1^5 x_2^5 x_3^5 - x_1^6 x_2^6 x_3^6}{(-1 + x_1^3 x_2^2 x_3^2)(-1 + x_1^2 x_2^3 x_3^2)(-1 + x_1^2 x_2^2 x_3^3)}\right\}$$

- The case $T = 3$, $k = 3$, $c = 1$ now works surprisingly fast:

In[3]:= OR[OSum[x$_1^{n_1}$ x$_2^{n_2}$ x$_3^{n_3}$, {3n$_1$+3n$_2$ ≥ 4n$_3$, 3n$_1$+3n$_3$ ≥ 4n$_2$, 3n$_2$+3n$_3$ ≥ 4n$_1$}, λ]] //
       Timing

Out[3]=

$$\{7.14 \text{ Second},$$
$$(-1 - x_1 x_2 x_3 - x_1^2 x_2^2 x_3 - x_1^2 x_2 x_3^2 - x_1 x_2^2 x_3^2 - x_1^2 x_2^2 x_3^2 - x_1^3 x_2^2 x_3^2 - x_1^2 x_2^3 x_3^2 -$$
$$x_1^3 x_2^3 x_3^2 - x_1^2 x_2^2 x_3^3 - x_1^3 x_2^2 x_3^3 - x_1^2 x_2^3 x_3^3 - x_1^4 x_2^3 x_3^3 - x_1^3 x_2^4 x_3^3 -$$
$$x_1^4 x_2^4 x_3^3 - x_1^3 x_2^3 x_3^4 - x_1^4 x_2^3 x_3^4 - x_1^3 x_2^4 x_3^4 - x_1^4 x_2^4 x_3^4 - x_1^5 x_2^4 x_3^3 - x_1^4 x_2^5 x_3^4 -$$
$$x_1^4 x_2^4 x_3^5 - x_1^5 x_2^5 x_3^3 - x_1^6 x_2^5 x_3^5 - x_1^5 x_2^6 x_3^5 - x_1^5 x_2^5 x_3^6 - x_1^6 x_2^6 x_3^6)/$$
$$((-1 + x_1^3 x_2^3 x_3)(-1 + x_1^3 x_2 x_3^3)(-1 + x_1 x_2^3 x_3^3))\}$$

- And also the case $T = 3$, $k = 1$, $c = 4$ can be handled now:

In[4]:= OR[OSum[x$_1^{n_1}$ x$_2^{n_2}$ x$_3^{n_3}$, {n$_1$+n$_2$ ≤ 5n$_3$, n$_1$+n$_3$ ≤ 5n$_2$, n$_2$+n$_3$ ≤ 5n$_1$}, λ]] //
       Timing

Out[4]=

$$\{643.86 \text{ Second},$$
$$(-1 - x_1 x_2 x_3 - x_1^2 x_2 x_3 - x_1^3 x_2 x_3 - x_1 x_2^2 x_3 - x_1^2 x_2^2 x_3 - x_1^3 x_2^2 x_3 - x_1 x_2^3 x_3 -$$
$$x_1^2 x_2^3 x_3 - x_1 x_2 x_3^2 - x_1^2 x_2 x_3^2 - x_1^3 x_2 x_3^2 - x_1 x_2^2 x_3^2 - x_1^2 x_2^2 x_3^2 - x_1^3 x_2^2 x_3^2 -$$
$$x_1^4 x_2^2 x_3^2 - x_1 x_2^3 x_3^2 - x_1^2 x_2^3 x_3^2 - x_1^3 x_2^3 x_3^2 - x_1^4 x_2^3 x_3^2 - x_1^2 x_2^4 x_3^2 - x_1^3 x_2^4 x_3^2 -$$
$$x_1^4 x_2^4 x_3^2 - x_1 x_2 x_3^3 - x_1^2 x_2 x_3^3 - x_1 x_2^2 x_3^3 - x_1^2 x_2^2 x_3^3 - x_1^3 x_2^2 x_3^3 - x_1^4 x_2^2 x_3^3 -$$
$$x_1^2 x_2^3 x_3^3 - x_1^3 x_2^3 x_3^3 - x_1^4 x_2^3 x_3^3 - x_1^2 x_2^4 x_3^3 - x_1^3 x_2^4 x_3^3 - x_1^4 x_2^4 x_3^3 - x_1^5 x_2^4 x_3^3 -$$
$$x_1^2 x_2^2 x_3^4 - x_1^3 x_2^2 x_3^4 - x_1^4 x_2^2 x_3^4 - x_1^2 x_2^3 x_3^4 - x_1^3 x_2^3 x_3^4 - x_1^4 x_2^3 x_3^4 - x_1^2 x_2^4 x_3^4 -$$
$$x_1^3 x_2^4 x_3^4 - x_1^4 x_2^4 x_3^4 - x_1^5 x_2^4 x_3^4 - x_1^4 x_2^5 x_3^4 - x_1^5 x_2^5 x_3^4 - x_1^5 x_2^3 x_3^5 - x_1^4 x_2^4 x_3^5 -$$
$$x_1^5 x_2^4 x_3^5 - x_1^3 x_2^5 x_3^5 - x_1^4 x_2^5 x_3^5 - x_1^5 x_2^5 x_3^5)/$$
$$((-1 + x_1^4 x_2 x_3)(-1 + x_1 x_2^4 x_3)(-1 + x_1 x_2 x_3^4))\}$$

4. The General Theorem

Our object here is to consider the non-negative integer solutions of

$$(3) \quad (k-c)(k(n_1 + n_2 + \cdots + n_T) - ((T-1)k + c)n_j) \geq 0, \qquad 1 \leq j \leq T,$$

where k and c are fixed unequal positive integers. Note that the only effect of multiplication by $(k-c)$ is to change the direction of the "$\geq$" depending on the sign of $(k-c)$. Furthermore we may assume that $\gcd(k,c) = 1$ because division of (3) by $\gcd(k,c)^2$ leaves the solutions unaltered.

DEFINITION 1. We denote by $\mathcal{P}_T$ the set of all T-tuples of non-negative integers $(n_1, \ldots, n_T)$ that satisfy (3).

DEFINITION 2. We define $A(q) = (\alpha_{i,j})_{1 \leq i,j \leq T}$ as the $(T \times T)$ matrix over rational numbers such that $\alpha_{i,i} = q$ and $\alpha_{i,j} = 1$ for $i \neq j$.

For the study of the set $\mathcal{P}_T$ we need various properties of the matrices $A(q)$, $q \in \mathbb{Q}$ being positive.

First of all, only $q = 1$ gives a singular matrix for which

$$A(1)^2 = T \cdot A(1).$$

PROPOSITION 1. *For $q \neq 1$ the matrix $A(q)$ is non-singular and its inverse matrix is*

$$(4) \qquad A(q)^{-1} = \frac{1}{(1-q)(T+q-1)} A(2-q-T).$$

PROOF. We rewrite (4) as $A(q)^{-1} = \gamma \cdot (\beta_{i,j})_{1 \leq i,j \leq T}$ with $\gamma = 1/((1-q)(T+q-1))$, $\beta_{i,i} = 2 - q - T$ and $\beta_{i,j} = 1$ for $i \neq j$. Let $A(q) = (\alpha_{i,j})_{1 \leq i,j \leq T}$ as in Definition 2. It suffices to show that

$$\sum_{h=1}^{T} \gamma \cdot \beta_{i,h} \cdot \alpha_{h,j} = \delta_{i,j} \qquad (1 \leq i, j \leq T),$$

where $\delta_{i,i} = 1$ and $\delta_{i,j} = 0$, if $i \neq j$. The sum equals

$$\frac{2-q-T}{(1-q)(T+q-1)} \alpha_{i,j} + \sum_{\substack{h=1 \\ h \neq i}}^{T} \frac{1}{(1-q)(T+q-1)} \alpha_{h,j}$$

which for $i \neq j$ reduces to

$$\frac{2-q-T}{(1-q)(T+q-1)} + \frac{q}{(1-q)(T+q-1)} + \frac{1}{(1-q)(T+q-1)} \sum_{\substack{h=1 \\ h \neq i, h \neq j}}^{T} 1 = 0,$$

whereas for $i = j$ it reduces to

$$\frac{2-q-T}{(1-q)(T+q-1)} q + \frac{1}{(1-q)(T+q-1)} \sum_{\substack{h=1 \\ h \neq i}}^{T} 1 = 1.$$

$\square$

REMARK. See also Exercise 8 of [**7**, p. 411] for an evaluation of the determinant of a more general matrix.

The singular and non-singular cases are coupled by the formula

$$(5) \qquad A(q) = A(1) - (1-q)I\,,$$

where $I = (\delta_{i,j})_{1 \le i,j \le T}$ denotes the identity matrix. Equation (5) together with Proposition 1 implies the analogous formula for $A(q)^{-1}$; namely, for $q \ne 1$

$$(6) \qquad A(q)^{-1} = \frac{1}{(1-q)(T+q-1)}\left(A(1) - (T+q-1)I\right).$$

Now we are ready to link these matrix considerations with our original problem.

PROPOSITION 2. *For $\boldsymbol{n} = (n_1,\ldots,n_T)$ the inequalities (3) can be rewritten in the form*

$$(7) \qquad A\!\left(\frac{c}{k}\right)^{-1}\boldsymbol{n} \ge \boldsymbol{0}\,,$$

where $\boldsymbol{0} = (0,\ldots,0) \in \mathbb{N}^T$. In other word, the set of all $\boldsymbol{n} \in \mathbb{N}^T$ satisfying (7) is $\mathcal{P}_T$.

PROOF. Obviously (3) is equivalent to

$$(8) \qquad (k-c)(k\,A(1) - ((T-1)k+c)I)\,\boldsymbol{n} \ge \boldsymbol{0}\,.$$

By (6) we see that

$$k\,A(1) - ((T-1)k+c)I = k\!\left(A(1) - \left(T+\frac{c}{k}-1\right)I\right)$$
$$= (k-c)\!\left(T+\frac{c}{k}-1\right)A\!\left(\frac{c}{k}\right)^{-1}.$$

Combining this with (8) gives (7), which completes the proof of Proposition 2. □

PROPOSITION 3. *Let $\boldsymbol{n},\boldsymbol{i} \in \mathbb{N}^T$ such that*

$$(9) \qquad \boldsymbol{n} = k\,A\!\left(\frac{c}{k}\right)\boldsymbol{m} + \boldsymbol{i}$$

for some $\boldsymbol{m} \in \mathbb{N}^T$. Then $\boldsymbol{i} \in \mathcal{P}_T$ implies $\boldsymbol{n} \in \mathcal{P}_T$.

PROOF. Multiplying both sides of (9) with $A(c/k)^{-1}$ gives

$$A\!\left(\frac{c}{k}\right)^{-1}\boldsymbol{n} = k\,\boldsymbol{m} + A\!\left(\frac{c}{k}\right)^{-1}\boldsymbol{i} \ge k\,\boldsymbol{m} \ge \boldsymbol{0}\,.$$

Applying Proposition 2 completes the proof. □

DEFINITION 3. Referring to Proposition 3, we say that $\boldsymbol{i} \in \mathcal{P}_T$ is a *reduction* of $\boldsymbol{n} \in \mathcal{P}_T$ if there exists an $\boldsymbol{m} \in \mathbb{N}^T$ such that

$$\boldsymbol{n} = k\,A\!\left(\frac{c}{k}\right)\boldsymbol{m} + \boldsymbol{i}\,.$$

We shall say that $\boldsymbol{n} \in \mathcal{P}_T$ is *irreducible* if it has no reduction except the trivial case where $\boldsymbol{m} = \boldsymbol{0}$.

Note that $n_1 + \cdots + n_T > i_1 + \cdots + i_T \ge 0$ if $\boldsymbol{i} = (i_1,\ldots,i_T)$ is a reduction of $\boldsymbol{n} = (n_1,\ldots,n_T)$; moreover, the reduction relation is transitive.

DEFINITION 4. We let $\mathcal{S}_T$ denote that subset of $\mathcal{P}_T$ consisting of those $(i_1,\ldots,i_T)$ such that: (1) $\max\limits_{1\le r,s \le T}|i_r - i_s| < |k-c|$ and (2) $\max\limits_{1 \le s \le T} i_s < k(T-1) + c$.

DEFINITION 5. We let $\mathcal{I}_T$ denote that subset of $\mathcal{S}_T$ consisting of all the irreducibles in $\mathcal{S}_T$.

REMARK. It is clear that $\mathcal{S}_T$ (and consequently $\mathcal{I}_T$) are finite sets. Indeed by condition (2) in Definition 4 we see that $\mathcal{S}_T$ certainly has at most $(k(T-1)+c)^T$ elements.

The next proposition is immediate from above. Nevertheless it will be convenient to state it explicitly.

PROPOSITION 4. *For any $\boldsymbol{l} = (l_1, \ldots, l_T)$, $\boldsymbol{m} = (m_1, \ldots, m_T) \in \mathbb{Q}^T$ such that*

$$\boldsymbol{l} = k\, A\!\left(\frac{c}{k}\right) \boldsymbol{m} \tag{10}$$

we have

$$\boldsymbol{m} = \frac{1}{(k-c)((T-1)k+c)}\left(k\, A(1) - ((T-1)k+c)I\right)\boldsymbol{l}. \tag{11}$$

In other words, for $1 \leq j \leq T$,

$$m_j = \frac{k \sum_{i=1}^T l_i - ((T-1)k+c)l_j}{(k-c)((T-1)k+c)}. \tag{12}$$

PROOF. From (10), $\boldsymbol{m} = k^{-1} A(c/k)^{-1}\boldsymbol{l}$ which by (6) equals the right hand side of (11). The reformulation of (11) to (12) is straightforward. □

PROPOSITION 5. *Every $\boldsymbol{n} \in \mathcal{P}_T$ has a reduction to an element of $\mathcal{S}_T$.*

PROOF. Define

$$\boldsymbol{N} := \frac{1}{k}\, A\!\left(\frac{c}{k}\right)^{-1} \boldsymbol{n}. \tag{13}$$

By Proposition 2, $\boldsymbol{N} \in \mathbb{Q}^T$ with non-negative entries. Define

$$\boldsymbol{\mu} := \boldsymbol{N} - \lfloor \boldsymbol{N} \rfloor,$$

where the floor function is taken component-wise. Finally we define

$$\boldsymbol{i} := k\, A\!\left(\frac{c}{k}\right) \boldsymbol{\mu}. \tag{14}$$

Obviously $\boldsymbol{i} \in \mathbb{Q}^T$ with non-negative entries; but in addition we have $\boldsymbol{i} \in \mathbb{N}^T$ since

$$\boldsymbol{i} = k\, A\!\left(\frac{c}{k}\right) \boldsymbol{N} - k\, A\!\left(\frac{c}{k}\right) \lfloor \boldsymbol{N} \rfloor = \boldsymbol{n} - k\, A\!\left(\frac{c}{k}\right) \lfloor \boldsymbol{N} \rfloor \tag{15}$$

by (13). Hence (15) gives

$$\boldsymbol{n} = k\, A\!\left(\frac{c}{k}\right) \boldsymbol{m} + \boldsymbol{i} \quad \text{with } \boldsymbol{m} = \lfloor \boldsymbol{N} \rfloor \in \mathbb{N}^T.$$

Therefore $\boldsymbol{i}$ is a reduction of $\boldsymbol{n}$ if $\boldsymbol{i} \in \mathcal{P}_T$. But this is easy to check since

$$A\!\left(\frac{c}{k}\right)^{-1} \boldsymbol{i} = A\!\left(\frac{c}{k}\right)^{-1} \boldsymbol{n} - k\boldsymbol{m} = k\boldsymbol{N} - k\lfloor \boldsymbol{N} \rfloor \geq \boldsymbol{0}$$

and $\boldsymbol{i} \in \mathcal{P}_T$ by Proposition 2.

Finally it remains to show that $\boldsymbol{i} \in \mathcal{S}_T$. Suppose $\boldsymbol{i} = (i_1, \ldots, i_T)$ and $\boldsymbol{\mu} = (\mu_1, \ldots, \mu_T)$, hence (14) implies

$$i_r - i_s = (c-k)\mu_r + (k-c)\mu_s,$$

and since $0 \leq \mu_r, \mu_s < 1$, we see immediately that

$$|i_r - i_s| < |k - c|.$$

Since each component μ_j of $\boldsymbol{\mu}$ lies in $[0, 1)$ we see, using (14) again, that

$$0 \leq i_j < k + \cdots + k + c + k + \cdots + k = (T-1)k + c.$$

This completes the proof of Proposition 5. $\square$

We have two further steps before we are ready for the generating function associated with $\mathcal{P}_T$.

PROPOSITION 6. *Every $\boldsymbol{n} \in \mathcal{P}_T$ has a reduction to an element of $\mathcal{I}_T$.*

PROOF. By Proposition 5, $\boldsymbol{n} = (n_1, \ldots, n_T)$ has a reduction $(i_1, \ldots, i_T) \in \mathcal{S}_T$. If $(i_1, \ldots, i_T)$ is irreducible we are done. If not, then there is $(m_1, \ldots, m_T) \in \mathcal{P}_T$ that is a non-trivial reduction of $(i_1, \ldots, i_T)$. By Definition 3, $\sum i_j > \sum m_j \geq 0$. If $(m_1, \ldots, m_T) \in \mathcal{S}_T$, then let $(i'_1, \ldots, i'_T) = (m_1, \ldots, m_T)$. If $(m_1, \ldots, m_T) \notin \mathcal{S}_T$, then by Proposition 5, it has a reduction $(i'_1, \ldots, i'_T) \in \mathcal{S}_T$. Hence since the reduction relation is clearly transitive, we see that $(i'_1, \ldots, i'_T) \in \mathcal{S}_T$ is a non-trivial reduction of $(i_1, \ldots, i_T)$. If $(i'_1, \ldots, i'_T)$ is irreducible then we are done because it too is a reduction of $(n_1, \ldots, n_T)$. Now we can repeat this construction producing a sequence of T-tuples in $\mathcal{S}_T$. For each entry the sum of the components forms a decreasing sequence of non-negative integers. Thus the sequence must terminate, and this is only possible provided the final entry we produce is irreducible. Consequently this last entry is in $\mathcal{I}_T$ and is a reduction of $(n_1, \ldots, n_T)$. $\square$

PROPOSITION 7. *The reduction of $\boldsymbol{n} \in \mathcal{P}_T$ lying in $\mathcal{I}_T$ is unique.*

PROOF. By Proposition 6 we know there is at least one reduction of $\boldsymbol{n} = (n_1, \ldots, n_T)$ lying in $\mathcal{I}_T$. Suppose there were two distinct ones: $(i_1, \ldots, i_T)$ and $(i'_1, \ldots, i'_T)$. So there are non-negative integers $m_1, \ldots, m_T, m'_1, \ldots, m'_T$ such that for all $1 \leq j \leq T$

$$n_j = km_1 + \cdots + km_{j-1} + cm_j + km_{j+1} + \cdots + km_T + i_j$$

and

$$n_j = km'_1 + \cdots + km'_{j-1} + cm'_j + km'_{j+1} + \cdots + km'_T + i'_j \,.$$

There must be at least one index r where $m_r > m'_r$. Otherwise

$$i_j = k(m'_1 - m_1) + \cdots + c(m'_j - m_j) + \cdots + k(m'_T - m_T) + i'_j$$

would be a non-trivial reduction of $(i_1, \ldots, i_T)$ (remember these are two distinct elements of $\mathcal{I}_T$) contradicting the irreducibility of $(i_1, \ldots, i_T)$.

By symmetry there must be at least one index s where $m'_s > m_s$. Therefore

(16) $$(m_r - m'_r) + (m'_s - m_s) \geq 2 \,,$$

by the integrality of the m_h and m'_h.

Now by Proposition 4,

$$m_r = \frac{k \sum_{j=1}^{T} (n_j - i_j) - ((T-1)k + c)(n_r - i_r)}{(k-c)((T-1)k + c)} \,.$$

Therefore

$$m_r - m_s = \frac{(n_s - i_s) - (n_r - i_r)}{(k-c)}$$

and similarly

$$m'_s - m'_r = \frac{(n_r - i'_r) - (n_s - i'_s)}{(k-c)} \,.$$

So by (16)

$$2 \leq |(m_r - m'_r) + (m'_s - m_s)|$$
$$= |(m_r - m_s) + (m'_s - m'_r)|$$
$$= \left| \frac{(n_s - i_s) - (n_r - i_r)}{k-c} + \frac{(n_r - i'_r) - (n_s - i'_s)}{k-c} \right|$$
$$= \left| \frac{i_r - i_s}{k-c} + \frac{i'_s - i'_r}{k-c} \right|$$
$$\leq \frac{|i_r - i_s|}{|k-c|} + \frac{|i'_s - i'_r|}{|k-c|}$$
$$< 1 + 1 = 2,$$

by Property (1) of Definition 4. This contradiction proves that there cannot be two distinct reductions of $(n_1, \ldots, n_T)$ within $\mathcal{I}_T$. $\square$

We are now prepared to obtain the T-variable generating function for $\mathcal{P}_T$.

THEOREM 1.
$$\sum_{(n_1,\ldots,n_T) \in \mathcal{P}_T} x_1^{n_1} x_2^{n_2} \cdots x_T^{n_T} = \frac{\sum_{(i_1,\ldots,i_T) \in \mathcal{I}_T} x_1^{i_1} x_2^{i_2} \cdots x_T^{i_T}}{\prod_{j=1}^T (1 - (x_1 x_2 \cdots x_T)^k x_j^{c-k})}.$$

PROOF. By Propositions 3 and 7,
$$\sum_{(n_1,\ldots,n_T) \in \mathcal{P}_T} x_1^{n_1} x_2^{n_2} \cdots x_T^{n_T}$$
$$= \sum_{m_1,m_2,\ldots,m_T \geq 0} \sum_{(i_1,\ldots,i_T) \in \mathcal{I}_T} \prod_{j=1}^T x_j^{km_1 + \cdots + cm_j + \cdots + km_T + i_j}$$
$$= \frac{\sum_{(i_1,\ldots,i_T) \in \mathcal{I}_T} x_1^{i_1} x_2^{i_2} \cdots x_T^{i_T}}{\prod_{j=1}^T (1 - (x_1 x_2 \cdots x_T)^k x_j^{c-k})}.$$

$\square$

COROLLARY 1. If k and c are positive integers with $|k - c| = 1$, then with $X = x_1 x_2 \cdots x_T$
$$\sum_{(n_1,\ldots,n_T) \in \mathcal{P}_T} x_1^{n_1} x_2^{n_2} \cdots x_T^{n_T} = \frac{1 - X^{k(T-1)+c}}{(1-X) \prod_{j=1}^T (1 - X^k x_j^{c-k})}.$$

PROOF. If $|k - c| = 1$, then the only candidates for membership in $\mathcal{I}_T$ are the T-tuples $(0, 0, \ldots, 0), (1, 1, \ldots, 1), \ldots, (k(T-1) + c - 1, \ldots, k(T-1) + c - 1)$. These each immediately are seen to satisfy (3) plus the two further conditions of Definition 4. Therefore these are all the elements of $\mathcal{S}_T$.

Now suppose that $(j, j, \ldots, j) \in \mathcal{S}_T$ is *not* irreducible. Then by Proposition 7 there must exist a unique reduction in $\mathcal{I}_T \subset \mathcal{S}_T$. This means there must be an irreducible $(h, h, \ldots, h) \in \mathcal{I}_T$ that is the reduction of $(j, j, \ldots, j)$. Therefore $j > h \geq 0$. Therefore from the definition of reduction, $(j - h, j - h, \ldots, j - h)$ has $(0, 0, \ldots, 0)$ as a reduction. Therefore by (12)

$$\frac{k \sum_{i=1}^T (j-h) - ((T-1)k + c)(j-h)}{(k-c)((T-1)k+c)}$$

must be a non-negative integer. Simplifying we see that
$$\frac{j-h}{((T-1)k+c)}$$
must be a non-negative integer, and this is impossible because $0 \leq h < j < (T-1)k+c$. Hence each entry of $\mathcal{S}_T$ is irreducible.

As a result, by Theorem 1
$$\sum_{(n_1,\ldots,n_T)\in\mathcal{P}_T} x_1^{n_1} x_2^{n_2} \cdots x_T^{n_T}$$
$$= \frac{1 + X + X^2 + \cdots + X^{(T-1)k+c-1}}{\prod_{j=1}^{T}(1 - X^k x_j^{c-k})}$$
$$= \frac{1 - X^{(T-1)k+c}}{(1-X)\prod_{j=1}^{T}(1 - X^k x_j^{c-k})}.$$

$\square$

5. Conclusion

There is obviously an important fact suggested by Theorem 1. Namely, it is immediate that if $k > c$, then
$$\sum_{(n_1,\ldots,n_T)\in\mathcal{P}_T} x_1^{n_1} x_2^{n_2} \cdots x_T^{n_T}$$
$$= \underset{\geq}{\Omega} \sum_{n_1\geq 0,\ldots,n_T\geq 0} x_1^{n_1} x_2^{n_2} \cdots x_T^{n_T} \lambda_1^{k\sum n_i - (k(T-1)+c)n_1} \cdots \lambda_T^{k\sum n_i - (k(T-1)+c)n_T}$$
$$= \underset{\geq}{\Omega} \frac{1}{\left(1 - x_1(\lambda_1\cdots\lambda_T)^k \lambda_1^{-(k(T-1)+c)}\right) \cdots \left(1 - x_T(\lambda_1\cdots\lambda_T)^k \lambda_T^{-(k(T-1)+c)}\right)}.$$

As we have seen in Section 3, the evaluation of this last expression as the right-hand side of the identity in Theorem 1 is quite memory consuming in Omega even for reasonably small values of T, k and c. It would be valuable if the algorithm for Omega could be improved so that it could do many other cases of Theorem 1.

On the theoretical side, the technique used to prove Theorem 1 may well have further application. Rather than find the full fundamental basis suggested by Hilbert's Syzygy Theorem, we found only a subset which were then used to produce an equivalence relation among all solutions (i.e. two elements of $\mathcal{P}_T$ are equivalent if they have the same reduction in $\mathcal{I}_T$). While this approach may not have the same generality as Hilbert's theorem, it may well be suited for specific combinational problems when the required generating function is particularly nice. Also bijective proofs of these theorems would be of interest.

Finally, we note that Theorem 1 generalizes the $S_{K,L}(x,y)$ of Section 2 only in the case $K = L$. Nonetheless, the final form of $S_{K,L}(x,y)$ suggests that there should be more general results in T dimensions than those in Theorem 1. For example we might consider (thanks to the referee) the non-negative integer solutions of the system
$$n_j \leq L_j n_{j+1}, \quad 1 \leq j \leq T-1; \qquad n_T \leq L_T n_1,$$
where $T, L_1, \ldots, L_T$ are integers each greater than 1. Empirical studies comparable to those in Section 3 suggest that this is a plausible subject for further study.

However, the methods of Section 4 are not immediately applicable to this alternative generalization.

References

[1] G.E. Andrews, *MacMahon's Partition Analysis II: Fundamental Theorems*, Ann. Comb. **4** (2000), 327–338.

[2] G.E. Andrews, P. Paule, and A. Riese, *MacMahon's Partition Analysis III: The Omega Package*, SFB Report **99-24**, J. Kepler University, Linz, 1999, (to appear in European J. Combin.).

[3] ———, *MacMahon's Partition Analysis VI: A New Reduction Algorithm*, SFB Report **01-4**, J. Kepler University, Linz, 2001, (to appear in Ann. Comb.).

[4] G.E. Andrews, P. Paule, A. Riese, and V. Strehl, *MacMahon's Partition Analysis V: Bijections, Recursions, and Magic Squares*, in "Algebraic Combinatorics and Applications" (A. Betten et al., eds.), pp. 1–39, Springer, Berlin, 2001.

[5] L.F. Klosinski, G.L. Alexanderson, and L.C. Larson, *The sixtieth William Lowell Putnam Mathematical Competition*, Amer. Math. Monthly **107** (2000), 721–732.

[6] P.A. MacMahon, *Combinatory Analysis*, Vols. 1 and 2, Cambridge Univ. Press, Cambridge, 1915 and 1916 (Reissued: Chelsea, New York, 1960).

[7] T. Muir, *A Treatise on the Theory of Determinants*, Dover, New York, 1960.

DEPARTMENT OF MATHEMATICS, THE PENNSYLVANIA STATE UNIVERSITY, UNIVERSITY PARK, PA 16802, USA
  E-mail address: andrews@math.psu.edu

RESEARCH INSTITUTE FOR SYMBOLIC COMPUTATION, JOHANNES KEPLER UNIVERSITY LINZ, A–4040 LINZ, AUSTRIA
  E-mail address: Peter.Paule@risc.uni-linz.ac.at

RESEARCH INSTITUTE FOR SYMBOLIC COMPUTATION, JOHANNES KEPLER UNIVERSITY LINZ, A–4040 LINZ, AUSTRIA
  E-mail address: Axel.Riese@risc.uni-linz.ac.at

Crystal bases and q-identities

Masato Okado, Anne Schilling, and Mark Shimozono

ABSTRACT. The relation of crystal bases with q-identities is discussed, and some new results on crystals and q-identities associated with the affine Lie algebra $C_n^{(1)}$ are presented.

1. Introduction

The purpose of this paper is two-fold. First, we would like to advocate the importance of crystal theory to the theory of q-series. In particular crystal base theory provides a unifying and general setting for a large class of q-identities. Second, as evidence, some new identities associated to the affine Lie algebra $C_n^{(1)}$ are presented. The emphasis here will not be on the completeness of the results since the field is evolving quite rapidly, but rather on the presentation of the main ideas and techniques used.

The Rogers–Ramanujan identities are undoubtably the most famous q-series identities. They are given by

$$(1.1) \qquad \sum_{n=0}^{\infty} \frac{q^{n^2}}{(q)_n} = \prod_{j=1}^{\infty} \frac{1}{(1-q^{5j-4})(1-q^{5j-1})}$$

$$(1.2) \qquad \sum_{n=0}^{\infty} \frac{q^{n(n+1)}}{(q)_n} = \prod_{j=1}^{\infty} \frac{1}{(1-q^{5j-3})(1-q^{5j-2})}$$

where $(q)_n = (1-q)(1-q^2)\cdots(1-q^n)$. We will view them here as identities of formal power series, meaning that in the expansion as series in the formal variable q the coefficients of q^N match on both sides for all $N \geq 0$. What contributes to their beauty is that these coefficients can be interpreted combinatorially. The coefficient of q^N on the left-hand side of (1.1) is the number of partitions of N for which the difference between any two parts is at least two. The coefficient of q^N of the right-hand side of (1.1) on the other hand is the number of partitions of N with parts congruent to 1 or 4 modulo 5. Similarly, the coefficients of q^N on the left and right side of (1.2) can be interpreted as the number of partitions of N for

---

2000 *Mathematics Subject Classification.* Primary 81R10 17B65 05A30; Secondary 82B20 82B23 05A17.

Key words and phrases. Crystal bases, q-series identities, quantum affine Lie algebras, fermionic formulas, q-analogues of tensor product multiplicities.

which the difference between any two parts is at least two and the smallest part is greater than one, and the number of partitions of N with parts congruent to 2 and 3 modulo 5, respectively.

Many of the ideas regarding crystals and q-identities can already be demonstrated in terms of the Rogers–Ramanujan identities. The point of focus here shall be the debut of the Rogers–Ramanujan identities on the mathematical physics stage, in particular their appearance in the hard hexagon model in a paper by Baxter [3] in 1981. In this setting the Rogers–Ramanujan identities can be viewed as two different evaluations of the generating function of certain paths which are coined bosonic and fermionic evaluations. Details are discussed in section 2. As it turns out the relation between the Rogers–Ramanujan identities and the hard hexagon model is only part of a much bigger picture. In terms of representation theory, the paths that occur in the hard hexagon model are elements of tensor products of crystals associated with the affine Lie algebra $\hat{\mathfrak{sl}}_2$. Crystal bases were introduced by Kashiwara [19, 20] and roughly speaking are bases of representations of quantum universal enveloping algebras $U_q(\mathfrak{g})$ as the parameter q (not to be confused with the q in the q-series!) tends to zero. Here $\mathfrak{g}$ is any symmetrizable Kac–Moody algebra. As in the Rogers–Ramanujan case, there are two different ways to evaluate generating functions of tensor products of crystals, thereby giving rise to q-identities. Hence crystal base theory provides a natural framework for q-identities. Crystal bases, path spaces and their generating functions are discussed in section 3. The two different ways to evaluate the paths generating functions are subject of sections 4 and 5, respectively. In particular in section 5.5 we present new fermionic formulas for level-restricted paths. We close in section 6 with some outstanding open problems.

Before indulging in the fascinating theory of crystal bases there is one important point that needs to be addressed. Applying Jacobi's triple product identity

$$\sum_{n=-\infty}^{\infty} z^n q^{n^2} = \prod_{n=0}^{\infty} (1-q^{2n+2})(1+zq^{2n+1})(1+z^{-1}q^{2n+1})$$

the right-hand sides of (1.1) and (1.2) can be rewritten as alternating sums yielding the identities

(1.3) $$\sum_{n=0}^{\infty} \frac{q^{n^2}}{(q)_n} = \frac{1}{(q)_\infty} \sum_{j=-\infty}^{\infty} (-1)^j q^{\frac{j}{2}(5j+1)}$$

(1.4) $$\sum_{n=0}^{\infty} \frac{q^{n(n+1)}}{(q)_n} = \frac{1}{(q)_\infty} \sum_{j=-\infty}^{\infty} (-1)^j q^{\frac{j}{2}(5j+3)}.$$

The two different evaluations of generating functions of crystal paths that were mentioned above really yield (polynomial) analogues of (1.3) and (1.4) rather than (1.1) and (1.2). Identities relating sums to alternating sums are more general than identities relating sums to products. Only in special cases can the alternating sums be evaluated as products, namely when Jacobi's triple product identity or more generally the Macdonald identities [32] can be applied.

Acknowledgements. We are grateful to Tim Baker for discussions. A.S. would like to thank the organizers Bruce Berndt and Ken Ono for the invitation to

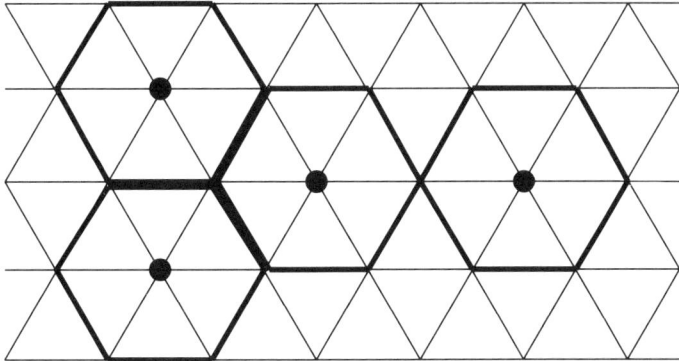

FIGURE 1. An allowed configuration of particles in the hard hexagon model.

present these results at the conference on q-Series with Applications to Combinatorics, Number Theory, and Physics held in Urbana-Champaign in October 2000. Thanks for this exciting conference!

2. The Rogers–Ramanujan identities and the Hard Hexagon model

The hard hexagon model is a two-dimensional lattice model of a gas of hard or non-overlapping particles. The particles are placed on a triangular lattice such that no two particles can occupy two adjacent sites. If one views each particle as the center of a hexagon, then the condition that no two particles can be adjacent translates into the condition that the hexagons cannot overlap. This explains the name of the model. An example of an allowed particle configuration is shown in Figure 1.

To pack the lattice densely all particles must lie on one of the three sublattices corresponding to the three corners of the triangles of the lattice. Hence one of the sublattices is distinguished from the others. At low particle density the probability for a particle to be on a particular site is equal for sites on all three sublattices (assuming either an infinite or sufficiently large lattice so that boundary effects can be neglected). Let ρ_a for $a = 1, 2, 3$ be the probability that there is a particle at a fixed site on sublattice $1, 2, 3$, respectively. If the boundary conditions of the model are such that at close packing all particles are on sublattice 1 then it is intuitively clear that the order parameter defined as $R = \rho_1 - \rho_2$ must undergo a phase transition. At low densities R is zero, but at some critical density R will become positive until at high densities it is one. Baxter [**3**] managed to determine the precise point at which the phase transition occurs exactly by using corner transfer matrices. In essence, the corner transfer matrix method reduces the two-dimensional problem to a one-dimensional problem. The precise details are beyond the scope of this paper and can be found in section 14 of Baxter's book [**4**].

The one-dimensional problem that Baxter encountered and which turns out to be of importance to the Rogers–Ramanujan identities is the following. Consider $L + 1$ points on a line labeled by $i = 0, 1, 2, \ldots, L$. Assign to each point a height variable σ_i which takes on the values 0 or 1. In addition the height variables satisfy the restrictions $\sigma_0 = \sigma_L = 0$ and $\sigma_i \sigma_{i+1} = 0$. An allowed configuration of height variables for a given length L is called a *path* of length L, and the set of all paths

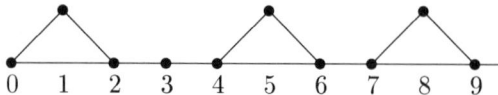

FIGURE 2. A path of length 9

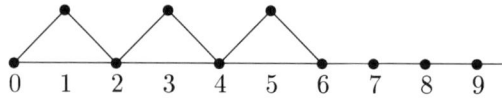

FIGURE 3. Ground state path with three particles

of length L is denoted by $\mathcal{D}_L$. One can illustrate a path graphically by drawing all points (i, σ_i) and connecting adjacent points by straight lines. An example for a path with $L = 9$ is given in Figure 2. The condition $\sigma_i \sigma_{i+1} = 0$ requires that the paths consist of a certain number of non-overlapping triangles (this condition comes directly from the condition in the two-dimensional hard hexagon model requiring that no two particles can be on adjacent sites). To each path $\sigma = (\sigma_0, \ldots, \sigma_L)$ one may assign an energy $E(\sigma)$ by summing up the positions of the peaks, that is

$$E(\sigma) = \sum_{j=1}^{L} j \sigma_j.$$

The energy of the path in Figure 2 is $E(\sigma) = 1 + 5 + 8 = 14$. The generating function of paths of length L which is also called a one dimensional configuration sum is defined as

(2.1) $$X(L) = \sum_{\sigma \in \mathcal{D}_L} q^{E(\sigma)}.$$

The path picture immediately implies that $X(L)$ satisfies the following initial conditions and recurrence which completely specify it

(2.2)
$$X(0) = X(1) = 1$$
$$X(L) = X(L-1) + q^{L-1} X(L-2).$$

The aim is to find explicit expressions for $X(L)$. We will describe two ways to obtain such an expression which will be related to the two sides of (1.3).

2.1. Fermionic formula. Interpret each peak in a path as a particle, that is, there is a particle at site i if $\sigma_i = 1$. Fix the number of particles to be n. The ground state path σ_G with minimal energy is the path with particles at positions $1, 3, \ldots, 2n-1$. The energy of σ_G is $E(\sigma_G) = 1 + 3 + \cdots + 2n - 1 = n^2$. An example of the ground state path with 3 particles is shown in Figure 3. All other paths with n particles can be obtained from σ_G by moving particles to the right in such a way that the particles never overtake each other. If the length of the path is L, the rightmost particle can move at most $L - 2n$ positions to the right. Hence the paths of length L with n particles are in one-to-one correspondence with partitions with at most n parts not exceeding $L - 2n$. If a path σ corresponds to partition λ then its energy is $E(\sigma) = E(\sigma_G) + |\lambda| = n^2 + |\lambda|$. The generating function of partitions with at most n parts not exceeding m is the q-binomial coefficient defined as

$$\begin{bmatrix} m+n \\ n \end{bmatrix} = \frac{(q)_{m+n}}{(q)_m (q)_n}$$

for $m, n \in \mathbb{N}$ and zero otherwise. This implies the following explicit expression for $X(L)$

(2.3) $$X(L) = \sum_{n=0}^{\infty} q^{n^2} \begin{bmatrix} L-n \\ n \end{bmatrix}.$$

Because of its interpretation in terms of non-overlapping particles this expression is called a quasiparticle or fermionic formula.

2.2. Bosonic formula. As opposed to (2.3) there is another expression for $X(L)$ given by

(2.4) $$X(L) = \sum_{j=-\infty}^{\infty} (-1)^j q^{\frac{j}{2}(5j+1)} \begin{bmatrix} L \\ \lfloor \frac{L-5j}{2} \rfloor \end{bmatrix}$$

where $\lfloor x \rfloor$ is the largest integer not exceeding x. It can be proven by showing that the right-hand side satisfies the recurrence and initial condition (2.2).

However, formula (2.4) can also be interpreted in terms of paths. For this purpose the above path picture needs to be slightly reformulated. The state diagram of the paths in the hard hexagon model is

This diagram is to be understood as follows. If the system is in the bottom state at position i, that is $\sigma_i = 0$, then at position $i+1$ it can either be in the bottom state again so that $\sigma_{i+1} = 0$ as indicated by the loop or it can be in the top state which means $\sigma_{i+1} = 1$ as indicated by the line. On the other hand, if the system is in the top state at position i then at the next position it has to be in the bottom state, meaning that $\sigma_i = 1$ implies $\sigma_{i+1} = 0$. This corresponds to condition on paths that $\sigma_i \sigma_{i+1} = 0$ for all i. Up to a $\mathbb{Z}_2$ symmetry this state diagram is isomorphic to

(2.5) $$\bigcirc \cong / \mathbb{Z}_2.$$

Under this correspondence the paths of the hard hexagon model become paths in a strip of height 4 with steps going up or down at each position. For example, with the convention that the paths start at height three, the path in Figure 2 becomes the path in Figure 4.

Let us now consider the following set of paths. Let $p = (p_1, p_2, \ldots, p_L)$ be a sequence of 1's and 2's and let $\lambda = (\lambda_1, \lambda_2)$ be the content of p where λ_i is the number of i's in p. Denote by $\mathcal{P}_{L,\lambda}$ the set of all $p = (p_1, \ldots, p_L)$ with content λ. Another way to represent p is by height variables $\sigma = (\sigma_0, \sigma_1, \ldots, \sigma_L)$ where by convention $\sigma_0 = 3$ and $\sigma_i = \sigma_{i-1} + 1$ if $p_i = 1$ and $\sigma_i = \sigma_{i-1} - 1$ if $p_i = 2$ for

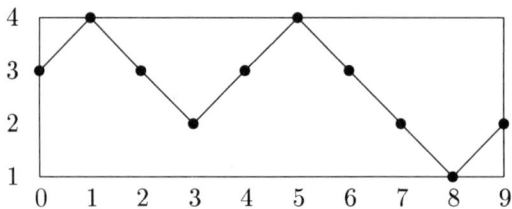

FIGURE 4. Path associated to the path in Figure 2 under the correspondence (2.5)

$1 \le i \le L$. We will use σ and p interchangeably. We are actually mostly concerned here with the set $\mathcal{P}_L := \mathcal{P}_{L,\lambda}$ where $\lambda = (\lfloor \frac{L}{2} \rfloor, \lfloor \frac{L+1}{2} \rfloor)$.

Consider the following subsets of $\mathcal{P}_L$. Let $\overline{\mathcal{P}}_L$ be the subset of $\mathcal{P}_L$ consisting of all paths in the strip of height four, that is all $\sigma \in \mathcal{P}_L$ such that $1 \le \sigma_i \le 4$ for all $1 \le i \le L$. Furthermore, let $\mathcal{P}_L^{\uparrow,j}$ be the set of all paths $\sigma \in \mathcal{P}_L$ such that there exist indices $1 \le i_1 < i_2 < \cdots < i_j \le L$ so that $\sigma_{i_1}, \sigma_{i_3}, \sigma_{i_5}, \ldots$ are greater than four and $\sigma_{i_2}, \sigma_{i_4}, \sigma_{i_6}, \ldots$ are less than one. Similarly let $\mathcal{P}_L^{\downarrow,j}$ be the set of all paths $\sigma \in \mathcal{P}_L$ such that there exist indices $1 \le i_1 < i_2 < \cdots < i_j \le L$ such that $\sigma_{i_1}, \sigma_{i_3}, \sigma_{i_5}, \ldots$ are less than one and $\sigma_{i_2}, \sigma_{i_4}, \sigma_{i_6}, \ldots$ are greater than four. By inclusion-exclusion we have

$$\overline{\mathcal{P}}_L = \left(\mathcal{P}_L \cup \bigcup_{j \ge 1}(\mathcal{P}_L^{\uparrow,2j} \cup \mathcal{P}_L^{\downarrow,2j})\right) \setminus \bigcup_{j \ge 1}(\mathcal{P}_L^{\uparrow,2j-1} \cup \mathcal{P}_L^{\downarrow,2j-1}).$$

Now define the energy function E for $\sigma \in \mathcal{P}_L$ as

$$E(\sigma) = \sum_{i=1}^{L-1} i \cdot h(\sigma_{i-1}, \sigma_i, \sigma_{i+1})$$

where h is the local energy function given by

$$h(\sigma_{i-1}, \sigma_i, \sigma_{i+1}) = \begin{cases} 1 & \text{if } \sigma_{i-1} = \sigma_i - 1 = \sigma_{i+1} \text{ and } \sigma_i > 3, \\ & \text{or } \sigma_{i-1} = \sigma_i + 1 = \sigma_{i+1} \text{ and } \sigma_i < 2, \\ 0 & \text{else.} \end{cases}$$

The term $q^{\frac{j}{2}(5j+1)}\begin{bmatrix} L \\ \lfloor \frac{L-5j}{2} \rfloor \end{bmatrix}$ in (2.4) can then be interpreted as the generating function of $\mathcal{P}_L^{\downarrow,j}$ given by $\sum_{\sigma \in \mathcal{P}_L^{\downarrow,j}} q^{E(\sigma)}$ if $j > 0$ and the generating function of $\mathcal{P}_L^{\uparrow,j}$ given by $\sum_{\sigma \in \mathcal{P}_L^{\uparrow,j}} q^{E(\sigma)}$ if $j < 0$. The term $j = 0$ in (2.4) is the generating function of $\mathcal{P}_L$.

In section 4 we will encounter more general inclusion-exclusion arguments in terms of operations on crystals.

2.3. Identities. Equations (2.3) and (2.4) yield the following polynomial identity

$$\sum_{n=0}^{\infty} q^{n^2} \begin{bmatrix} L-n \\ n \end{bmatrix} = \sum_{j=-\infty}^{\infty} (-1)^j q^{\frac{j}{2}(5j+1)} \begin{bmatrix} L \\ \lfloor \frac{L-5j}{2} \rfloor \end{bmatrix}.$$

Using $\lim_{L \to \infty} \begin{bmatrix} L \\ m \end{bmatrix} = \frac{1}{(q)_m}$ this identity implies (1.3) in the limit $L \to \infty$.

All discussion so far focused on (1.3). The second Rogers–Ramanujan identity (1.4) can in fact be treated in a very similar fashion. Define the set $\mathcal{D}'_L$ analogous to $\mathcal{D}_L$ with the only difference that now $\sigma_0 = 1$ instead of 0. The generating function is defined to be $X'(L) = \sum_{\sigma \in \mathcal{D}'_L} q^{E(\sigma)}$. Analogous arguments to those in sections 2.1 and 2.2 yield the following two expressions for $X'(L)$

$$X'(L) = \sum_{n=0}^{\infty} q^{n(n+1)} \begin{bmatrix} L-n-1 \\ n \end{bmatrix} = \sum_{j=-\infty}^{\infty} (-1)^j q^{\frac{j}{2}(5j+3)} \begin{bmatrix} L \\ \lfloor \frac{L-5j-1}{2} \rfloor \end{bmatrix}.$$

In the limit $L \to \infty$ this polynomial identity implies (1.4).

3. Crystal bases

In this section we review the main results about crystal bases and explain how they provide a general setting for the definition of one-dimensional configuration sums.

The quantized universal enveloping algebra $U_q(\mathfrak{g})$ associated with a symmetrizable Kac–Moody Lie algebra $\mathfrak{g}$ was introduced independently by Drinfeld [7] and Jimbo [14] in their study of two dimensional solvable lattice models in statistical mechanics. The parameter q corresponds to the temperature of the underlying model. Kashiwara [18] showed that at zero temperature or $q = 0$ the representations of $U_q(\mathfrak{g})$ have bases, which he coined crystal bases, with a beautiful combinatorial structure and favorable properties such as uniqueness and stability under tensor products. The existence and uniqueness of crystal bases for integrable highest weight modules for an arbitrary symmetrizable Kac–Moody algebra was given in [19].

In this setting, the paths are tensor products of finite-dimensional crystals and the energy function is determined by the affine crystal action. At this point it is worth mentioning that the energy statistics defined in this section does not specialize to the statistics of the Rogers–Ramanujan paths of section 2. The reason is that here we will be dealing with crystals of integrable modules whereas the Rogers–Ramanujan identities come from admissible modules as introduced by Kac and Wakimoto [16, 17]. As will be discussed in section 6, it is still an open problem to define the energy statistics for general admissible modules.

3.1. Axiomatic definition of crystals. Let $\mathfrak{g}$ be a symmetrizable Kac–Moody algebra, P the weight lattice, I the index set for the vertices of the Dynkin diagram of $\mathfrak{g}$, $\{\alpha_i \in P \mid i \in I\}$ the simple roots, and $\{h_i \in P^* = \text{Hom}_{\mathbb{Z}}(P, \mathbb{Z}) \mid i \in I\}$ the simple coroots. Let $U_q(\mathfrak{g})$ be the quantized universal enveloping algebra of $\mathfrak{g}$. A $U_q(\mathfrak{g})$-crystal is a nonempty set B equipped with maps $\text{wt} : B \to P$ and $e_i, f_i : B \to B \cup \{\emptyset\}$ for all $i \in I$, satisfying

(3.1) $\qquad\qquad f_i(b) = b' \Leftrightarrow e_i(b') = b$ if $b, b' \in B$

(3.2) $\qquad\qquad \text{wt}(f_i(b)) = \text{wt}(b) - \alpha_i$ if $f_i(b) \in B$

(3.3) $\qquad\qquad \langle h_i, \text{wt}(b) \rangle = \varphi_i(b) - \epsilon_i(b)$

where $\langle \cdot, \cdot \rangle$ is the natural pairing. Here for $b \in B$,

(3.4) $\qquad\quad \epsilon_i(b) = \max\{n \geq 0 \mid e_i^n(b) \neq \emptyset\}$
$\qquad\qquad \varphi_i(b) = \max\{n \geq 0 \mid f_i^n(b) \neq \emptyset\}.$

(It is assumed that $\varphi_i(b), \epsilon_i(b) < \infty$ for all $i \in I$ and $b \in B$.)

A $U_q(\mathfrak{g})$-crystal B can be viewed as a directed edge-colored graph (the crystal graph) whose vertices are the elements of B, with a directed edge from b to b' labeled $i \in I$, if and only if $f_i(b) = b'$. The element b is said to be highest weight if $e_i(b) = \emptyset$ for all $i \in I$. Let K be a subset of I. Then the K-component of the crystal B is the graph obtained by only considering edges colored by $i \in K$.

We also define the crystal reflection operator $s_i : B \to B$ by

$$s_i(b) = \begin{cases} f_i^{\varphi_i(b)-\epsilon_i(b)}(b) & \text{if } \varphi_i(b) > \epsilon_i(b) \\ b & \text{if } \varphi_i(b) = \epsilon_i(b) \\ e_i^{\epsilon_i(b)-\varphi_i(b)}(b) & \text{if } \varphi_i(b) < \epsilon_i(b). \end{cases}$$

It is obvious that s_i is an involution.

3.2. Tensor products of crystals. Given two crystals B and B', there is also a crystal obtained by taking the tensor product $B \otimes B'$. As a set $B \otimes B'$ is just given by the Cartesian product of the sets B and B'. The weight function wt for $b \otimes b' \in B \otimes B'$ is $\text{wt}(b \otimes b') = \text{wt}(b) + \text{wt}(b')$ and the raising and lowering operators e_i and f_i act as follows

(3.5)
$$e_i(b \otimes b') = \begin{cases} e_i b \otimes b' & \text{if } \epsilon_i(b) > \phi_i(b'), \\ b \otimes e_i b' & \text{otherwise,} \end{cases}$$

$$f_i(b \otimes b') = \begin{cases} f_i b \otimes b' & \text{if } \epsilon_i(b) \geq \phi_i(b'), \\ b \otimes f_i b' & \text{otherwise.} \end{cases}$$

The reader is warned that this convention is different from Kashiwara's convention. The order of the tensor factors is interchanged.

3.3. Finite and infinite crystals. Let us fix some notation. From now on let $\mathfrak{g}$ be a simple complex Lie algebra and $\hat{\mathfrak{g}}$ be the associated untwisted affine algebra. That is, let $\hat{\mathfrak{g}}' = \mathfrak{g} \otimes \mathbb{C}[t, t^{-1}] \oplus \mathbb{C}c$ be the central extension of the loop algebra of $\mathfrak{g}$ and $\hat{\mathfrak{g}} = \hat{\mathfrak{g}}' \oplus \mathbb{C}d$ where d is the degree derivation. Let J (resp. $I = J \cup \{0\}$) be a set indexing the vertices of the Dynkin diagram of $\mathfrak{g}$ (resp. $\hat{\mathfrak{g}}$ and $\hat{\mathfrak{g}}'$). A classical weight is a weight with respect to the algebra $\mathfrak{g}$. Denote by $U_q(\mathfrak{g})$, $U_q'(\hat{\mathfrak{g}})$, and $U_q(\hat{\mathfrak{g}})$ the quantized universal enveloping algebras of $\mathfrak{g}$, $\hat{\mathfrak{g}}'$, and $\hat{\mathfrak{g}}$ respectively.

There are two main categories of crystals. The first one contains the crystal bases of irreducible integrable $U_q(\hat{\mathfrak{g}})$-modules $V(\Lambda)$ of highest weight $\Lambda \in P^+$ where P^+ denotes the set of dominant weights of $\hat{\mathfrak{g}}$. These crystals are infinite-dimensional. The second one contains finite-dimensional crystals corresponding to $U_q'(\hat{\mathfrak{g}})$. Since the set B is finite in this case these crystals are called finite crystals. The level of a finite crystal B is defined as

$$\text{lev} B = \min\{\langle c, \epsilon(b) \rangle \mid b \in B\}$$

where $\epsilon(b) = \sum_{i \in I} \epsilon_i(b) \Lambda_i$ and $\{\Lambda_i \mid i \in I\}$ is the $\mathbb{Z}$-basis of the weight lattice of $\hat{\mathfrak{g}}'$. A $U_q'(\hat{\mathfrak{g}})$-crystal can be viewed as a $U_q(\mathfrak{g})$-crystal by restricting to the J-component.

The crystal of each integrable highest weight $U_q(\hat{\mathfrak{g}})$-module can be realized in terms of a semi-infinite tensor product of perfect crystals [24, 25, 23]. Perfect crystals are finite crystals with some additional properties (for details see [24, Definition 4.6.1]). At least one perfect crystal for each integrable $U_q(\hat{\mathfrak{g}})$-module for $\hat{\mathfrak{g}} = A_n^{(1)}, B_n^{(1)}, C_n^{(1)}, D_n^{(1)}, A_n^{(2)}$ and $D_{n+1}^{(2)}$ was given in [25].

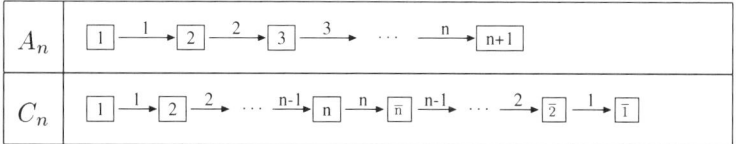

TABLE 1. Classical crystals $B(\Lambda_1)$

Kashiwara and Nakashima [27] constructed the finite $U_q(\mathfrak{g})$-crystals for $\mathfrak{g} = A_n$, B_n, C_n and D_n explicitly in terms of tableaux. The cases $\mathfrak{g} = A_n$ and C_n are discussed in more detail in the next subsection.

3.4. Finite crystals of type A_n and C_n. The finite crystals associated with $\mathfrak{g} = A_n$ and C_n are presented in more detail since they will be our main examples throughout the paper. Let Λ_i for $i \in J$ be the fundamental weights of $\mathfrak{g}$. For later purposes it will be convenient to give the root and weight structure of A_n and C_n explicitly. Let $(\cdot|\cdot)$ be the standard bilinear form normalized such that $(\alpha_i \mid \alpha_i) = 2$ for the long roots α_i.

EXAMPLE 3.1. For either $\mathfrak{g} = A_n$ or $\mathfrak{g} = C_n$ the index set for the simple roots is $J = \{1, 2, \ldots, n\}$.

For $\mathfrak{g} = A_n$ the weight lattice is embedded in the subspace of $\mathbb{R}^{n+1}$ orthogonal to the vector $e = \sum_{j=1}^{n+1} \epsilon_j$, where ϵ_i is the i-th standard basis vector of $\mathbb{R}^{n+1}$. The simple roots are $\alpha_i = \epsilon_i - \epsilon_{i+1}$ for all $i \in J$. The fundamental weights are $\Lambda_i = \sum_{j=1}^{i} \epsilon_j - \frac{i}{n+1} e$ for $i \in J$. The half-sum of positive roots is $\rho = (n, n-1, \ldots, 1, 0) - \frac{n}{2} e$. The scalar product is the standard one: $(\alpha|\beta) = \alpha \cdot \beta$.

For $\mathfrak{g} = C_n$ the simple roots are given by the short roots $\alpha_i = \epsilon_i - \epsilon_{i+1}$ for $1 \le i < n$ and the long root $\alpha_n = 2\epsilon_n$ where ϵ_i is the i-th unit vector in $\mathbb{R}^n$. The fundamental weights are $\Lambda_i = \epsilon_1 + \cdots + \epsilon_i$ for $i \in J$. Half the sum of all positive roots is $\rho = (n, n-1, \ldots, 1)$. The bilinear form is given by $(\alpha|\beta) = \alpha \cdot \beta/2$.

For both type A_n and C_n we will identify dominant weights, that is weights of the form $\Lambda = \Lambda_{k_1} + \cdots + \Lambda_{k_m}$, with partitions. A partition $\lambda = (\lambda_1, \ldots, \lambda_{n+1})$ (with $\lambda_{n+1} = 0$ for type C_n) corresponds to the weight $\Lambda = \sum_{i \in J} (\lambda_i - \lambda_{i+1}) \Lambda_i$.

For each dominant weight Λ there is a finite classical crystal $B(\Lambda)$ [27]. The crystals $B(\Lambda_1)$ for type A_n and C_n are given in Table 1 where the arrow $\xrightarrow{i}$ stands for f_i. The finite crystals $B(\Lambda_k)$ for $k \in J$ can be obtained in the following way. Let $u_{\Lambda_k} = \boxed{k} \otimes \cdots \otimes \boxed{2} \otimes \boxed{1}$ be the unique highest weight vector of weight Λ_k in $B(\Lambda_1)^{\otimes k}$. Then $B(\Lambda_k)$ is the connected component of $B(\Lambda_1)^{\otimes k}$ containing u_{Λ_k}. In [27] the elements in this connected component were identified with certain one-column tableaux. The finite crystal $B(\Lambda)$ for a dominant weight $\Lambda = \Lambda_{k_1} + \cdots + \Lambda_{k_m}$ with $k_1 \ge k_2 \ge \cdots \ge k_m$ is isomorphic to the connected component in $B(\Lambda_{k_1}) \otimes \cdots \otimes B(\Lambda_{k_m})$ which contains the highest weight element $u_{\Lambda_{k_1}} \otimes \cdots \otimes u_{\Lambda_{k_m}}$. Combining the two embeddings it follows that $B(\Lambda)$ is isomorphic to a certain connected component in $B(\Lambda_1)^{\otimes M}$ where $M = k_1 + \cdots + k_m$.

For type A_n the elements in $B(\Lambda)$ can be identified with semi-standard Young tableaux of shape λ where λ is the partition corresponding to the weight Λ. The paths encountered in section 2 are sequences of 1's and 2's. Viewed as single box

Young tableaux over the alphabet $\{1,2\}$, these are exactly the elements of the crystal $B(\Lambda_1)$ of type A_1.

3.5. Simple crystals. Simple crystals were introduced by Akasaka and Kashiwara [1]. As will be explained in the next section they have an isomorphism and energy function which are required for the definition of the one-dimensional configuration sums.

Let $\hat{W}$ be the Weyl group of the affine Kac–Moody algebra $\hat{\mathfrak{g}}$ generated by the simple reflections r_i for $i \in I$ defined as

$$(3.6) \qquad r_i(\beta) = \beta - \langle h_i, \beta \rangle \alpha_i$$

where β is a root. Let B be the crystal graph of an integrable $U_q(\hat{\mathfrak{g}})$-module. Say that $b \in B$ is an extremal vector of weight $\Lambda \in P$ provided that $\text{wt}(b) = \Lambda$ and there exists a family of elements $\{b_w \mid w \in \hat{W}\} \subset B$ such that

1. $b_w = b$ for $w = e$.
2. If $\langle h_i, w\Lambda \rangle \geq 0$ then $e_i(b) = 0$ and $f_i^{\langle h_i, w\Lambda \rangle}(b_w) = b_{r_i w}$.
3. If $\langle h_i, w\Lambda \rangle \leq 0$ then $f_i(b) = 0$ and $e_i^{\langle h_i, w\Lambda \rangle}(b_w) = b_{r_i w}$.

DEFINITION 3.2. Say that a $U_q'(\hat{\mathfrak{g}})$-crystal B is *simple* if

1. B is the crystal base of a finite-dimensional integrable $U_q'(\hat{\mathfrak{g}})$-module.
2. There is a dominant weight Λ with respect to the weight lattice of the classical algebra $\mathfrak{g}$ such that B has a unique vector (denoted $u(B)$) of weight Λ, and the weight of any extremal vector of B is contained in $W\Lambda$. Here W is the Weyl group corresponding to the classical algebra $\mathfrak{g}$.

THEOREM 3.3. [1]

1. *Simple crystals are connected as graphs.*
2. *The tensor product of simple crystals is simple.*

Let B be a simple $U_q'(\hat{\mathfrak{g}})$-crystal, equipped with a function $D = D_B : B \to \mathbb{Z}$, called its intrinsic energy, which is required to be constant on J-components and defined up to a global additive constant. Call the pair (B, D) a graded simple $U_q'(\hat{\mathfrak{g}})$-crystal. We normalize the intrinsic energy function by the requirement that

$$D_B(u(B)) = 0.$$

3.6. Finite dimensional affine crystals. Recently, new families of crystals of the finite dimensional representations of $U_q(\hat{\mathfrak{g}})$ were conjectured [13, 21] where $\hat{\mathfrak{g}}$ is a untwisted affine Lie algebra.

CONJECTURE 3.4. [13, 21] For each $r \in J$ and $s \geq 1$, there exists an irreducible finite-dimensional integrable $U_q'(\hat{\mathfrak{g}})$-module $W_s^{(r)}$ with simple crystal base $B^{r,s}$ generated a unique extremal vector $u(B^{r,s})$ of weight $s\Lambda_r$, and a prescribed $U_q(\mathfrak{g})$-crystal decomposition of the form $B^{r,s} \cong B(s\Lambda_r) \oplus B$, where B is a direct sum of $U_q(\mathfrak{g})$-crystals of the form $B(\Lambda)$ where Λ is a classical dominant weight and $s\Lambda_r \triangleright \lambda$. Here $\Lambda' \trianglerighteq \Lambda$ if and only if $\Lambda' - \Lambda \in \bigoplus_{i \in J} \mathbb{N}\alpha_i$. Moreover there is a prescribed intrinsic energy function $D = D_{B^{r,s}} : B^{r,s} \to \mathbb{Z}$, that is constant on J-components, such that $0 = D(u) > D(b)$ where u is the J-highest weight vector of weight $s\Lambda_r$ in $B^{r,s}$, and b is any element not in the J-component of u.

3.7. Combinatorial R-matrix and energy function.

Suppose B_1 and B_2 are simple $U_q'(\mathfrak{g})$-crystals. Then there is a unique isomorphism of $U_q'(\mathfrak{g})$-crystals $\sigma : B_2 \otimes B_1 \to B_1 \otimes B_2$. There is also a function $H = H_{B_2, B_1} : B_2 \otimes B_1 \to \mathbb{Z}$, called local energy function which is unique up to global additive constant, such that for all $b_2 \otimes b_1 \in B_2 \otimes B_1$ with $b_1' \otimes b_2' = \sigma(b_2 \otimes b_1)$

$$(3.7) \quad H(e_i(b_2 \otimes b_1)) = H(b_2 \otimes b_1) + \begin{cases} -1 & \text{if } i = 0, \, \epsilon_0(b_2) > \varphi_0(b_1), \, \epsilon_0(b_1') > \varphi_0(b_2') \\ 1 & \text{if } i = 0, \, \epsilon_0(b_2) \le \varphi_0(b_1), \, \epsilon_0(b_1') \le \varphi_0(b_2') \\ 0 & \text{otherwise.} \end{cases}$$

The pair (σ, H) is called the combinatorial R-matrix. It is convenient to normalize the local energy function H by requiring that

$$(3.8) \quad H(u(B_2) \otimes u(B_1)) = 0.$$

With this convention it follows by definition that

$$(3.9) \quad H_{B_1, B_2} \circ \sigma_{B_2, B_1} = H_{B_2, B_1}$$

as operators on $B_2 \otimes B_1$.

Let (B_j, D_j) be graded simple $U_q'(\hat{\mathfrak{g}})$-crystals for $1 \le j \le L$ and set $B = B_L \otimes \cdots \otimes B_1$. Following [**34**] define the energy function $E_B : B \to \mathbb{Z}$ by

$$(3.10) \quad E_B = \sum_{1 \le i < j \le L} H_i \sigma_{i+1} \sigma_{i+2} \cdots \sigma_{j-1}$$

where H_i (resp. σ_i) is the local energy function (resp. isomorphism) acting on the i-th and $(i+1)$-st tensor factor. By the normalization assumption (3.8) it follows that

$$(3.11) \quad E_B(u(B)) = 0.$$

As shown in [**35**], the intrinsic energy D_B for the L-fold tensor product $B = B_L \otimes \cdots \otimes B_1$ is given by

$$(3.12) \quad D_B = E_B + \sum_{j=1}^{L} D_j \sigma_1 \sigma_2 \cdots \sigma_{j-1}$$

where D_j acts on the rightmost tensor factor which is B_j.

The coenergy and intrinsic coenergy are defined as

$$\overline{E}_B = -E_B \quad \text{and} \quad \overline{D}_B = -D_B.$$

3.8. One-dimensional configuration sums.

There are three different sets of paths that we consider. Let $B = B_L \otimes \cdots \otimes B_1$ where all B_i are simple crystals. For a classical weight Λ the set of unrestricted paths is defined as

$$(3.13) \quad \mathcal{P}(B, \Lambda) = \{b \in B \mid \mathrm{wt}(b) = \Lambda\}.$$

For a dominant classical weight Λ the set of classically restricted paths is

$$(3.14) \quad \mathcal{P}'(B, \Lambda) = \{b \in B \mid \mathrm{wt}(b) = \Lambda \text{ and } e_i b = 0 \text{ for all } i \in J\}$$

and the set of level-restricted paths for $\ell \in \mathbb{N}$ is

$$(3.15) \quad \mathcal{P}^\ell(B, \Lambda) = \{b \in B \mid \mathrm{wt}(b) = \Lambda, \, e_i b = 0 \text{ for all } i \in J \text{ and } e_0^{\ell+1} b = 0\}.$$

The corresponding one-dimensional configuration sums are the generating functions of these sets of paths with energy/coenergy statistics. The one-dimensional sums

(3.16)
$$S(B,\Lambda) = \sum_{b\in\mathcal{P}(B,\Lambda)} q^{D_B(b)}$$
$$\overline{S}(B,\Lambda) = \sum_{b\in\mathcal{P}(B,\Lambda)} q^{\overline{D}_B(b)}$$

are called supernomials, whereas

(3.17)
$$X(B,\Lambda) = \sum_{b\in\mathcal{P}'(B,\Lambda)} q^{D_B(b)}$$
$$\overline{X}(B,\Lambda) = \sum_{b\in\mathcal{P}'(B,\Lambda)} q^{\overline{D}_B(b)}$$

are the classically restricted configuration sums or generalized Kostka polynomials, and

(3.18)
$$X^\ell(B,\Lambda) = \sum_{b\in\mathcal{P}^\ell(B,\Lambda)} q^{D_B(b)}$$
$$\overline{X}^\ell(B,\Lambda) = \sum_{b\in\mathcal{P}^\ell(B,\Lambda)} q^{\overline{D}_B(b)}$$

are the level-ℓ restricted configuration sums or ℓ-generalized Kostka polynomials.

The classically restricted configurations sums (3.17) are graded tensor product multiplicities. The level-restricted configuration sums (3.18) are graded level ℓ fusion coefficients. Let $B_{\Lambda'}$ denote the crystal corresponding to the affine irreducible highest weight representation $V(\Lambda')$. By the Verlinde formula [40], the fusion coefficient is the coefficient of $B_{\Lambda'}$ of weight $\Lambda' = \Lambda + \ell\Lambda_0$ in the decomposition of $B \otimes B_{\ell\Lambda_0}$. The affine highest weight vectors of weight Λ', whose number is the above multiplicity, are the summands of $\overline{X}^\ell(B,\Lambda)$.

It will be shown in sections 4 and 5 that there are two different evaluations of $\overline{X}(B,\Lambda)$ and $\overline{X}^\ell(B,\Lambda)$ which give rise to q-identities.

4. Bosonic evaluation

Here we present the bosonic evaluation of $\overline{X}(B,\Lambda)$ and $\overline{X}^\ell(B,\Lambda)$ as defined in (3.17) and (3.18). Similarly to the bosonic evaluation of $X(L)$ of section 2.2, the bosonic evaluation of $\overline{X}(B,\Lambda)$ and $\overline{X}^\ell(B,\Lambda)$ can be obtained by inclusion-exclusion arguments as shown in [36]. We discuss the main ideas and techniques of sign-reversing involutions.

4.1. Classically-restricted case. Let $B = B_L \otimes \cdots \otimes B_1$ be a $U_q(\mathfrak{g})$-crystal and let Λ be a classical weight. The Weyl group W of $\mathfrak{g}$ which is generated by the simple reflections r_i as in (3.6) with i restricted to $i \in J$. The bosonic expression for the generating function of classically-restricted paths $\overline{X}(B,\Lambda)$ as defined in (3.17) is given by

(4.1)
$$\overline{X}(B,\Lambda) = \sum_{\omega\in W} (-1)^\omega \, \overline{S}(B,\omega(\Lambda+\rho)-\rho)$$

where $(-1)^\omega$ is the sign of ω and ρ is half the sum of all positive roots. This formula follows directly from Weyl's character formula.

As a warm-up for the level-restricted case, we would like to briefly sketch for types A_n and C_n how (4.1) can be derived via a sign-reversing involution.

EXAMPLE 4.1. For $\mathfrak{g} = A_n$ the Weyl group W is generated by the reflections $r_1, \ldots, r_n$ which act on $\lambda \in \mathbb{Z}^{n+1}$ as follows

$$r_i(\lambda) = (\lambda_1, \ldots, \lambda_{i+1}, \lambda_i, \ldots, \lambda_{n+1}).$$

Hence on $\mathbb{Z}^{n+1}$ the Weyl group acts as the symmetric group S_{n+1}.

For $\mathfrak{g} = C_n$ the Weyl group is generated by

$$r_i(\lambda) = (\lambda_1, \ldots, \lambda_{i+1}, \lambda_i, \ldots, \lambda_n) \quad \text{for } 1 \le i < n$$
$$r_n(\lambda) = (\lambda_1, \ldots, \lambda_{n-1}, -\lambda_n)$$

so that W on $\mathbb{Z}^n$ acts by all permutations and sign changes.

Set

$$\mathcal{S} = \{(\omega, b) \in W \times B \mid \omega(\text{wt}(b) + \rho) = \Lambda + \rho\}$$

On the set $\mathcal{S}$ we define an involution $\Phi : \mathcal{S} \to \mathcal{S}$ with the properties that the fixed points are the pairs $(1, b)$ with $b \in \mathcal{P}'(B, \Lambda)$ which recall are the paths underlying $\overline{X}(B, \Lambda)$ (see (3.17)). Furthermore, if $\Phi(\omega, b) = (\omega', b')$ with $(\omega, b) \ne (\omega', b')$ then ω and ω' have opposite signs.

Let $\mathcal{S}_i$ for $i \in J$ be the set of pairs $(\omega, b) \in \mathcal{S}$ such that $\epsilon_i(b) > 0$. Define $\Phi_i : \mathcal{S}_i \to \mathcal{S}_i$ by $\Phi_i(\omega, b) = (\omega r_i, s_i e_i(b))$. Define the set $\mathcal{S}' = \mathcal{S} - \{(1, b) \mid b \in \mathcal{P}'(B, \Lambda)\}$ so that $\mathcal{S}' = \bigcup_{i \in J} \mathcal{S}_i$. Then Φ is given by

$$\Phi(\omega, b) = \begin{cases} (\omega, b) & \text{if } (\omega, b) \in \mathcal{S} \setminus \mathcal{S}' \\ \Phi_i(\omega, b) & \text{if } (\omega, b) \in \mathcal{S}' \text{ and } i = v(\omega, b) \end{cases}$$

where v is some functions $v : \mathcal{S}' \to J$. To show that Φ indeed exists and is an involution it needs to be shown that if $v(\omega, b) = i$ then

(4.2) $\qquad (\omega, b) \in \mathcal{S}_i \quad \text{and} \quad v(\Phi_i(\omega, b)) = i.$

As mentioned in section 3.4, the crystal $B = B_L \otimes \cdots \otimes B_1$ of type A_n or C_n can be embedded into $B(\Lambda_1)^{\otimes M}$ for some M. Let $p = p_M \otimes \cdots \otimes p_1$ be the image of $b \in B$ under this embedding. Then v can be defined as follows. Let k be minimal such that $\epsilon_i(p_k \otimes \cdots \otimes p_1) > 0$ for some $i \in J$. Then it is clear from Table 1 that there is a unique i satisfying $\epsilon_i(p_k \otimes \cdots \otimes p_1) > 0$. Set $v(\omega, b) = i$. It follows from equation (3.5) and Table 1 that the first k tensor factors of p stay invariant under Φ_i since there are no strings of length greater than one. This ensures (4.2). Hence inclusion-exclusion implies (4.1).

4.2. Level-restricted case. The bosonic expression for the level-restricted generating function $\overline{X}^\ell(B, \Lambda)$ defined in (3.18) can also be found by a sign-reversing involution. The difference is that one needs to consider elements ω in the affine Weyl group $\hat{W}$ which is generated by r_i with $i \in I$ (rather than $i \in J$).

EXAMPLE 4.2. For $\hat{\mathfrak{g}} = A_n^{(1)}$ the affine Weyl group $\hat{W}$ is generated by the reflections $r_1, \ldots, r_n$ as in example 4.1 and

$$r_0(\lambda) = (\lambda_{n+1} + \ell + n + 1, \lambda_2, \ldots, \lambda_n, \lambda_1 - (\ell + n + 1)).$$

For $\hat{\mathfrak{g}} = C_n^{(1)}$ the affine Weyl group $\hat{W}$ is generated by the reflections $r_1, \ldots, r_n$ as in example 4.1 and
$$r_0(\lambda) = (-\lambda_1 + 2(\ell + n + 1), \lambda_2, \ldots, \lambda_n).$$

The affine Weyl group is isomorphic to $\hat{W} \cong T \rtimes W$ where T is the set of certain translations t_α indexed by $\alpha \in M$ for a particular set M (for more details see [15, Section 6]). Then it was shown in [36] that [1]

$$(4.3) \quad \overline{X}^\ell(B, \Lambda) = \sum_{\omega \in W} \sum_{\alpha \in M} (-1)^\omega q^{\frac{1}{2}a_0(\alpha|\alpha)(\ell+h^\vee) - a_0(\rho+\Lambda|\alpha)}$$
$$\times \overline{S}(B, \omega(\Lambda + \rho - (\ell + h^\vee)\alpha) - \rho)$$

where $h^\vee$ is the dual Coxeter number of $\hat{\mathfrak{g}}$ and a_0 is the label of the zeroth node in the Dynkin diagram of $\hat{\mathfrak{g}}$.

EXAMPLE 4.3. Let us give (4.3) more explicitly for $\hat{\mathfrak{g}} = A_n^{(1)}$ and $C_n^{(1)}$. In both cases the dual Coxeter number is $h^\vee = n + 1$ and $a_0 = 1$. For type $A_n^{(1)}$ the set M is given by all $\beta \in \mathbb{Z}^{n+1}$ such that $|\beta| := \beta_1 + \cdots + \beta_{n+1} = 0$ so that

$$\overline{X}^\ell(B, \lambda) = \sum_{\omega \in W} \sum_{\substack{\beta \in \mathbb{Z}^{n+1} \\ |\beta|=0}} (-1)^\omega q^{\frac{1}{2}\beta \cdot \beta(\ell+n+1) - (\rho+\lambda)\cdot\beta}$$
$$\times \overline{S}(B, \omega(\lambda + \rho - (\ell + n + 1)\beta) - \rho)$$

where $W = S_{n+1}$ is the set of permutations.

For type $C_n^{(1)}$ we have $M = 2\mathbb{Z}^n$. Hence

$$\overline{X}^\ell(B, \lambda) = \sum_{\omega \in W} \sum_{\beta \in 2\mathbb{Z}^n} (-1)^\omega q^{\frac{1}{4}\beta \cdot \beta(\ell+n+1) - \frac{1}{2}(\rho+\lambda)\cdot\beta}$$
$$\times \overline{S}(B, \omega(\lambda + \rho - (\ell + n + 1)\beta) - \rho)$$

where the Weyl group W in this case is the set of all permutations and sign changes. The extra factor of $1/2$ in the exponent of q comes from the normalization of $(\cdot|\cdot)$ as alluded to in example 3.1.

The arguments in [36] involve a sign-reversing involution. Similarly to the classically restricted case set

$$\mathcal{S} = \{(\omega, b) \in \hat{W} \times B \mid \omega(\text{wt}(b) + \rho) = \Lambda + \rho\}.$$

For $i \in J$ define $\mathcal{S}_i$ and Φ_i as before. In addition, let $\mathcal{S}_0$ be the subset of all pairs (ω, b) in $\mathcal{S}$ such that $\epsilon_0(b) > \ell$ and define $\Phi_0(\omega, b) = (\omega r_0, s_0 e_0^{\ell+1} b)$. One can find a sign-reversing involution Φ with fixed point set being the set of level-ℓ restricted paths $\mathcal{P}^\ell(B, \Lambda)$ by again specifying a function $v : \mathcal{S} - \{(1, b) \mid b \in \mathcal{P}^\ell(B, \Lambda)\} \to I$ satisfying (4.2). The existence of such a function v was proven in [36] for a large class of crystals.

---

[1]The arguments in [36] require that B is a tensor product of almost perfect crystals and that the energy function obeys certain properties. For the examples of type $A_n^{(1)}$ with $B_i = B^{r_i, s_i}$ and $C_n^{(1)}$ with $B_i = B^{r_i, 1}$, for which we will consider fermionic formulas in the next section, these conditions are all satisfied.

4.3. Supernomial coefficients.

The bosonic formulas (4.1) and (4.3) involve the supernomial coefficients defined in (3.16). To obtain truely explicit expressions it is still necessary to give formulas for the supernomial coefficients. These are not yet known in general. Here we give a few examples for which formulas exist.

EXAMPLE 4.4. The supernomial coefficients for type $A_n^{(1)}$ for single columns were given in [12, 22]. Let $B = B^{\mu_L,1} \otimes \cdots \otimes B^{\mu_1,1}$ be the tensor product of crystals of type $A_n^{(1)}$ corresponding to the partition $\mu = (\mu_1, \ldots, \mu_L)$. Furthermore, let $\lambda \in \mathbb{N}^{n+1}$ be a composition. Then

$$(4.4) \qquad \overline{S}(B,\lambda) = \sum_{\nu} \prod_{\substack{1 \leq a \leq n \\ 1 \leq i \leq \mu_1}} \begin{bmatrix} \nu_i^{(a+1)} - \nu_{i+1}^{(a+1)} \\ \nu_i^{(a)} - \nu_{i+1}^{(a+1)} \end{bmatrix}$$

where the sum is over all sequences of partitions $\nu = (\nu^{(1)}, \ldots, \nu^{(n)})$ such that

$$\emptyset = \nu^{(0)} \subset \nu^{(1)} \subset \cdots \subset \nu^{(n+1)} = \mu^t$$

$\nu^{(a)}/\nu^{(a-1)}$ is a horizontal λ_a-strip.

A horizontal p-strip is a skew shape with p boxes such that each column contains at most one box. For $\mu = (1^{|\lambda|})$ equation (4.4) reduces to the q-multinomial coefficient

$$(4.5) \qquad \begin{bmatrix} L \\ \lambda_1, \ldots, \lambda_{n+1} \end{bmatrix} = \frac{(q)_L}{(q)_{\lambda_1} \cdots (q)_{\lambda_{n+1}}} \qquad \text{if } |\lambda| = L$$

and zero otherwise.

EXAMPLE 4.5. The supernomials for type $A_n^{(1)}$ for single rows were also given in [12, 22]. Let $\mu = (\mu_1, \ldots, \mu_L)$ be a partition, $B = B^{1,\mu_L} \otimes \cdots \otimes B^{1,\mu_1}$ and $\lambda \in \mathbb{N}^{n+1}$ a composition. Then

$$(4.6) \qquad \overline{S}(B,\lambda) = \sum_{\nu} q^{\phi(\nu)} \prod_{\substack{1 \leq a \leq n \\ 1 \leq i \leq \mu_1}} \begin{bmatrix} \nu_i^{(a+1)} - \nu_{i+1}^{(a)} \\ \nu_i^{(a)} - \nu_{i+1}^{(a)} \end{bmatrix}$$

where the sum is over all sequences of partitions $\nu = (\nu^{(1)}, \ldots, \nu^{(n)})$ such that

$$\emptyset = \nu^{(0)} \subset \nu^{(1)} \subset \cdots \subset \nu^{(n+1)} = \mu^t$$

$|\nu^{(a)}| = \lambda_1 + \cdots + \lambda_a$ for all $1 \leq a \leq n$.

Here $\phi(\nu)$ is defined as

$$\phi(\nu) = \sum_{\substack{1 \leq a \leq n \\ 1 \leq i \leq \mu_1}} \nu_{i+1}^{(a)}(\nu_i^{(a+1)} - \nu_i^{(a)}).$$

As in the previous example this reduces to the q-multinomial coefficient (4.5) if $\mu = (1^{|\lambda|})$. For type A_1 this formula coincides with that in [38].

EXAMPLE 4.6. The supernomials of type C_n for single boxes can be obtained by the following arguments. Let $B = (B^{1,1})^{\otimes L}$ and $\lambda \in \mathbb{Z}^n$ with $\|\lambda\| := |\lambda_1| + \cdots + |\lambda_n| \leq L$. This means that there are at least λ_i letters i if $\lambda_i \geq 0$ and at least λ_i

letter $\bar{\imath}$ if $\lambda_i < 0$. If $\|\lambda\| < L$ then there have to be $(L - \|\lambda\|)/2$ pairs of barred and unbarred letters in order to have weight λ. Hence we have

$$(4.7) \qquad \overline{S}(B,\lambda) = \sum_{\substack{\mu \in \mathbb{N}^n \\ 2|\mu|=L-\|\lambda\|}} \begin{bmatrix} L \\ |\lambda_1|+\mu_1,\ldots,|\lambda_n|+\mu_n,\mu_1,\ldots,\mu_n \end{bmatrix}$$

where $\begin{bmatrix} L \\ p_1,\ldots,p_k \end{bmatrix}$ is the q-multinomial as defined in (4.5).

Explicit formulas also exist for level one cases for $B_n^{(1)}$, $D_n^{(1)}$ [8] and $A_{2n-1}^{(2)}$, $A_{2n}^{(2)}$, $D_{n+1}^{(2)}$ [26].

5. Fermionic evaluation

The derivation of fermionic evaluations of the classically- and level-restricted configuration sums (3.17) and (3.18) is in general much more intricate than for the hard-hexagon model. There exists a vast literature on conjectures and proofs of fermionic formulas, most of which deal with the case $\hat{\mathfrak{g}} = A_1^{(1)}$ or $A_n^{(1)}$. A relatively complete list of references can be found in [13]. Fermionic formulas for all untwisted quantum affine algebras were recently conjectured in [13]. For type $A_n^{(1)}$ these are proven for the classically-restricted case in [31] and the level-restricted case in [37]. For type $C_n^{(1)}$ the classically-restricted formulas are proven in [35]. We will present these results here and also derive the fermionic level-restricted formulas for type $C_n^{(1)}$ in section 5.5.

Interestingly, Kirillov and Reshetikhin [30] conjectured that the coefficients of the decomposition of the representations of $U_q'(\hat{\mathfrak{g}})$ naturally associated with multiples of the fundamental weights into direct sums of irreducible representations of $U_q(\mathfrak{g})$ are given by the fermionic formulas at $q = 1$. Chari [6] proved this conjecture for a single tensor factor for $\mathfrak{g}$ a simple Lie algebra of classical type and also for some exceptional cases.

5.1. Classically-restricted case.
We will state here the fermionic formulas conjectured in [13]. Let $\hat{\mathfrak{g}} = X_n^{(1)}$ with $X = A, B, C, D$ or E for $n = 6, 7, 8$ or F for $n = 4$ or G for $n = 2$. Let $B = \bigotimes_{a=1}^n \bigotimes_{i \geq 1} (B^{a,i})^{\otimes L_i^{(a)}}$ where $L_i^{(a)} \in \mathbb{N}$ for all $i \geq 1$ and $1 \leq a \leq n$ and only finitely many $L_i^{(a)}$ are nonzero. Define the following polynomial in q depending on B and a dominant weight Λ

$$(5.1) \qquad \overline{F}(B,\Lambda) = \sum_{\{m\}} q^{cc(\{m\})} \prod_{a=1}^{n} \prod_{i \geq 1} \begin{bmatrix} m_i^{(a)} + p_i^{(a)} \\ m_i^{(a)} \end{bmatrix}$$

where the sum is over all $\{m_i^{(a)} \in \mathbb{N} \mid 1 \leq a \leq n, i \geq 1\}$ subject to the constraints

$$(5.2) \qquad \sum_{a=1}^{n} \sum_{i \geq 1} i m_i^{(a)} \alpha_a = \sum_{a=1}^{n} \sum_{i \geq 1} i L_i^{(a)} \Lambda_a - \Lambda.$$

The variables $p_i^{(a)}$ and the exponent $cc(\{m\})$ are defined as

$$(5.3) \qquad p_i^{(a)} = \sum_{j \geq 1} L_j^{(a)} \min(i,j) - \sum_{b=1}^n (\alpha_a|\alpha_b) \sum_{k \geq 1} \min(t_b i, t_a k) m_k^{(b)}$$

$$(5.4) \qquad cc(\{m\}) = \frac{1}{2} \sum_{a,b=1}^n (\alpha_a|\alpha_b) \sum_{j,k \geq 1} \min(t_b j, t_a k) m_j^{(a)} m_k^{(b)}$$

where $t_a = \frac{2}{(\alpha_a|\alpha_a)}$. Recall that $(\cdot|\cdot)$ is normalized such that $(\alpha_a|\alpha_a) = 2$ if α_a is a long root. Then it was conjectured [13] that

$$\overline{X}(B,\Lambda) = \overline{F}(B,\Lambda).$$

For type $A_n^{(1)}$ and general B this is proven in [31] and for type $C_n^{(1)}$ and $B = \bigotimes_{a=1}^n (B^{a,1})^{\otimes L_1^{(a)}}$ a proof is given in [35]. Parts of the proofs given in [31, 35] are quite technical, but we would like to highlight the general ideas of the proof which also give more insight into the fermionic formulas.

5.2. Rigged configurations. Rigged configurations provide a combinatorial interpretation of the fermionic formula (5.1). We will focus first on type $A_n^{(1)}$.

The sum over the variables $\{m_i^{(a)}\}$ in (5.1) subject to the restriction (5.2) can be interpreted as follows. Let $\nu = (\nu^{(1)}, \ldots, \nu^{(n)})$ be a sequence of partitions with constraints on their sizes given by

$$(5.5) \qquad |\nu^{(a)}| = -\sum_{j=1}^a \lambda_j + \sum_{i \geq 1} \sum_{b=1}^n i\, L_i^{(b)} \min(a,b)$$

where λ is the partition corresponding to the weight Λ. Then (5.5) and (5.2) are equivalent provided that $m_i^{(a)}$ is interpreted as the number of parts of $\nu^{(a)}$ of size i. In terms of ν the definitions (5.3) and (5.4) read

$$P_i^{(a)}(\nu) = Q_i(\nu^{(a-1)}) - 2Q_i(\nu^{(a)}) + Q_i(\nu^{(a+1)}) + \sum_{j \geq 1} L_j^{(a)} Q_i(j)$$

$$cc(\nu) = \sum_{a=1}^n \sum_{i \geq 1} \alpha_i^{(a)} (\alpha_i^{(a)} - \alpha_i^{(a+1)})$$

where $Q_i(\mu)$ is the number of boxes in the first i columns of the partition μ, $\alpha_i^{(a)}$ is the size of the i-th column of $\nu^{(a)}$ and $p_i^{(a)} = P_i^{(a)}(\nu)$.

To interpret (5.1) combinatorially one uses the fact the q-binomial coefficient $\begin{bmatrix} m+p \\ m \end{bmatrix}$ is the generating function of partitions in a box of size $m \times p$. More precisely, these are the partitions with at most m parts each not exceeding p. Hence (5.1) can be restated as

$$\overline{F}(B,\Lambda) = \sum_{(\nu,J) \in \mathrm{RC}(B,\Lambda)} q^{cc(\nu,J)}$$

where $\mathrm{RC}(B,\Lambda)$ is the set of all (ν,J) where ν is a sequence of partitions satisfying (5.5) and J is a double sequence of partitions

$$J = \{J^{(a,i)}\}_{\substack{1 \leq a \leq n \\ i \geq 1}}.$$

The partition $J^{(a,i)}$ has to fit in a box of size $m_i^{(a)}(\nu) \times P_i^{(a)}(\nu)$ where $m_i^{(a)}(\nu) = m_i^{(a)}$ is the number of parts of size i in $\nu^{(a)}$. In particular this requires that $P_i^{(a)}(\nu) \geq 0$ for all $i \geq 1$ and $1 \leq a \leq n$. The exponent is defined as

$$cc(\nu, J) = cc(\nu) + \sum_{a=1}^{n} \sum_{i \geq 1} |J^{(a,i)}|.$$

The elements in $\mathrm{RC}(B, \Lambda)$ are called rigged configurations.

Originally, rigged configurations were introduced in papers by Kerov, Kirillov and Reshetikhin [28, 30] in their study of the XXX model using Bethe Ansatz techniques. Rigged configurations index the solutions to the Bethe Ansatz equations.

In [31] the fermionic formula was proven by showing that there is a statistic preserving bijection between classically restricted paths and rigged configurations. It should be noted that for type $A_n^{(1)}$ the intrinsic energy is equal to the energy, $\overline{D}_B = \overline{E}_B$.

THEOREM 5.1. [31] *For* $B = \bigotimes_{a=1}^{n} \bigotimes_{i \geq 1} (B^{a,i})^{\otimes L_i^{(a)}}$ *a crystal of type* $A_n^{(1)}$, *there is a bijection* $\phi : \mathcal{P}'(B, \Lambda) \to \mathrm{RC}(B, \Lambda)$ *such that for* $b \in \mathcal{P}'(B, \Lambda)$ *we have* $\overline{E}_B(b) = cc(\theta \circ \phi(b))$. *Here* $\theta : \mathrm{RC}(B, \Lambda) \to \mathrm{RC}(B, \Lambda)$ *maps* (ν, J) *to* $(\nu, \tilde{J})$ *where* $\tilde{J}$ *is obtained from* J *by complementing each* $J^{(a,i)}$ *in the box of dimensions* $m_i^{(a)}(\nu) \times P_i^{(a)}(\nu)$.

The bijection ϕ is given explicitly in [31], [35, Section 5.4].

The fermionic formula for type $C_n^{(1)}$ crystals of the form $B_C = \bigotimes_{a=1}^{n} (B_C^{a,1})^{\otimes L_1^{(a)}}$ was proven in [35] using an embedding of type $C_n^{(1)}$ crystals into type $A_{2n-1}^{(1)}$ crystals. Let

$$\Psi(B_C^{r,1}) = \begin{cases} B_A^{2n-r,1} \otimes B_A^{r,1} & \text{if } 1 \leq r < n \\ B_A^{n,2} & \text{if } r = n. \end{cases}$$

Baker [2] showed that there is an embedding $\Psi^{r,1} : B_C^{r,1} \to \Psi(B_C^{r,1})$. It can be defined by requiring that the $U_q(C_n)$-highest weight vector $u_C^{r,1}$ in $B_C^{r,1}$ is mapped to $u_A^{2n-r,1} \otimes u_A^{r,1}$ where $u_A^{r,s}$ is the $U_q(A_{2n-1})$-highest weight vector in $B_A^{r,s}$, and

$$\Psi^{r,1} \circ f_i^C = f_{2n-i}^A \circ f_i^A \circ \Psi^{r,1}$$
$$\Psi^{r,1} \circ e_i^C = e_{2n-i}^A \circ e_i^A \circ \Psi^{r,1}.$$

For a tensor product, define $\Psi_L : B_C^{r_L,1} \otimes \cdots \otimes B_C^{r_1,1} \to \Psi(B_C^{r_L,1}) \otimes \cdots \otimes \Psi(B_C^{r_1,1})$ by $\Psi_L = \Psi^{r_L,1} \otimes \cdots \otimes \Psi^{r_1,1}$.

THEOREM 5.2. [35] *Let* $B_C = B_C^{r_L,1} \otimes \cdots \otimes B_C^{r_1,1}$. *The image* $\mathrm{Im}(\phi \circ \Psi_L)$ *of* $\phi \circ \Psi_L : \mathcal{P}'(B_C, \cdot) \to \mathrm{RC}(\Psi(B_C), \cdot)$ *is characterized by the set of rigged configurations* (ν, J) *satisfying*:

1. $(\nu, J)^{(k)} = (\nu, J)^{(2n-k)}$.
2. *All parts of* $\nu^{(n)}$ *are even*.
3. *All riggings in* $(\nu, J)^{(n)}$ *are even*.

This characterization of the image of $\phi \circ \Psi_L$ suggests the following definition of type C rigged configurations. Let λ be a partition and $B_C = \bigotimes_{a=1}^{n} (B_C^{a,1})^{\otimes L_1^{(a)}}$, and let $\nu = (\nu^{(1)}, \ldots, \nu^{(n)})$ be a sequence of partitions with the properties

(5.6)
$$|\nu^{(a)}| = -\sum_{j=1}^{a} \lambda_j + \sum_{b=1}^{n} L_1^{(b)} \min(a,b) \quad \text{for } 1 \leq a \leq n$$

$\nu^{(n)}$ has only even parts.

Define the vacancy numbers as

(5.7)
$$P_i^{(a)}(\nu) = Q_i(\nu^{(a-1)}) - 2Q_i(\nu^{(a)}) + Q_i(\nu^{(a+1)}) + L_1^{(a)} \quad \text{for } 1 \leq a < n,$$
$$P_i^{(n)}(\nu) = Q_i(\nu^{(n-1)}) - Q_i(\nu^{(n)}) + \frac{1}{2}L_1^{(n)}Q_i(2).$$

The set of rigged configurations of type C corresponding to a weight Λ with associated partition λ and crystal B_C, denoted by $\mathrm{RC}_C(B_C, \Lambda)$, is given by (ν, J) where ν is a sequence of partitions satisfying (5.6) and J is a double sequence of partitions

$$J = \{J^{(a,i)}\}_{\substack{1 \leq a \leq n \\ i \geq 1}}$$

where $J^{(a,i)}$ is a partition in a box of size $m_i^{(a)}(\nu) \times P_i^{(a)}(\nu)$ with $P_i^{(a)}(\nu)$ as in (5.7) and $m_i^{(a)}(\nu)$ the number of parts of $\nu^{(a)}$.

It is shown in [**35**] that $\overline{D}_A \circ \Psi_L = 2\overline{D}_C$. Hence using Theorem 5.2 the statistics of type C rigged configurations becomes

$$cc_C(\nu, J) = cc_C(\nu) + \sum_{a=1}^{n} \sum_{i \geq 1} |J^{(a,i)}|$$

where $\quad cc_C(\nu) = \sum_{i \geq 1} \Big(\sum_{a=1}^{n-1} \alpha_i^{(a)}(\alpha_i^{(a)} - \alpha_i^{(a+1)}) + \frac{1}{2}\alpha_i^{(n)\,2} \Big)$

which implies that

$$\overline{X}(B_C, \Lambda) = \sum_{(\nu, J) \in \mathrm{RC}_C(B_C, \Lambda)} q^{cc_C(\nu, J)}.$$

It is also not so hard to show that

$$\overline{F}(B_C, \Lambda) = \sum_{(\nu, J) \in \mathrm{RC}_C(B_C, \Lambda)} q^{cc_C(\nu, J)}$$

by identifying

$$p_i^{(a)} = \begin{cases} P_i^{(a)}(\nu) & \text{for } 1 \leq a < n \\ P_{2i}^{(n)}(\nu) & \text{for } a = n \end{cases}$$

$$m_i^{(a)} = \begin{cases} m_i(\nu^{(a)}) & \text{for } 1 \leq a < n \\ m_{2i}(\nu^{(n)}) & \text{for } a = n. \end{cases}$$

This proves that $\overline{X}(B_C, \Lambda) = \overline{F}(B_C, \Lambda)$.

5.3. Level-restricted case.
Fermionic formulas for the level-restricted one-dimensional configuration sums $\overline{X}^\ell(B,\Lambda)$ were conjectured in [13] for all $\hat{\mathfrak{g}}$ as in section 5.1 and special weight $\Lambda = 0$. Let $\ell \in \mathbb{N}$ and define the following polynomial in q depending on B

$$\overline{F}^\ell(B) = \sum_{\{m\}} q^{cc^\ell(\{m\})} \prod_{(a,i) \in H^\ell} \begin{bmatrix} m_i^{(a)} + p_i^{(a)} \\ m_i^{(a)} \end{bmatrix} \tag{5.8}$$

where $H^\ell = \{(a,i) \mid 1 \le a \le n, 1 \le j \le t_a \ell\}$ and the sum is over all $\{m_i^{(a)} \in \mathbb{N} \mid (a,i) \in H^\ell\}$ subject to the constraints

$$\sum_{(a,i)\in H^\ell} i\, m_i^{(a)} \alpha_a = \sum_{(a,i)\in H^\ell} i\, L_i^{(a)} \Lambda_a. \tag{5.9}$$

The variables $p_i^{(a)}$ and the exponent $cc^\ell(\{m\})$ are defined as

$$p_i^{(a)} = \sum_{j=1}^{t_a \ell} L_j^{(a)} \min(i,j) - \sum_{(b,k)\in H^\ell} (\alpha_a|\alpha_b) \min(t_b i, t_a k) m_k^{(b)} \tag{5.10}$$

$$cc^\ell(\{m\}) = \frac{1}{2} \sum_{(a,j),(b,k)\in H^\ell} (\alpha_a|\alpha_b) \min(t_b j, t_a k) m_j^{(a)} m_k^{(b)}. \tag{5.11}$$

Then it was conjectured [13] that

$$\overline{X}^\ell(B,0) = \overline{F}^\ell(B). \tag{5.12}$$

5.4. Level-restricted case: type $A_n^{(1)}$.
For type $A_n^{(1)}$ the conjecture (5.12) and its generalization to arbitrary weights Λ was proven in [37]. These formulas can again be understood in terms of rigged configurations. We will explain this here since it will enable us to derive the level-restricted fermionic formulas for type $C_n^{(1)}$ in the next section.

Let $\lambda = (\lambda_1, \ldots, \lambda_{n+1})$ be the partition corresponding to the dominant integral weight Λ. A partition λ is restricted of level ℓ if $\lambda_1 - \lambda_{n+1} \le \ell$. Define $\tilde{\ell} = \ell - (\lambda_1 - \lambda_{n+1})$, which is nonnegative by assumption. Set $\lambda' = (\lambda_1 - \lambda_{n+1}, \ldots, \lambda_n - \lambda_{n+1})^t$ (where ^{t} stands for transpose) and denote the set of all column-strict tableaux of shape λ' over the alphabet $\{1, 2, \ldots, \lambda_1 - \lambda_{n+1}\}$ by $\mathrm{CST}(\lambda')$. Define a table of modified vacancy numbers depending on a sequence of partitions ν and $t \in \mathrm{CST}(\lambda')$ by

$$P_i^{(a)}(\nu,t) = P_i^{(a)}(\nu) - \sum_{j=1}^{\lambda_a - \lambda_{n+1}} \chi(i \ge \tilde{\ell} + t_{j,a}) + \sum_{j=1}^{\lambda_{a+1} - \lambda_{n+1}} \chi(i \ge \tilde{\ell} + t_{j,a+1}) \tag{5.13}$$

where $t_{j,a}$ denotes the entry in t in row j and column a and $\chi(S) = 1$ if the statement S is true and $\chi(S) = 0$ otherwise.

DEFINITION 5.3. Let $B = \bigotimes_{a=1}^n \bigotimes_{i=1}^\ell (B^{a,i})^{\otimes L_i^{(a)}}$ be a crystal of type $A_n^{(1)}$ and Λ a dominant integral weight with corresponding partition λ. Say that $(\nu, J) \in \mathrm{RC}(B,\Lambda)$ is restricted of level ℓ provided that

1. $\nu_1^{(a)} \le \ell$ for all a.
2. There exists a tableau $t \in \mathrm{CST}(\lambda')$, such that for every $i, a \ge 1$, the largest part of $J^{(a,i)}$ does not exceed $P_i^{(a)}(\nu, t)$.

Denote by $\mathrm{RC}^\ell(B,\Lambda)$ the set of $(\nu,J) \in \mathrm{RC}(B,\Lambda)$ that are restricted of level ℓ.

Note in particular that the second condition requires that $P_i^{(a)}(\nu,t) \geq 0$ for all $i, a \geq 1$.

EXAMPLE 5.4. Let $\lambda = (m^{n+1})$ be rectangular with $n+1$ rows. Then $\lambda' = \emptyset$ and $P_i^{(a)}(\nu,\emptyset) = P_i^{(a)}(\nu)$ for all $i,a \geq 1$ so that the modified vacancy numbers are equal to the vacancy numbers.

THEOREM 5.5. [**37**, Theorem 8.2] *The bijection* $\phi: \mathcal{P}'(B,\Lambda) \to \mathrm{RC}(B,\Lambda)$ *restricts to a well-defined bijection* $\phi: \mathcal{P}^\ell(B,\Lambda) \to \mathrm{RC}^\ell(B,\Lambda)$.

Since $\overline{E}_B = cc \circ \theta \circ \phi$ by Theorem 5.1 it follows from Theorem 5.5 that for type $A_n^{(1)}$

$$\overline{X}^\ell(B,\Lambda) = \sum_{(\nu,J) \in \mathrm{RC}^\ell(B,\Lambda)} q^{cc(\theta(\nu,J))}. \tag{5.14}$$

Note that $cc(\theta(\nu,J)) = cc(\nu) + \sum_{a=1}^n \sum_{i=1}^\ell P_i^{(a)}(\nu) m_i^{(a)}(\nu) - \sum_{a=1}^n \sum_{i=1}^\ell |J^{(a,i)}|$. Let $\mathrm{SCST}(\lambda')$ be the set of all nonempty subsets of $\mathrm{CST}(\lambda')$. Since the q-binomial $\begin{bmatrix} m+p \\ m \end{bmatrix}$ is the generating function of partitions with at most m parts each not exceeding p it follows by inclusion-exclusion that

$$\overline{X}^\ell(B,\Lambda) = \sum_{S \in \mathrm{SCST}(\lambda')} (-1)^{|S|+1} \sum_{\{m\}} q^{c^\ell(\{m\})} \prod_{(a,i) \in H^\ell} \begin{bmatrix} m_i^{(a)} + p_i^{(a)}(S) \\ m_i^{(a)} \end{bmatrix}_{1/q} \tag{5.15}$$

where the sum is over all $\{m_i^{(a)} \in \mathbb{N} \mid (a,i) \in H^\ell\}$ subject to the constraints

$$\sum_{(a,i) \in H^\ell} i m_i^{(a)} \alpha_a = \sum_{(a,i) \in H^\ell} i L_i^{(a)} \Lambda_a - \Lambda.$$

Also $p_i^{(a)}(S)$ and $c^\ell(\{m\})$ are defined as

$$p_i^{(a)}(S) = p_i^{(a)} + \min_{t \in S}\left\{ -\sum_{j=1}^{\lambda_a - \lambda_{n+1}} \chi(i \geq \tilde{\ell} + t_{j,a}) + \sum_{j=1}^{\lambda_{a+1} - \lambda_{n+1}} \chi(i \geq \tilde{\ell} + t_{j,a+1}) \right\}$$

$$c^\ell(\{m\}) = cc^\ell(\{m\}) + \sum_{(a,i) \in H^\ell} p_i^{(a)} m_i^{(a)}$$

with $p_i^{(a)}$ as in (5.10) and $cc^\ell(\{m\})$ as in (5.11). $\begin{bmatrix} m+p \\ m \end{bmatrix}_{1/q}$ is the q-binomial with q replaced by $1/q$. In particular if $\lambda = (m^{n+1})$ as in Example 5.4 so that $\Lambda = 0$ and $p_i^{(a)}(\nu, \{\emptyset\}) = p_i^{(a)}$ the fermionic form (5.15) reduces to (5.8) since $\begin{bmatrix} m+p \\ m \end{bmatrix}_{1/q} = q^{-mp}\begin{bmatrix} m+p \\ m \end{bmatrix}$. Further details can be found in [**37**].

It should be remarked that even though (5.15) contains explicit signs, it is clear from the equivalent combinatorial formula (5.14) that it is a nonnegative polynomial in q.

5.5. Level-restricted case: type $C_n^{(1)}$. In this section we will show how the level-restricted fermionic formulas for type $C_n^{(1)}$ can be obtained from Theorems 5.5 and 5.2.

Under the embedding Ψ a dominant weight $\Lambda_C = \sum_{k=1}^m \Lambda_{i_k}^C$ of type $C_n^{(1)}$ becomes the weight $\Lambda_A = \sum_{k=1}^m (\Lambda_{i_k}^A + \Lambda_{2n-i_k}^A)$ of type $A_{2n-1}^{(1)}$ where all $1 \le i_k \le n$. In terms of the corresponding partitions λ^A and λ^C this implies that

$$\lambda_a^A - \lambda_{2n}^A = \lambda_1^C + \lambda_a^C \qquad \text{for } 1 \le a \le n$$
$$\lambda_a^A - \lambda_{2n}^A = \lambda_1^C - \lambda_{2n+1-a}^C \qquad \text{for } n < a \le 2n.$$

Let $B_C = \bigotimes_{a=1}^n (B_C^{a,1})^{\otimes L_1^{(a)}}$. Hence, under the embedding $\Psi_L : B_C \to \Psi(B_C)$, the conditions for level-restriction for rigged configurations of type A as given in Definition 5.3 become the following.

For a partition λ^C define $(\lambda^C)' = (2\lambda_1^C, \lambda_1^C + \lambda_2^C, \ldots, \lambda_1^C + \lambda_n^C, \lambda_1^C - \lambda_n^C, \lambda_1^C - \lambda_{n-1}^C, \ldots, \lambda_1^C - \lambda_1^C)^t$. Let $\mathrm{CST}((\lambda^C)')$ be the set of all semi-standard tableaux of shape $(\lambda^C)'$ in the alphabet $\{1, 2, \ldots, 2\lambda_1^C\}$. For $t \in \mathrm{CST}((\lambda^C)')$ set

$$f_i^{(a)}(t) = \begin{cases} -\sum_{j=1}^{\lambda_1^C + \lambda_a^C} \chi(i \ge 2\ell - 2\lambda_1^C + t_{j,a}) \\ \quad + \sum_{j=1}^{\lambda_1^C + \lambda_{a+1}^C} \chi(i \ge 2\ell - 2\lambda_1^C + t_{j,a+1}) & \text{for } 1 \le a < n \\ -\sum_{j=1}^{\lambda_1^C + \lambda_n^C} \chi(i \ge 2\ell - 2\lambda_1^C + t_{j,n}) \\ \quad + \sum_{j=1}^{\lambda_1^C - \lambda_n^C} \chi(i \ge 2\ell - 2\lambda_1^C + t_{j,n+1}) & \text{for } a = n \\ -\sum_{j=1}^{\lambda_1^C - \lambda_{2n+1-a}^C} \chi(i \ge 2\ell - 2\lambda_1^C + t_{j,a}) \\ \quad + \sum_{j=1}^{\lambda_1^C - \lambda_{2n-a}^C} \chi(i \ge 2\ell - 2\lambda_1^C + t_{j,a+1}) & \text{for } n < a < 2n. \end{cases}$$

Define modified vacancy numbers as

$$(5.16) \qquad P_i^{(a)}(\nu, t) = P_i^{(a)}(\nu) + \begin{cases} \min\{f_i^{(a)}(t), f_i^{(2n-a)}(t)\} & \text{for } 1 \le a < n \\ \frac{1}{2} f_i^{(n)}(t) & \text{for } a = n \end{cases}$$

with $P_i^{(a)}(\nu)$ as defined in (5.7).

DEFINITION 5.6. Let Λ_C be a dominant weight and λ^C the corresponding partition. Let $B_C = \bigotimes (B_C^{a,1})^{\otimes L_1^{(a)}}$ be a crystal of type $C_n^{(1)}$. Say that $(\nu, J) \in \mathrm{RC}_C(B_C, \Lambda_C)$ is restricted of level ℓ provided that

1. $\nu_1^{(a)} \le 2\ell$ for all a.
2. There exists a tableau $t \in \mathrm{CST}((\lambda^C)')$, such that for every $i, a \ge 1$, the largest part of $J^{(a,i)}$ does not exceed $P_i^{(a)}(\nu, t)$ defined in (5.16).

Denote by $\mathrm{RC}_C^\ell(B_C, \Lambda_C)$ the set of $(\nu, J) \in \mathrm{RC}_C(B_C, \Lambda_C)$ that are restricted of level ℓ.

It follows that

$$\overline{X}^\ell(B_C, \Lambda_C) = \sum_{(\nu, J) \in \mathrm{RC}_C^\ell(B_C, \Lambda_C)} q^{cc_C(\theta(\nu, J))}.$$

Let $\mathrm{SCST}((\lambda^C)')$ be the set of all nonempty subsets of $\mathrm{CST}((\lambda^C)')$. By the same arguments as in the type A we find

$$(5.17) \quad \overline{X}^\ell(B_C, \Lambda_C) = \sum_{S \in \mathrm{SCST}((\lambda^C)')} (-1)^{|S|+1} \sum_{\{m\}} q^{c^\ell(\{m\})} \prod_{(a,i) \in H^\ell} \begin{bmatrix} m_i^{(a)} + p_i^{(a)}(S) \\ m_i^{(a)} \end{bmatrix}_{1/q}$$

where the sum is over all $\{m_i^{(a)} \in \mathbb{N} \mid (a,i) \in H^\ell\}$ subject to the constraints

$$\sum_{(a,i) \in H^\ell} im_i^{(a)} \alpha_a = \sum_{(a,1) \in H^\ell} L_1^{(a)} \Lambda_a^C - \Lambda_C.$$

The variable $p_i^{(a)}(S)$ is defined as

$$p_i^{(a)}(S) = p_i^{(a)} + \begin{cases} \min_{t \in S}\{f_i^{(a)}(t), f_i^{(2n-a)}(t)\} & \text{for } 1 \leq a < n \\ \frac{1}{2}\min_{t \in S}\{f_i^{(n)}(t)\} & \text{for } a = n \end{cases}$$

$$c^\ell(\{m\}) = cc^\ell(\{m\}) + \sum_{(a,i) \in H^\ell} p_i^{(a)} m_i^{(a)}$$

with $p_i^{(a)}$ as in (5.3) and $cc^\ell(\{m\})$ as in (5.4). In particular if $\lambda^C = \emptyset$ so that $\Lambda_C = 0$ the fermionic form (5.17) reduces to (5.8).

6. Summary and open problems

Equating the bosonic and fermionic evaluations for $\overline{X}(B, \Lambda)$ and $\overline{X}^\ell(B, \Lambda)$ as given in sections 4 and 5 yields polynomial identities in q. One may take limits of these identities in various ways to obtain q-series identities. We refer the interested reader to [12, 13, 37, 38] for details.

It is still an outstanding problem to prove the conjectured fermionic formulas (5.1) and (5.8) for general $\hat{\mathfrak{g}}$. There is strong evidence that the rigged configuration approach will still be applicable in this case. Part of this program also requires the proof of the existence and structure of the crystals $B^{r,s}$ for general $\hat{\mathfrak{g}}$ as mentioned in Conjecture 3.4.

As mentioned in the beginning of section 3, the energy function introduced in section 3 does not specialize to the statistics on path of section 2 which yield the Rogers-Ramanujan identities. This is due to the fact that the Rogers-Ramanujan identities correspond to a non-unitary physical model (more precisely the minimal model $M(2,5)$) whereas the crystal base theory introduced in section 3 is related to unitary physical models. In representation theoretic terms, this is reflected in the difference between integrable versus admissible modules. It is still an open problem to define the (intrinsic) energy for general admissible modules. For $\hat{\mathfrak{g}} = \widehat{\mathfrak{sl}}_2$ results are available [9, 5, 10, 11]. However for general $\hat{\mathfrak{g}}$ details are not yet known in general. Some results in this direction can be found in [33].

References

1. T. Akasaka and M. Kashiwara, *Finite-dimensional representations of quantum affine algebras*, Publ. RIMS, Kyoto Univ. **33** (1997) 839–867.
2. T. Baker, *Zero actions and energy functions for perfect crystals*, Publ. RIMS, Kyoto Univ. **36** (2000) 533–572.
3. R. J. Baxter, *Rogers-Ramanujan identities in the hard hexagon model*, J. Stat. Phys. **26** (1981) 427–452.
4. R. J. Baxter, Exactly Solved Models in Statistical Mechanics, Academic Press, 1982.
5. A. Berkovich, B. M. McCoy and A. Schilling, *Rogers-Schur-Ramanujan type identities for the $M(p,p')$ minimal models of conformal field theory*, Commun. Math. Phys. **191** (1998) 325–395.
6. V. Chari, *On the fermionc formula and the Kirillov-Reshetikhin conjecture*, preprint math.QA/0006090.
7. V. G. Drinfeld, *Hopf algebra and the Yang–Baxter equation*, Soviet. Math. Dokl. **32** (1985) 254–258.

8. E. Date, M. Jimbo, A. Kuniba, T. Miwa and M. Okado, *One dimensional configuration sums in vertex models and affine Lie algebra characters*, Lett. Math. Phys. **17** (1989) 69–77.
9. P. J. Forrester and R. J. Baxter, *Further exact solutions of the eight-vertex SOS model and generalizations of the Rogers–Ramanujan identities*, J. Statist. Phys. **38** (1985) 435–472.
10. O. Foda and T. A. Welsh, *On the combinatorics of Forrester-Baxter models*, in: Physical combinatorics (Kyoto, 1999), 49–103, Progr. Math., **191**, Birkhuser Boston, Boston, MA, 2000.
11. O. Foda, K. S. Lee, Y. Pugai and T. A. Welsh, *Path generating transforms*, in: q-series from a contemporary perspective, Contemp. Math. **254** 157–186.
12. G. Hatayama, A. N. Kirillov, A. Kuniba, M. Okado, T. Takagi and Y. Yamada, *Character formulae of $\widehat{sl}_n$-modules and inhomogeneous paths*, Nucl. Phys. **B536** (1999) 575–616.
13. G. Hatayama, A. Kuniba, M. Okado, T. Takagi and Y. Yamada, *Remarks on fermionic formula*, in: Recent developments in quantum affine algebras and related topics, Contemp. Math. **248** (1999) 243–291.
14. M. Jimbo, *A q-difference analogue of $U(\mathcal{G})$ and the Yang–Baxter equation*, Lett. Math. Phys. **10** (1985) 63–69.
15. V. G. Kac, *Infinite dimensional Lie algebras*, Cambridge University Press, 1990, third edition.
16. V. G. Kac and M. Wakimoto, *Modular invariant representations of infinite-dimensional Lie algebras and superalgebras*, Proc. Nat. Acad. Sci. U.S.A. **85** (1988), no. 14, 4956–4960.
17. V. G. Kac and M. Wakimoto, *Classification of modular invariant representations of affine algebras*, Adv. Ser. Math. Phys. **7**, 138–177 (World Sci. Publishing, Teaneck, 1989).
18. M. Kashiwara, *Crystalizing the q-analogue of universal enveloping algebras*, Commun. Math. Phys. **133** (1990) 249–260.
19. M. Kashiwara, *On crystal bases of the q-analogue of universal enveloping algebras*, Duke Math. J. **63** (1991) 465–516.
20. M. Kashiwara, *On crystal bases*, in: Representations of groups (Banff, AB, 1994), 155–197, CMS Conf. Proc., 16, Amer. Math. Soc., Providence, RI, 1995.
21. M. Kashiwara, *On level zero representations of quantized affine algebras*, preprint math.QA/0010293.
22. A. N. Kirillov, *New combinatorial formula for modified Hall-Littlewood polynomials*, in: q-series from a contemporary perspective, Contemp. Math. **254** (2000) 283–333.
23. S.-J. Kang, M. Kashiwara and K. C. Misra, *Crystal bases of Verma modules for quantum affine Lie algebras*, Comp. Math. **92** (1994) 299–325.
24. S.-J. Kang, M. Kashiwara, K. C. Misra, T. Miwa, T. Nakashima and A. Nakayashiki, *Affine crystals and vertex models*, Int. J. Mod. Phys. **7**, Suppl. 1A (1992) 449–484.
25. S.-J. Kang, M. Kashiwara, K. C. Misra, T. Miwa, T. Nakashima and A. Nakayashiki, *Perfect crystals of quantum affine Lie algebras*, Duke Math. J. **68** (1992) 499–607.
26. A. Kuniba, K. C. Misra, M. Okado, T. Takagi and J. Uchiyama, *Characters of Demazure modules and solvable models*, Nucl. Phys. **B510** [PM] (1998) 555-576.
27. M. Kashiwara and T. Nakashima, *Crystal graphs for representations of the q-analogue of classical Lie algebras*, J. Alg. **165** (1994) 295–345.
28. S. V. Kerov, A. N. Kirillov and N. Y. Reshetikhin, *Combinatorics, the Bethe ansatz and representations of the symmetric group*, J. Soviet Math. **41** (1988), no. 2, 916–924.
29. A. N. Kirillov and N. Y. Reshetikhin, *The Bethe ansatz and the combinatorics of Young tableaux*, J. Soviet Math. **41** (1988), no. 2, 925–955.
30. A. N. Kirillov and N. Y. Reshetikhin, *Representations of Yangians and multiplicities of the inclusion of the irreducible components of the tensor product of representations of simple Lie algebras*, J. Soviet Math. **52** (1990), no. 3, 3156–3164.
31. A. N. Kirillov, A. Schilling and M. Shimozono, *A bijection between Littlewood-Richardson tableaux and rigged configurations*, to appear in Selecta Mathematica (N.S.) (math.CO/9901037).
32. I. G. Macdonald, *Affine roots systems and Dedekind's η-function*, Inv. Math. **15** (1972) 91–143.
33. T. Nakanishi, *Non-unitary minimal models and RSOS models*, Nucl. Phys. B **334** (1990) 745–766.
34. A. Nakayashiki and Y. Yamada, *Kostka polynomials and energy functions in solvable lattice models*, Selecta Math. (N.S.) **3** (1997) 547–599.

35. M. Okado, A. Schilling and M. Shimozono, *Virtual crystals and fermionic formulas of type $D_{n+1}^{(2)}$, $A_{2n}^{(2)}$, and $C_n^{(1)}$*, preprint math.QA/0105017.
36. A. Schilling and M. Shimozono, *Bosonic formula for level-restricted paths*, Adv. Studies in Pure Mathematics **28**, Combinatorial Methods in Representation Theory (2000) 305–325.
37. A. Schilling and M. Shimozono, *Fermionic formulas for level-restricted generalized Kostka polynomials and coset branching functions*, to appear in Commun. Math. Phys. (math.QA/0001114).
38. A. Schilling and S. O. Warnaar, *Supernomial coefficients, polynomial identities and q-series*, The Ramanujan Journal **2** (1998) 459–494.
39. A. Schilling and S. O. Warnaar, *Inhomogeneous lattice paths, generalized Kostka polynomials and A_{n-1} supernomials*, Commun. Math. Phys. **202** (1999) 359–401.
40. E. Verlinde, *Fusion rules and modular transformations in 2D conformal field theory*, Nucl. Phys. B **300** (1988) 360–376.

DEPARTMENT OF INFORMATICS AND MATHEMATICAL SCIENCE, GRADUATE SCHOOL OF ENGINEERING SCIENCE, OSAKA UNIVERSITY, TOYONAKA, OSAKA 560-8531, JAPAN
  E-mail address: okado@sigmath.es.osaka-u.ac.jp

DEPARTMENT OF MATHEMATICS 2-279, M.I.T., CAMBRIDGE, MA 02139, U.S.A. AND DEPARTMENT OF MATHEMATICS, UNIVERSITY OF CALIFORNIA, ONE SHIELDS AVENUE, DAVIS, CA 95616-8633, U.S.A.
  E-mail address: anne@math.mit.edu, anne@math.ucdavis.edu

DEPARTMENT OF MATHEMATICS, 460 MCBRYDE HALL, VIRGINIA TECH, BLACKSBURG, VA 24061-0123, U.S.A
  E-mail address: mshimo@math.vt.edu

The Bailey-Rogers-Ramanujan group

D. Stanton

ABSTRACT. A certain group of upper triangular 2×2 matrices is explicitly defined via generators. Any element of this group has an associated multisum identity of Rogers-Ramanujan type. Several infinite families of identities are given as examples. Different expressions for an element in the generators can yield distinct identities. An application to the Borwein polynomials is given.

1. Introduction. The Rogers-Ramanujan identities have many proofs [5]. One idea which has been fruitful [3], [16], [17] is the concept of a Bailey pair. This technique allows for iteration to objects called Bailey chains [3], and results in multisum generalizations of the Rogers-Ramanujan theorems to arbitrary modulus. The purpose of this paper is to define a group of 2×2 rational matrices, which contains the standard iteration of Bailey chains. Any element of this group has a corresponding identity of Rogers-Ramanujan type, in fact there may be many such identities.

We review Bailey pairs and give the relevant transformations for Bailey pairs in §2. These transformations are written as 2×2 matrices in §3, where the Bailey-Rogers-Ramanujan group is defined in Definition 2. A Rogers-Ramanujan type identity is given for an element of the group in Theorem 1 in §4. Examples of the identities are given in §5 and §6, and an application to the Borwein polynomials is given in §7.

We use standard notation for q-series found in [2], [12], and we shall also use the Jacobi triple product identity

$$(1.1) \qquad \sum_{n=-\infty}^{\infty} q^{n^2} x^n = (q^2, -qx, -q/x; q^2)_\infty.$$

2. Bailey pairs. In this section we review Bailey pairs, and give the versions of the transformations on pairs which are needed for the Bailey-Rogers-Ramanujan group.

Recall [3] the definition of a Bailey pair.

1991 *Mathematics Subject Classification.* Primary 11P57, 33D65; Secondary 05A19.
Key words and phrases. Rogers-Ramanujan identities.
Partially supported by NSF grant DMS 99-70627.

DEFINITION 1. *A pair of sequences* $(\alpha_n(a,q), \beta_n(a,q))$ *is called a Bailey pair with parameters* (a,q) *if*

$$\beta_n(a,q) = \sum_{r=0}^{n} \frac{\alpha_r(a,q)}{(q;q)_{n-r}(aq;q)_{n+r}}$$

for all $n \geq 0$.

The first example of a Bailey pair, which will be used throughout this paper, is the unit Bailey pair

(UBP) $\quad \beta_n^{(0)}(a,q) = \begin{cases} 1, & \text{if } n = 0 \\ 0, & \text{if } n > 0, \end{cases} \quad \alpha_n^{(0)}(a,q) = \frac{(a;q)_n}{(q;q)_n} \frac{(1-aq^{2n})}{(1-a)} (-1)^n q^{\binom{n}{2}}.$

Bailey's lemma [3],[17] takes a Bailey pair $(\alpha_n(a,q), \beta_n(a,q))$ and produces another Bailey pair $(\alpha'_n(a,q), \beta'_n(a,q))$ with parameters (a,q). One limiting case of Bailey's lemma is denoted here by (S1)

(S1)
$$\alpha'_r(a,q) = a^r q^{r^2} \alpha_r(a,q),$$
$$\beta'_n(a,q) = \sum_{k=0}^{n} \frac{a^k q^{k^2}}{(q;q)_{n-k}} \beta_k(a,q).$$

If we start with (UBP), apply (S1) twice, we have

(2.1)
$$\beta_n^{(2)}(a,q) = \sum_{r=0}^{n} \frac{a^{2r} q^{2r^2} \alpha_r^{(0)}(a,q)}{(q;q)_{n-r}(aq;q)_{n+r}}$$
$$= \sum_{s=0}^{n} \frac{a^s q^{s^2}}{(q;q)_{n-s}(q;q)_s}.$$

The Rogers-Ramanujan identities modulo 5 occur if $a=1$ and $n \to \infty$ in (2.1).

Another limiting case of Bailey's lemma is

(S2)
$$\alpha'_r(a,q) = a^{r/2} q^{r^2/2} \alpha_r(a,q),$$
$$\beta'_n(a,q) = \sum_{k=0}^{n} \frac{(-\sqrt{aq};q)_k}{(q;q)_{n-k}(-\sqrt{aq};q)_n} a^{k/2} q^{k^2/2} \beta_k(a,q).$$

It is clear from the action on $\alpha_n(a,q)$ that applying (S2) twice is equivalent to applying (S1) once.

Some other transformations of Bailey pairs which changed the base q were given in [11]. The following three choices, denoted (D1), (D2), and (D3), all have $(\alpha'_n(a,q), \beta'_n(a,q))$ as a Bailey pair with parameters (a,q).

(D1)
$$\alpha'_r(a,q) = \alpha_r(a^2, q^2),$$
$$\beta'_n(a,q) = \sum_{k=0}^{n} \frac{(-aq;q)_{2k}}{(q^2;q^2)_{n-k}} q^{n-k} \beta_k(a^2, q^2),$$

(D2)
$$\alpha'_r(a,q) = a^{-r} q^{-r^2} \alpha_r(a^2, q^2),$$
$$\beta'_n(a,q) = \sum_{k=0}^{n} \frac{(-aq;q)_{2k}}{(q^2;q^2)_{n-k}} q^{k^2+k-2kn-n} (-1)^{n-k} a^{-n} \beta_k(a^2, q^2),$$

and

(D3)
$$\alpha'_r(a,q) = a^{-r/2}q^{-r^2/2}\alpha_r(a^2,q^2),$$
$$\beta'_n(a,q) = \sum_{k=0}^{n} \frac{(-aq;q)_{2k}(q^{-1/2-k}a^{-1/2}, q^{k+3/2}a^{1/2};q)_{n-k}}{(aq^{2k+1};q^2)_{n-k}(q^2;q^2)_{n-k}}$$
$$\times q^{-\binom{k}{2}}(aq)^{-k/2}\beta_k(a^2,q^2).$$

The inverse versions of (D1)-(D3), denoted (E1)-(E3), follow from Theorem 2.2 of [11]. To avoid fractional powers we choose to write (E1)-(E3) in such a way that $(\alpha'_n(a,q), \beta'_n(a,q))$ is a Bailey pair with parameters (a^4, q^4) in each of these three cases.

(E1)
$$\alpha'_r(a^4,q^4) = \alpha_r(a^2,q^2),$$
$$\beta'_n(a^4,q^4) = \sum_{k=0}^{n} \frac{(-1)^{n-k}q^{2(n-k)^2}}{(-a^2q^2;q^2)_{2n}(q^4;q^4)_{n-k}}\beta_k(a^2,q^2),$$

(E2)
$$\alpha'_r(a^4,q^4) = a^{2r}q^{2r^2}\alpha_r(a^2,q^2),$$
$$\beta'_n(a^4,q^4) = \sum_{k=0}^{n} \frac{a^{2k}q^{2k^2}}{(-a^2q^2;q^2)_{2n}(q^4;q^4)_{n-k}}\beta_k(a^2,q^2),$$

(E3)
$$\alpha'_r(a^4,q^4) = a^r q^{r^2}\alpha_r(a^2,q^2),$$
$$\beta'_n(a^4,q^4) = \sum_{k=0}^{n} \frac{(aq;q^2)_{2n-k}(-aq;q^2)_k a^k q^{k^2}}{(-a^2q^2;q^2)_{2n}(q^4;q^4)_{n-k}(a^2q^2;q^4)_n}\beta_k(a^2,q^2).$$

For changing the base q to q^3 we have one possibility and its inverse, denoted (T1) and (T2). In (T1) (α'_n, β'_n) has parameters (a^3, q^3), while in (T2) it has parameters (a,q).

(T1)
$$\alpha'_r(a^3,q^3) = a^r q^{r^2}\alpha_r(a,q),$$
$$\beta'_n(a^3,q^3) = \sum_{k=0}^{n} \frac{(aq;q)_{3n-k} a^k q^{k^2}}{(a^3q^3;q^3)_{2n}(q^3;q^3)_{n-k}}\beta_k(a,q),$$

(T2)
$$\alpha'_r(a,q) = a^{-r}q^{-r^2}\alpha_r(a^3,q^3),$$
$$\beta'_n(a,q) = \sum_{k=0}^{n} \frac{(aq^{2n+1};q^{-1})_{3k}(a^3q^3;q^3)_{2(n-k)}}{(aq;q)_{2n}(q^3;q^3)_k}$$
$$\times (-1)^k q^{3\binom{k}{2}-n^2} a^{-n}\beta_{n-k}(a^3,q^3).$$

3. 2×2 matrices. In this section we realize the operations of §2 on Bailey pairs as 2×2 matrices. These are the generators of the Bailey-Rogers-Ramanujan group in Definition 2.

We are concerned with iterating the transformations (S), (D), (E), and (T) of §2. Our initial choice is always the unit Bailey pair (UBP) with $a = 1$. We shall also assume that each iteration yields a Bailey pair with parameters $(1, q)$. We have

$$\alpha'_r(1,q) = q^{Ar^2}\alpha_r^{(0)}(1,q^B).$$

for some rational numbers A and B. Thus we need only keep track A and B while carrying out the iteration.

We encode a transformation
$$\alpha'_r(1,q) = q^{Ar^2}\alpha_r(1,q^B),$$
where $(\alpha'_n(1,q), \beta'_n(1,q))$ has parameters $(1,q)$, by the 2×2 matrix
$$\begin{bmatrix} 1 & A \\ 0 & B \end{bmatrix}.$$

We next check that matrix multiplication on the right does correspond to the composition of transformations. If
$$\alpha''_r(1,q) = q^{Cr^2}\alpha'_r(1,q^D) = q^{Cr^2+ADr^2}\alpha_r(1,q^{BD}),$$
the corresponding matrix is
$$\begin{bmatrix} 1 & C+AD \\ 0 & BD \end{bmatrix} = \begin{bmatrix} 1 & A \\ 0 & B \end{bmatrix}\begin{bmatrix} 1 & C \\ 0 & D \end{bmatrix}.$$

With this notation we see that
$$(S1) = \begin{bmatrix} 1 & 1 \\ 0 & 1 \end{bmatrix}, \quad (S2) = \begin{bmatrix} 1 & 1/2 \\ 0 & 1 \end{bmatrix}, \quad (D1) = \begin{bmatrix} 1 & 0 \\ 0 & 2 \end{bmatrix}, \quad (D2) = \begin{bmatrix} 1 & -1 \\ 0 & 2 \end{bmatrix},$$
$$(D3) = \begin{bmatrix} 1 & -1/2 \\ 0 & 2 \end{bmatrix}, \quad (E1) = \begin{bmatrix} 1 & 0 \\ 0 & 1/2 \end{bmatrix}, \quad (E2) = \begin{bmatrix} 1 & 1/2 \\ 0 & 1/2 \end{bmatrix},$$
$$(E3) = \begin{bmatrix} 1 & 1/4 \\ 0 & 1/2 \end{bmatrix}, \quad (T1) = \begin{bmatrix} 1 & 1/3 \\ 0 & 1/3 \end{bmatrix}, \quad (T2) = \begin{bmatrix} 1 & -1 \\ 0 & 3 \end{bmatrix}.$$

DEFINITION 2. *The Bailey-Rogers-Ramanujan group is the subgroup of 2×2 upper triangular rational matrices generated by*
$$\{(S1), (S2), (D1), (D2), (D3), (E1), (E2), (E3), (T1), (T2)\}.$$

Even though (S1), (E1), (E2), (E3) and (T2) are unnecessary as generators, it will be convenient in the following sections to keep their designation as generators. There are other relations amongst the generators, for example
$$(S1)(D1) = (D1)(S1), \quad (E2)(D3) = (S2).$$

4. The Rogers-Ramanujan type identities. Let
$$g = w_1 w_2 \cdots w_{k+1}$$
be an element of the Bailey-Rogers-Ramanujan group, with each w_i a generator from Definition 2. Suppose that the corresponding Bailey pairs are
$$(\alpha_n^{(0)}, \beta_n^{(0)}), (\alpha_n^{(1)}, \beta_n^{(1)}), \cdots, (\alpha_n^{(k+1)}, \beta_n^{(k+1)}).$$

Let the corresponding relations for w_{i+1} between $\beta_n^{(i)}$ and $\beta_n^{(i+1)}$ be expressed as
$$\beta_n^{(i+1)} = \sum_{s_i=0}^{n} M_{n,s_i}^{(i)} \beta_{s_i}^{(i)}, \quad 0 \leq i \leq k.$$

We say that $M^{(i)}$ is the infinite lower triangular matrix corresponding to w_{i+1}. For example, if $w_1 = (S1)$, then

$$M^{(0)}_{nk} = q^{k^2}/(q;q)_{n-k}.$$

THEOREM 1. *If $w = w_1 w_2 \cdots w_{k+1}$ is an element of the Bailey-Rogers- Ramanujan group, the corresponding finite Rogers-Ramanujan identity is given by*

$$\beta_n^{(k+1)} = \sum_{n \geq s_k \geq \cdots \geq s_1 \geq 0} M^{(k)}_{n,s_k} \cdots M^{(1)}_{s_2,s_1} M^{(0)}_{s_1,0} =$$

$$= \sum_{r=0}^{n} \frac{\alpha_r^{(k+1)}}{(q;q)_{n+r}(q;q)_{n-r}},$$

where $M^{(i)}$ is the infinite lower triangular matrix corresponding to w_{i+1}.

PROOF. The first equality is the expression for $\beta_n^{(k+1)}$ as a $(k+1)$-fold sum over $\beta_s^{(0)}$. This sum reduces to a k-fold sum because of the (UBP) condition. The right side expresses the fact that $(\alpha_n^{(k+1)}, \beta_n^{(k+1)})$ is a Bailey pair with parameters $(1, q)$. □

Next we consider the $n \to \infty$ limit of Theorem 1. Let

$$w = w_1 w_2 \cdots w_{k+1} = \begin{bmatrix} 1 & A \\ 0 & B \end{bmatrix}.$$

The right-side of Theorem 1, using the (UBP) and the Jacobi triple product identity (1.1), approaches

(4.1) $$\frac{(q^{2A+B}, q^{A+B}, q^A; q^{2A+B})_\infty}{(q;q)_\infty^2}.$$

The left-side will have a termwise limit if

$$\lim_{n \to \infty} M^{(k)}_{n,s_k} = M^{(k)}_{\infty,s_k}$$

exists. This is the case if $w_{k+1} = (S1), (S2), (D3), (E2), (E3),$ or $(T1)$. If $w_{k+1} = (D1)$ or $(E1)$, we see that

$$\lim_{n \to \infty} M^{(k)}_{n,n-s_k}$$

exists, so that the limit may be taken as long as

$$\lim_{n \to \infty} M^{(k-1)}_{n-s_k, s_{k-1}}$$

exists. It will exist if $w_k = (S1), (S2), (D3), (E2), (E3),$ or $(T1)$, otherwise we may need to replace s_{k-1} by $n - s_{k-1}$ and continue.

We also see that $w_k = (D2)$ or $(T2)$ will lead to interesting identities, even though the termwise limit does not exist. Each term will be a Laurent series, yet the sum has a limit with no negative powers of q as $n \to \infty$. Several such examples are given in §5.

5. Single sum identities. In this section we record which single sum Rogers-Ramanujan type identities appear from words of length 2 in Theorem 1. In each case we have taken the $n \to \infty$ limit in Theorem 1 and multiplied by the infinite product occurring in $M_{\infty,s_k}^{(k)}$. Each of these identities has many multisum generalizations by considering longer words, a few are given in §6.

Rogers-Ramanujan identity (Slater (18)):

$$(S1)(S1) = \begin{bmatrix} 1 & 2 \\ 0 & 1 \end{bmatrix},$$

$$\sum_{s=0}^{\infty} \frac{q^{s^2}}{(q;q)_s} = \frac{(q^5, q^2, q^3; q^5)_\infty}{(q;q)_\infty}.$$

Bailey's mod 9 identity (Slater (42)):

$$(T1)(S1) = \begin{bmatrix} 1 & 4/3 \\ 0 & 1/3 \end{bmatrix},$$

$$\sum_{s=0}^{\infty} \frac{q^{3s^2}(q;q)_{3s}}{(q^3;q^3)_{2s}(q^3;q^3)_s} = \frac{(q^9, q^4, q^5; q^9)_\infty}{(q^3;q^3)_\infty}.$$

Rogers' mod 7 identity (Slater (33)):

$$(E2)(S1) = \begin{bmatrix} 1 & 3/2 \\ 0 & 1/2 \end{bmatrix},$$

$$\sum_{s=0}^{\infty} \frac{q^{2s^2}}{(-q;q)_{2s}(q^2;q^2)_s} = \frac{(q^7, q^4, q^3; q^7)_\infty}{(q^2;q^2)_\infty}.$$

Rogers' mod 5 identity, (Slater (19)):

$$(E1)(S1) = \begin{bmatrix} 1 & 1 \\ 0 & 1/2 \end{bmatrix},$$

$$\sum_{s=0}^{\infty} \frac{(-1)^s q^{3s^2}}{(-q;q)_{2s}(q^2;q^2)_s} = \frac{(q^5, q^2, q^3; q^5)_\infty}{(q^2;q^2)_\infty}.$$

Rogers' mod 5 identity (Slater (20)):

$$(E2)(S2) = \begin{bmatrix} 1 & 1 \\ 0 & 1/2 \end{bmatrix},$$

$$\sum_{s=0}^{\infty} \frac{q^{s^2}}{(q^4;q^4)_s} = \frac{(q^5, q^2, q^3; q^5)_\infty (-q;q^2)_\infty}{(q^2;q^2)_\infty}.$$

Slater's identity (36):

$$(S1)(S2) = \begin{bmatrix} 1 & 3/2 \\ 0 & 1 \end{bmatrix},$$

$$\sum_{s=0}^{\infty} \frac{(-q;q^2)_s}{(q^2;q^2)_s} q^{s^2} = \frac{1}{(q^1, q^4, q^7; q^8)_\infty}.$$

Slater's identity (39):

$$(S2)(S1) = \begin{bmatrix} 1 & 3/2 \\ 0 & 1 \end{bmatrix},$$

$$\sum_{s=0}^{\infty} \frac{q^{2s^2}}{(q^2;q^2)_s(-q;q^2)_s} = \frac{(q^8,q^3,q^5;q^8)_\infty}{(q^2;q^2)_\infty}.$$

Slater's identity (53):
$$(E3)(S1) = \begin{bmatrix} 1 & 5/4 \\ 0 & 1/2 \end{bmatrix},$$
$$\sum_{s=0}^{\infty} \frac{q^{4s^2}(q;q^2)_{2s}}{(q^4;q^4)_{2s}} = \frac{(q^{12},q^5,q^7;q^{12})_\infty}{(q^4;q^4)_\infty}.$$

mod 4 identity:
$$(D2)(S2) = \begin{bmatrix} 1 & -1/2 \\ 0 & 2 \end{bmatrix},$$
$$\sum_{s=0}^{\infty} \frac{(-q;q^2)_s}{(q^4;q^4)_s}(-1)^{s+1}q^{(s-1)^2} = \frac{(q;q)_\infty}{(q^4;q^4)_\infty}.$$

mod 6 identity:
$$(D3)(S1) = \begin{bmatrix} 1 & 1/2 \\ 0 & 2 \end{bmatrix},$$
$$\sum_{s=0}^{\infty} \frac{(q^{-1};q^2)_s(q^3;q^2)_s}{(q^2;q^2)_{2s}}q^{2s^2} = \frac{(q^1,q^5;q^6)_\infty}{(q^2,q^4;q^6)_\infty}.$$

mod 8 identity:
$$(E3)(S2) = \begin{bmatrix} 1 & 3/4 \\ 0 & 1/2 \end{bmatrix},$$
$$\sum_{s=0}^{\infty} \frac{(q;q^2)_{2s}(-q^2;q^4)_s}{(q^4;q^4)_{2s}}q^{2s^2} = \frac{(q^3,q^5;q^8)_\infty}{(q^2,q^6;q^8)_\infty}.$$

mod 12 identity:
$$(D1)(S2) = \begin{bmatrix} 1 & 1/2 \\ 0 & 2 \end{bmatrix},$$
$$\sum_{s=0}^{\infty} \frac{(-q;q^2)_s}{(q^4;q^4)_s}q^{s^2+2s} = \frac{(q^6;q^{12})_\infty}{(q^3,q^4,q^8,q^9;q^{12})_\infty}.$$

mod 12 identity:
$$(T1)(S2) = \begin{bmatrix} 1 & 5/6 \\ 0 & 1/3 \end{bmatrix},$$
$$\sum_{s=0}^{\infty} \frac{(-q^3;q^6)_s(q^2;q^2)_{3s}}{(q^6;q^6)_{2s}(q^6;q^6)_s}q^{3s^2} = \frac{(q^{12},q^5,q^7;q^{12})_\infty(-q^3;q^6)_\infty}{(q^6;q^6)_\infty}.$$

mod 4 identity:
$$(D1)(T2) = \begin{bmatrix} 1 & -1 \\ 0 & 6 \end{bmatrix},$$
$$\lim_{n\to\infty}\sum_{s=0}^{n} \frac{(q^{2n+1};q^{-1})_{3s}(q^3;q^6)_{n-s}}{(q^3;q^3)_s}(-1)^{s+1}q^{3\binom{s}{2}-n^2+3(n-s)+1} = \frac{1}{(q^2;q^4)_\infty}.$$

mod 2 identity:
$$(E3)(T2) = \begin{bmatrix} 1 & -1/4 \\ 0 & 3/2 \end{bmatrix},$$
$$\lim_{n\to\infty}\sum_{s=0}^{n} \frac{(q^{8n+4};q^{-4})_{3s}(q^3;q^6)_{2(n-s)}}{(q^{12};q^{12})_s}(-1)^{s+1}q^{12\binom{s}{2}-4n^2+1} = (q;q^2)_\infty.$$

mod 10 identity:
$$(S1)(D3) = \begin{bmatrix} 1 & 3/2 \\ 0 & 2 \end{bmatrix},$$

$$\lim_{n\to\infty} \sum_{s=0}^{n} \frac{(-q^2;q^2)_{2s}(q^{-1-2s},q^{3+2s};q^2)_{n-s}}{(q^{4s+2};q^4)_{n-s}(q^4;q^4)_s(q^4;q^4)_{n-s}} q^{-s^2} = \frac{(q^{10},q^7,q^3;q^{10})_\infty}{(q^2;q^2)_\infty^2}.$$

mod 7 identity:
$$(S1)(T2) = \begin{bmatrix} 1 & 2 \\ 0 & 3 \end{bmatrix},$$

$$\lim_{n\to\infty} \sum_{s=0}^{n} \frac{(q^{2n+1};q^{-1})_{3s}(q^3;q^3)_{2(n-s)}}{(q^3;q^3)_s(q^3;q^3)_{n-s}} (-1)^s q^{3\binom{s}{2}-n^2} = \frac{(q^7,q^2,q^5;q^7)_\infty}{(q;q)_\infty}.$$

mod 8 identity:
$$(S2)(T2) = \begin{bmatrix} 1 & 1/2 \\ 0 & 3 \end{bmatrix},$$

$$\lim_{n\to\infty} \sum_{s=0}^{n} \frac{(q^{4n+2};q^{-2})_{3s}(q^6;q^6)_{2(n-s)}(-1)^s q^{6\binom{s}{2}-2n^2}}{(q^6;q^6)_s(-q^3;q^6)_{n-s}(q^6;q^6)_{n-s}} = \frac{(q^8,q^7,q^1;q^8)_\infty}{(q^2;q^2)_\infty}.$$

mod 5 identity:
$$(E2)(T2) = \begin{bmatrix} 1 & 1/2 \\ 0 & 3/2 \end{bmatrix},$$

$$\lim_{n\to\infty} \sum_{s=0}^{n} \frac{(q^{4n+2};q^{-2})_{3s}(q^6;q^6)_{2(n-s)}(-1)^s q^{6\binom{s}{2}-2n^2}}{(q^6;q^6)_s(-q^3;q^3)_{2(n-s)}(q^6;q^6)_{n-s}} = \frac{(q^5,q^4,q^1;q^5)_\infty}{(q^2;q^2)_\infty}.$$

mod 2 identity:
$$(D3)(T2) = \begin{bmatrix} 1 & -5/2 \\ 0 & 6 \end{bmatrix},$$

$$\lim_{n\to\infty} \sum_{s=0}^{n} \frac{(q^{4n+2};q^{-2})_{3s}(q^{-3},q^9;q^6)_{n-s}}{(q^6;q^6)_s} (-1)^{s+1} q^{6\binom{s}{2}-2n^2+9} = (q;q^2)_\infty^2.$$

mod 2 identity:
$$(D3)(D3) = \begin{bmatrix} 1 & -3/2 \\ 0 & 4 \end{bmatrix},$$

$$\lim_{n\to\infty} \sum_{s=0}^{n} \frac{(-q^2;q^2)_{2s}(q^{-1-2s},q^{2s+3};q^2)_{n-s}(q^{-2},q^6;q^4)_s}{(q^{4s+2};q^4)_{n-s}(q^4;q^4)_{n-s}(q^4;q^4)_{2s}} q^{-s^2+4} = \frac{(q;q^2)_\infty^2}{(q^2;q^2)_\infty}.$$

mod 2 identity:
$$(D1)(D2) = \begin{bmatrix} 1 & -1 \\ 0 & 4 \end{bmatrix},$$

$$\lim_{n\to\infty} \sum_{s=0}^{n} \frac{(-q;q)_{2s}(-1)^{n-s+1} q^{s^2+3s-2sn-n+1}}{(q^2;q^2)_{n-s}(q^4;q^4)_s} = \frac{1}{(q^2;q^2)_\infty}.$$

mod 2 identity:
$$(S1)(D2) = \begin{bmatrix} 1 & 1 \\ 0 & 2 \end{bmatrix},$$

$$\lim_{n\to\infty} \sum_{s=0}^{n} \frac{(-q;q)_{2s}(-1)^{n-s} q^{s^2+s-2sn-n}}{(q^2;q^2)_{n-s}(q^2;q^2)_s} = \frac{(-q^2;q^2)_\infty}{(q;q)_\infty}.$$

mod 6 identity:

$$(D1)(D3) = \begin{bmatrix} 1 & -1/2 \\ 0 & 4 \end{bmatrix},$$

$$\lim_{n\to\infty} \sum_{s=0}^{n} \frac{(-q^2;q^2)_{2s}(q^{-1-2s},q^{2s+3};q^2)_{n-s}q^{-s^2+4s+1}}{(q^{4s+2};q^4)_{n-s}(q^4;q^4)_{n-s}(q^8;q^8)_s} = \frac{-(q^6,q^1,q^5;q^6)_\infty}{(q^2;q^2)_\infty^2}.$$

mod 6 identity:

$$(S2)(D3) = \begin{bmatrix} 1 & 1/2 \\ 0 & 2 \end{bmatrix},$$

$$\lim_{n\to\infty} \sum_{s=0}^{n} \frac{(-q^2;q^2)_{2s}(q^{-1-2s},q^{2s+3};q^2)_{n-s}q^{-s^2}}{(q^{4s+2};q^4)_{n-s}(q^4;q^4)_{n-s}(q^4;q^4)_s(-q^2;q^4)_s} = \frac{(q^6,q^1,q^5;q^6)_\infty}{(q^2;q^2)_\infty^2}.$$

mod 6 identity:

$$(T1)(D3) = \begin{bmatrix} 1 & 1/6 \\ 0 & 2/3 \end{bmatrix},$$

$$\lim_{n\to\infty} \sum_{s=0}^{n} \frac{(q^{-3-6s},q^{6s+9};q^6)_{n-s}(q^4,q^8;q^{12})_s q^{-3s^2}}{(q^{12s+6};q^{12})_{n-s}(q^{12};q^{12})_{n-s}(q^6;q^6)_{2s}} = \frac{(q^1,q^5;q^6)_\infty}{(q^6;q^6)_\infty}.$$

6. Multisum identities. In this section we give several specific examples of multisum identities which correspond to group elements via Theorem 1.

Andrews-Gordon identities, ($k \geq 1$): Since

$$(S1)^{k+1} = \begin{bmatrix} 1 & k+1 \\ 0 & 1 \end{bmatrix},$$

the result from (4.1) is

$$\sum_{s_1 \geq s_2 \geq \cdots \geq s_k \geq 0} \frac{q^{s_1^2+\cdots+s_k^2}}{(q;q)_{s_1-s_2}(q;q)_{s_2-s_3}\cdots(q;q)_{s_k}} = \frac{(q^{2k+3},q^{k+2},q^{k+1};q^{2k+3})_\infty}{(q;q)_\infty}.$$

Bressoud identities, ($k \geq 2$): Choose

$$(D1)(S1)^{k-1} = \begin{bmatrix} 1 & k-1 \\ 0 & 2 \end{bmatrix},$$

the result from (4.1) is

$$\sum_{s_1 \geq s_2 \geq \cdots \geq s_{k-1} \geq 0} \frac{q^{s_1^2+\cdots+s_{k-1}^2+s_{k-1}}}{(q;q)_{s_1-s_2}(q;q)_{s_2-s_3}\cdots(q;q)_{s_{k-2}-s_{k-1}}(q^2;q^2)_{s_{k-1}}}$$
$$= \frac{(q^{2k},q^{k-1},q^{k+1};q^{2k})_\infty}{(q;q)_\infty}.$$

mod $2^k + 2i$ identities, ($k+i \geq 2$): Choose

$$(D1)^k(S1)^i = \begin{bmatrix} 1 & i \\ 0 & 2^k \end{bmatrix},$$

$$\sum_{s_{k+i-1}\geq\cdots\geq s_1\geq s_0=0} \frac{q^{s_{k+i-1}^2+\cdots+s_k^2}}{\prod_{j=k}^{k+i-2}(q;q)_{s_{j+1}-s_j}} \prod_{j=0}^{k-1} \frac{q^{2^j(s_{k-j}-s_{k-j-1})}(-q^{2^j};q^{2^j})_{2s_{k-j-1}}}{(q^{2^{j+1}};q^{2^{j+1}})_{s_{k-j}-s_{k-j-1}}}$$

$$= \frac{(q^{2^k+2i},q^{2^k+i},q^i;q^{2^k+2i})_\infty}{(q;q)_\infty}.$$

If $k = 0$ these are the Andrews-Gordon identities, while for $k = 1$ they are the Bressoud identities. The $i = 2$ case was previously given in [**11, Corollary 4.4**].

mod $i2^{k+1} + 1$ identities, $(k + i \geq 2)$: Choose

$$(E1)^k(S1)^i = \begin{bmatrix} 1 & i \\ 0 & 2^{-k} \end{bmatrix},$$

$$\sum_{s_{k+i-1}\geq\cdots\geq s_1\geq s_0=0} \frac{q^{2^k(s_{k+i-1}^2+\cdots+s_k^2)}}{\prod_{j=k}^{k+i-2}(q^{2^k};q^{2^k})_{s_{j+1}-s_j}} \prod_{j=0}^{k-1} \frac{(-1)^{s_{j+1}-s_j}q^{2^j(s_{j+1}-s_j)^2}}{(q^{2^{j+1}};q^{2^{j+1}})_{s_{j+1}-s_j}(-q^{2^j};q^{2^j})_{2s_{j+1}}}$$

$$= \frac{(q^{i2^{k+1}+1},q^{i2^k+1},q^{i2^k};q^{i2^{k+1}+1})_\infty}{(q^{2^k};q^{2^k})_\infty}.$$

If $k = 0$ these are again the Andrews-Gordon identities, for $k = i = 1$ they are Rogers' identities for modulus 5.

mod $(i + 1)2^{k+1} - 1$ identities, $(k + i \geq 2)$: Choose

$$(E2)^k(S1)^i = \begin{bmatrix} 1 & 1+i-2^{-k} \\ 0 & 2^{-k} \end{bmatrix},$$

$$\sum_{s_{k+i-1}\geq\cdots\geq s_1\geq s_0=0} \frac{q^{2^k(s_{k+i-1}^2+\cdots+s_k^2)}}{\prod_{j=k}^{k+i-2}(q^{2^k};q^{2^k})_{s_{j+1}-s_j}} \prod_{j=0}^{k-1} \frac{q^{2^j s_j^2}}{(q^{2^{j+1}};q^{2^{j+1}})_{s_{j+1}-s_j}(-q^{2^j};q^{2^j})_{2s_{j+1}}}$$

$$= \frac{(q^{(i+1)2^{k+1}-1},q^{(i+1)2^k},q^{(i+1)2^k-1};q^{(i+1)2^{k+1}-1})_\infty}{(q^{2^k};q^{2^k})_\infty}.$$

If $k = 0$ these are again the Andrews-Gordon identities, for $k = i = 1$ they are Rogers' identities for modulus 7.

mod $(2i + 1)3^k$ identities, $(k + i \geq 2)$: Choose

$$(T1)^k(S1)^i = \begin{bmatrix} 1 & i+(1-3^{-i})/2 \\ 0 & 2^k 3^{-i} \end{bmatrix},$$

$$\sum_{s_{k+i-1}\geq\cdots\geq s_1\geq s_0=0} \frac{q^{3^k(s_{k+i-1}^2+\cdots+s_k^2)}}{\prod_{j=k}^{k+i-2}(q^{3^k};q^{3^k})_{s_{j+1}-s_j}}$$

$$\times \prod_{j=0}^{k-1} \frac{(q^{3^j};q^{3^j})_{3s_{j+1}-s_j} q^{3^j s_j^2}}{(q^{3^{j+1}};q^{3^{j+1}})_{2s_{j+1}}(q^{3^{j+1}};q^{3^{j+1}})_{s_{j+1}-s_j}}$$

$$= \frac{(q^{(2i+1)3^k},q^{(2i+1)3^k/2-1/2},q^{(2i+1)3^k/2+1/2};q^{(2i+1)3^k})_\infty}{(q^{3^k};q^{3^k})_\infty}.$$

If $k = i = 1$ this is Bailey's mod 9 identity.

mod $3^i + 2^k - 1$ identities ($i \geq 1, k \geq 0$): Choose

$$(D1)^k(T1)^i = \begin{bmatrix} 1 & (1-3^{-i})/2 \\ 0 & 2^k 3^{-i} \end{bmatrix},$$

$$\sum_{s_{k+i-1} \geq \cdots \geq s_1 \geq s_0 = 0} q^{3^{i-1} s_{k+i-1}^2} \prod_{j=k}^{k+i-2} \frac{q^{3^{j-k} s_j^2}(q^{3^{j-k}}; q^{3^{j-k}})_{3s_{j+1}-s_j}}{(q^{3^{j-k+1}}; q^{3^{j-k+1}})_{2s_{j+1}} (q^{3^{j-k+1}}; q^{3^{j-k+1}})_{s_{j+1}-s_j}}$$

$$\times \prod_{j=0}^{k-1} \frac{q^{2^j(s_{k-j}-s_{k-j-1})}(-q^{2^j}; q^{2^j})_{2s_{k-j-1}}}{(q^{2^{j+1}}; q^{2^{j+1}})_{s_{k-j}-s_{k-j-1}}}$$

$$= \frac{(q^{3^i+2^k-1}, q^{2^k+(3^i-1)/2}, q^{(3^i-1)/2}; q^{3^i+2^k-1})_\infty}{(q^{3^{i-1}}; q^{3^{i-1}})_\infty}.$$

If $k = i = 1$, this is the special case of the q-binomial theorem, which says that partitions of N into odd parts are equinumerous with partitions of N into distinct parts.

We next give examples of mod 11 identities which are double sums. Any word $w = w_1 w_2 w_3$

$$w = \begin{bmatrix} 1 & A \\ 0 & B \end{bmatrix},$$

with $(2A + B)/B = 11$ will give such an identity. A Mathematica run finds all 16 such words. These 16 words give 6 distinct identities. We list these 6 identities along with a representative word.

$$(S1)(T1)(S1) = \begin{bmatrix} 1 & 5/3 \\ 0 & 1/3 \end{bmatrix},$$

(6.1) $$\sum_{s_1, s_2 \geq 0} \frac{q^{3s_1^2 + s_2^2}(q; q)_{3s_1 - s_2}}{(q^3; q^3)_{2s_1}(q^3; q^3)_{s_1 - s_2}(q; q)_{s_2}} = \frac{(q^5, q^6, q^{11}; q^{11})_\infty}{(q^3; q^3)_\infty},$$

$$(E2)(E1)(S1) = \begin{bmatrix} 1 & 5/4 \\ 0 & 1/4 \end{bmatrix},$$

(6.2) $$\sum_{s_1, s_2 \geq 0} \frac{q^{4s_1^2 + 2(s_1 - s_2)^2}(-1)^{s_1 - s_2}}{(-q^2; q^2)_{2s_1}(q^4; q^4)_{s_1 - s_2}(-q; q)_{2s_2}(q^2; q^2)_{s_2}} = \frac{(q^5, q^6, q^{11}; q^{11})_\infty}{(q^4; q^4)_\infty},$$

$$(E2)(E2)(S2) = \begin{bmatrix} 1 & 5/4 \\ 0 & 1/4 \end{bmatrix},$$

(6.3) $$\sum_{s_1, s_2 \geq 0} \frac{q^{2s_2^2 + 2s_1^2}(-q^2; q^4)_{s_1}}{(-q^2; q^2)_{2s_1}(q^4; q^4)_{s_1 - s_2}(-q; q)_{2s_2}(q^2; q^2)_{s_2}} = \frac{(q^5, q^6, q^{11}; q^{11})_\infty}{(q^8; q^8)_\infty (q^2; q^4)_\infty},$$

$$(E2)(S1)(S1) = \begin{bmatrix} 1 & 5/2 \\ 0 & 1/2 \end{bmatrix},$$

(6.4) $$\sum_{s_1, s_2 \geq 0} \frac{q^{2s_2^2 + 2s_1^2}}{(q^2; q^2)_{s_1 - s_2}(-q; q)_{2s_2}(q^2; q^2)_{s_2}} = \frac{(q^5, q^6, q^{11}; q^{11})_\infty}{(q^2; q^2)_\infty},$$

$$(E1)(E3)(S1) = \begin{bmatrix} 1 & 5/4 \\ 0 & 1/4 \end{bmatrix},$$

(6.5)
$$\sum_{s_1,s_2 \geq 0} \frac{q^{2s_2^2+4s_1^2}(-1)^{s_2}(-q;q^2)_{s_2}(q;q^2)_{2s_1-s_2}}{(-q^2;q^2)_{2s_1}(q^2;q^4)_{s_1}(q^4;q^4)_{s_1-s_2}(-q;q)_{2s_2}(q^2;q^2)_{s_2}} = \frac{(q^5,q^6,q^{11};q^{11})_\infty}{(q^4;q^4)_\infty},$$

$$(E1)(T1)(S2) = \begin{bmatrix} 1 & 5/6 \\ 0 & 1/6 \end{bmatrix},$$

(6.6)
$$\sum_{s_1,s_2 \geq 0} \frac{q^{3s_1^2+3s_2^2}(-1)^{s_2}(-q^3;q^6)_{s_1}(q^2;q^2)_{3s_1-s_2}}{(q^6;q^6)_{2s_1}(q^6;q^6)_{s_1-s_2}(-q;q)_{2s_2}(q^2;q^2)_{s_2}} = \frac{(q^5,q^6,q^{11};q^{11})_\infty}{(q^{12};q^{12})_\infty (q^3;q^6)_\infty}.$$

Note that the 2nd, 3rd, and 5th identities are distinct even though they correspond to the same group element. Perhaps the easiest version of this is $(S1)(S2) = (S2)(S1)$, see §5.

These six identities, particularly (6.2) and (6.5), are reminiscent of, but not the same as, Andrews' mod 11 identities in [**4**], one of which is

$$\sum_{n,j=0}^\infty \frac{(q;q)_{4n+2j}(-1)^j q^{4n^2+12nj+8j^2+j}}{(q^4;q^4)_n (q^2;q^2)_j (q^4;q^4)_{2n+2j}} = \frac{(q^5,q^6,q^{11};q^{11})_\infty}{(q^4;q^4)_\infty}.$$

Perhaps the most exotic double sum which appears corresponds to

$$(T1)(T1)(S1) = \begin{bmatrix} 1 & 13/9 \\ 0 & 1/9 \end{bmatrix},$$

$$\sum_{s_1,s_2 \geq 0} \frac{q^{9s_2^2+3s_1^2}(q^3;q^3)_{3s_2-s_1}(q;q)_{3s_1}}{(q^9;q^9)_{2s_2}(q^9;q^9)_{s_2-s_1}(q^3;q^3)_{2s_1}(q^3;q^3)_{s_1}} = \frac{(q^{13},q^{14},q^{27};q^{27})_\infty}{(q^9;q^9)_\infty}.$$

There are 348 words $w = w_1 w_2 w_3 w_4$ of length 4 with corresponding integer values of $11B/(2A+B)$, these could lead to 202 possible triple sum identities modulo multiples of 11.

7. Borwein polynomials. In this section we use the group element

$$(T1) = \begin{bmatrix} 1 & 1/3 \\ 0 & 1/3 \end{bmatrix}$$

to find alternative forms of the Borwein polynomials.

These polynomials were defined by Andrews [**6**] as

$$A_n(q) = \sum_{k=-\infty}^\infty \begin{bmatrix} 2n \\ n-3k \end{bmatrix}_q (-1)^k q^{k(9k-1)/2}$$

$$B_n(q) = \sum_{k=-\infty}^\infty \begin{bmatrix} 2n \\ n-3k-1 \end{bmatrix}_q (-1)^k q^{k(9k+5)/2}$$

$$C_n(q) = \sum_{k=-\infty}^\infty \begin{bmatrix} 2n \\ n-3k+1 \end{bmatrix}_q (-1)^k q^{k(9k-7)/2}.$$

mod $3^i + 2^k - 1$ identities ($i \geq 1, k \geq 0$): Choose

$$(D1)^k(T1)^i = \begin{bmatrix} 1 & (1-3^{-i})/2 \\ 0 & 2^k 3^{-i} \end{bmatrix},$$

$$\sum_{s_{k+i-1} \geq \cdots \geq s_1 \geq s_0 = 0} q^{3^{i-1}s_{k+i-1}^2} \prod_{j=k}^{k+i-2} \frac{q^{3^{j-k}s_j^2}(q^{3^{j-k}};q^{3^{j-k}})_{3s_{j+1}-s_j}}{(q^{3^{j-k+1}};q^{3^{j-k+1}})_{2s_{j+1}}(q^{3^{j-k+1}};q^{3^{j-k+1}})_{s_{j+1}-s_j}}$$

$$\times \prod_{j=0}^{k-1} \frac{q^{2^j(s_{k-j}-s_{k-j-1})}(-q^{2^j};q^{2^j})_{2s_{k-j-1}}}{(q^{2^{j+1}};q^{2^{j+1}})_{s_{k-j}-s_{k-j-1}}}$$

$$= \frac{(q^{3^i+2^k-1}, q^{2^k+(3^i-1)/2}, q^{(3^i-1)/2}; q^{3^i+2^k-1})_\infty}{(q^{3^{i-1}};q^{3^{i-1}})_\infty}.$$

If $k = i = 1$, this is the special case of the q-binomial theorem, which says that partitions of N into odd parts are equinumerous with partitions of N into distinct parts.

We next give examples of mod 11 identities which are double sums. Any word $w = w_1 w_2 w_3$

$$w = \begin{bmatrix} 1 & A \\ 0 & B \end{bmatrix},$$

with $(2A + B)/B = 11$ will give such an identity. A Mathematica run finds all 16 such words. These 16 words give 6 distinct identities. We list these 6 identities along with a representative word.

$$(S1)(T1)(S1) = \begin{bmatrix} 1 & 5/3 \\ 0 & 1/3 \end{bmatrix},$$

(6.1) $$\sum_{s_1,s_2 \geq 0} \frac{q^{3s_1^2+s_2^2}(q;q)_{3s_1-s_2}}{(q^3;q^3)_{2s_1}(q^3;q^3)_{s_1-s_2}(q;q)_{s_2}} = \frac{(q^5,q^6,q^{11};q^{11})_\infty}{(q^3;q^3)_\infty},$$

$$(E2)(E1)(S1) = \begin{bmatrix} 1 & 5/4 \\ 0 & 1/4 \end{bmatrix},$$

(6.2) $$\sum_{s_1,s_2 \geq 0} \frac{q^{4s_1^2+2(s_1-s_2)^2}(-1)^{s_1-s_2}}{(-q^2;q^2)_{2s_1}(q^4;q^4)_{s_1-s_2}(-q;q)_{2s_2}(q^2;q^2)_{s_2}} = \frac{(q^5,q^6,q^{11};q^{11})_\infty}{(q^4;q^4)_\infty},$$

$$(E2)(E2)(S2) = \begin{bmatrix} 1 & 5/4 \\ 0 & 1/4 \end{bmatrix},$$

(6.3) $$\sum_{s_1,s_2 \geq 0} \frac{q^{2s_2^2+2s_1^2}(-q^2;q^4)_{s_1}}{(-q^2;q^2)_{2s_1}(q^4;q^4)_{s_1-s_2}(-q;q)_{2s_2}(q^2;q^2)_{s_2}} = \frac{(q^5,q^6,q^{11};q^{11})_\infty}{(q^8;q^8)_\infty(q^2;q^4)_\infty},$$

$$(E2)(S1)(S1) = \begin{bmatrix} 1 & 5/2 \\ 0 & 1/2 \end{bmatrix},$$

(6.4) $$\sum_{s_1,s_2 \geq 0} \frac{q^{2s_2^2+2s_1^2}}{(q^2;q^2)_{s_1-s_2}(-q;q)_{2s_2}(q^2;q^2)_{s_2}} = \frac{(q^5,q^6,q^{11};q^{11})_\infty}{(q^2;q^2)_\infty},$$

$$(E1)(E3)(S1) = \begin{bmatrix} 1 & 5/4 \\ 0 & 1/4 \end{bmatrix},$$

(6.5)
$$\sum_{s_1,s_2\geq 0} \frac{q^{2s_2^2+4s_1^2}(-1)^{s_2}(-q;q^2)_{s_2}(q;q^2)_{2s_1-s_2}}{(-q^2;q^2)_{2s_1}(q^2;q^4)_{s_1}(q^4;q^4)_{s_1-s_2}(-q;q)_{2s_2}(q^2;q^2)_{s_2}} = \frac{(q^5,q^6,q^{11};q^{11})_\infty}{(q^4;q^4)_\infty},$$

$$(E1)(T1)(S2) = \begin{bmatrix} 1 & 5/6 \\ 0 & 1/6 \end{bmatrix},$$

(6.6)
$$\sum_{s_1,s_2\geq 0} \frac{q^{3s_1^2+3s_2^2}(-1)^{s_2}(-q^3;q^6)_{s_1}(q^2;q^2)_{3s_1-s_2}}{(q^6;q^6)_{2s_1}(q^6;q^6)_{s_1-s_2}(-q;q)_{2s_2}(q^2;q^2)_{s_2}} = \frac{(q^5,q^6,q^{11};q^{11})_\infty}{(q^{12};q^{12})_\infty(q^3;q^6)_\infty}.$$

Note that the 2nd, 3rd, and 5th identities are distinct even though they correspond to the same group element. Perhaps the easiest version of this is $(S1)(S2) = (S2)(S1)$, see §5.

These six identities, particularly (6.2) and (6.5), are reminiscent of, but not the same as, Andrews' mod 11 identities in [**4**], one of which is

$$\sum_{n,j=0}^{\infty} \frac{(q;q)_{4n+2j}(-1)^j q^{4n^2+12nj+8j^2+j}}{(q^4;q^4)_n(q^2;q^2)_j(q^4;q^4)_{2n+2j}} = \frac{(q^5,q^6,q^{11};q^{11})_\infty}{(q^4;q^4)_\infty}.$$

Perhaps the most exotic double sum which appears corresponds to

$$(T1)(T1)(S1) = \begin{bmatrix} 1 & 13/9 \\ 0 & 1/9 \end{bmatrix},$$

$$\sum_{s_1,s_2\geq 0} \frac{q^{9s_2^2+3s_1^2}(q^3;q^3)_{3s_2-s_1}(q;q)_{3s_1}}{(q^9;q^9)_{2s_2}(q^9;q^9)_{s_2-s_1}(q^3;q^3)_{2s_1}(q^3;q^3)_{s_1}} = \frac{(q^{13},q^{14},q^{27};q^{27})_\infty}{(q^9;q^9)_\infty}.$$

There are 348 words $w = w_1w_2w_3w_4$ of length 4 with corresponding integer values of $11B/(2A+B)$, these could lead to 202 possible triple sum identities modulo multiples of 11.

7. Borwein polynomials.

In this section we use the group element

$$(T1) = \begin{bmatrix} 1 & 1/3 \\ 0 & 1/3 \end{bmatrix}$$

to find alternative forms of the Borwein polynomials.

These polynomials were defined by Andrews [**6**] as

$$A_n(q) = \sum_{k=-\infty}^{\infty} \begin{bmatrix} 2n \\ n-3k \end{bmatrix}_q (-1)^k q^{k(9k-1)/2}$$

$$B_n(q) = \sum_{k=-\infty}^{\infty} \begin{bmatrix} 2n \\ n-3k-1 \end{bmatrix}_q (-1)^k q^{k(9k+5)/2}$$

$$C_n(q) = \sum_{k=-\infty}^{\infty} \begin{bmatrix} 2n \\ n-3k+1 \end{bmatrix}_q (-1)^k q^{k(9k-7)/2}.$$

A conjecture of P. Borwein is equivalent (see [**6**], [**9**], [**15**]) to the conjecture that $A_n(q)$, $B_n(q)$, and $C_n(q)$ have non-negative coefficients as polynomials in q. We give in Theorem 2 an alternative form of these polynomials.

First we review some the hook difference polynomials, which may be defined by [**7**]

$$D_{K,i}(N,M;\alpha,\beta)(q) = \sum_{\lambda=-\infty}^{\infty} q^{\lambda(K\lambda+i)(\alpha+\beta)-K\beta\lambda} \begin{bmatrix} N+M \\ N-K\lambda \end{bmatrix}_q$$

(7.1)
$$- \sum_{\lambda=-\infty}^{\infty} q^{\lambda(K\lambda-i)(\alpha+\beta)-K\beta\lambda+\beta i} \begin{bmatrix} N+M \\ N-K\lambda+i \end{bmatrix}_q.$$

Note that

$$D_{6,3}(N,M;\alpha,\beta,q) = \sum_{k=-\infty}^{\infty} \begin{bmatrix} N+M \\ N-3k \end{bmatrix}_q (-1)^k q^{3k^2(\alpha+\beta)/2 + 3k(\alpha-\beta)/2},$$

$$D_{6,3}(N,M;\alpha,\beta,q) = D_{6,3}(M,N;\beta,\alpha,q)$$
$$D_{6,3}(N,M;\alpha,\beta,q) = q^{MN} D_{6,3}(M,N; 3+M-N-\alpha, 3-M+N-\beta, q^{-1})$$

so that

$$A_n(q) = D_{6,3}(n,n; 4/3, 5/3, q) = D_{6,3}(n,n; 5/3, 4/3, q),$$
$$B_n(q) = D_{6,3}(n-1, n+1; 7/3, 2/3, q) = D_{6,3}(n+1, n-1; 2/3, 7/3, q),$$
$$C_n(q) = D_{6,3}(n+1, n-1; 1/3, 8/3, q) = q^{n^2-1} D_{6,3}(n+1, n-1; 2/3, 7/3, q^{-1}).$$

It is known [**7**] that if α and β are positive integers satisfying

$$\alpha + \beta < K, \quad -i + \beta \leq N - M \leq K - i - \alpha,$$

then the $D_{K,i}(N,M;\alpha,\beta)(q)$ has non-negative coefficients. The next proposition realizes fractional values as an element of a Bailey pair.

PROPOSITION 1. *For (T1),*
1. *if $\beta_n = (q;q)_{2n}^{-1} D_{6,3}(n,n;\alpha,\beta,q)$,*
   then $\beta'_n = (q^3;q^3)_{2n}^{-1} D_{6,3}(n,n; 1+\alpha/3, 1+\beta/3, q^3)$,
2. *if $\beta_n = (q;q)_{2n}^{-1} D_{6,3}(n-1, n+1; \alpha, \beta, q)$,*
   then $\beta'_n = q(q^3;q^3)_{2n}^{-1} D_{6,3}(n-1, n+1; (5+\alpha)/3, (1+\beta)/3, q^3)$.

PROOF. If $a = 1$ in a Bailey pair, we have

$$\beta_n(1,q) = (q;q)_{2n}^{-1} \sum_{k=0}^{n} \begin{bmatrix} 2n \\ n-k \end{bmatrix}_q \alpha_k(1,q).$$

For part (1) choose

$$\alpha_k = \begin{cases} 1 & \text{if } k=0, \\ (-1)^K q^{3K^2(\alpha+\beta)/2} (q^{3K(\alpha-\beta)/2} + q^{-3K(\alpha-\beta)/2}), & \text{if } k = 3K > 0, \\ 0 & \text{otherwise.} \end{cases}$$

so that $\beta_n(1,q) = (q;q)_{2n}^{-1} D_{6,3}(n,n;\alpha,\beta,q)$. Applying (T1) we see that

$$\beta'_n = (q^3;q^3)_{2n}^{-1} D_{6,3}(n,n;\alpha',\beta',q^3),$$

where
$$3(\alpha+\beta)/2+9 = 9(\alpha'+\beta')/2, \quad 3(\alpha-\beta)/2 = 9(\alpha'-\beta')/2.$$
The solution is $\alpha' = 1 + \alpha/3$, $\beta' = 1 + \beta/3$.

For part (2), choose non-zero values
$$\alpha_{3k+1} = (-1)^k q^{3k^2(\alpha+\beta)/2 + 3k(\alpha-\beta)/2},$$
$$\alpha_{3k-1} = (-1)^k q^{3k^2(\alpha+\beta)/2 - 3k(\alpha-\beta)/2},$$
and apply (T1). □

Now we use the known values [**13, Proposition 2**]
$$D_{6,3}(n,n;1,2,q) = (1+q^n)\frac{(q^3;q^3)_{n-1}}{(q;q)_{n-1}},$$
$$D_{6,3}(n+1,n-1;1,2,q) = D_{6,3}(n-1,n+1;2,1,q) = \frac{(q^3;q^3)_{n-1}}{(q;q)_{n-1}},$$
to obtain from Proposition 1 the following theorem.

THEOREM 2. *We have*
$$A_n(q^3) = \frac{(q;q)_{3n}}{(q^3;q^3)_n} + \sum_{k=1}^{n} \frac{(q;q)_{3n-3k}}{(q^3;q^3)_{n-k}} q^{k^2} \begin{bmatrix} 3n-k \\ 2k \end{bmatrix}_q \frac{(q^3;q^3)_{k-1}}{(q;q)_{k-1}}(1+q^k),$$
$$qB_n(q^3) = \sum_{k=1}^{n} \frac{(q;q)_{3n-3k}}{(q^3;q^3)_{n-k}} q^{k^2} \begin{bmatrix} 3n-k \\ 2k \end{bmatrix}_q \frac{(q^3;q^3)_{k-1}}{(q;q)_{k-1}},$$
$$q^2 C_n(q^3) = \sum_{k=1}^{n} \frac{(q;q)_{3n-3k}}{(q^3;q^3)_{n-k}} q^{k^2+k} \begin{bmatrix} 3n-k \\ 2k \end{bmatrix}_q \frac{(q^3;q^3)_{k-1}}{(q;q)_{k-1}}.$$

Finding the $n \to \infty$ limits of Theorem 2 gives the mod 27 identities in Slater [**14**]: (93), (91), and (90) respectively.

Unfortunately Theorem 2 does not establish positivity for the polynomials, but new recurrences do follow from Theorem 2 using Axel Riese's q-Zeil package. If $\gamma_n = qB_n(q^3)$ or $q^2 C_n(q^3)$ then we have
$$\gamma_n = -q^{-15}(q^8 + q^{6n} + q^{4+3n})(q^{10} + q^{6n} + q^{5+3n})\gamma_{n-2}$$
$$+ q^{-6}(q^6 + q^9 + q^{6n} + q^{6n+3} + q^{4+3n} + q^{5+3n})\gamma_{n-1}$$
$$- q^{6n-7}(1+q+q^2)(q;q)_{3n-6}/(q^3;q^3)_{n-2}, \quad n \geq 2.$$

8. Remarks. One may ask where the second Rogers-Ramanujan identity appears from the Bailey-Rogers-Ramanujan group. The group may be extended by a simple transformation on Bailey pairs (see [**11, Proposition 4.1**]) which puts
$$\beta'_n(1,q) = q^n \beta_n(1,q).$$

The full Andrews-Gordon identities then appear. However once an element of type (D), (E) or (T) is used in our word the base q has changed, and only certain linear exponents may be inserted. These, in turn, allow special sets of excluded bases on the product sides. An example of these choices is given in [**11, Corollary 4.4**].

Some of the sums involving $\lim_{n\to\infty}$ in §5 have striking finite forms. For example, the finite sum factors to

$$\frac{(q,q^5;q^6)_n}{(q^6;q^6)_{2n}} \quad \text{for (T1)(D3)},$$

$$(q;q^2)_{2n-1}(1-q^{4n+1}) \quad \text{for (E3)(T2)},$$

$$(q;q^2)_{n-3}(q;q^2)_{n+3} \quad \text{for (D3)(T2)},$$

$$\frac{(q;q^2)_{n+2}(q;q^2)_{n-2}}{(q^2;q^2)_{2n}} \quad \text{for (D3)(D3)},$$

$$\frac{1-q^{2n}-q^{2n+1}}{(q^2;q^2)_{2n}} \quad \text{for (D1)(D2)}.$$

Perhaps these elements of the group are particularly useful.

Another large set of infinite families of multisum Rogers-Ramanujan identities has been given by Warnaar [15].

A combinatorial interpretation of the Rogers-Ramanujan identity which corresponds to a general element of the Bailey-Rogers-Ramanujan group is not known. Some preliminary work on the mod $2^k + 2i$ identities for $i = 2$ has been done by Bressoud [10].

Acknowledgment. The author would like to thank David Bressoud, Tina Garrett, and Mourad Ismail for their contributions to this work.

References

1. A. Agarwal, G. Andrews, and D. Bressoud, *The Bailey lattice*, J. Indian Math. Soc. **51** (1987), 57–73.
2. G. Andrews, *The Theory of Partitions*, Cambridge University Press, New York, 1976.
3. ———, *q-series: their development and application in analysis, number theory, combinatorics, physics, and computer algebra*, CBMS Regional Conference Series in Mathematics, 66, AMS, Providence, R.I., 1986.
4. ———, *On Rogers-Ramanujan type identities related to the modulus 11*, Proc. Lon. Math. Soc. **30** (1975), 330-346.
5. ———, *On the proofs of the Rogers-Ramanujan identities*, IMA Volumes in Mathematics and its Applications, vol. 18, 1989, pp. 1-14.
6. ———, *On a conjecture of Peter Borwein*, J. Symbolic Comput. **20**, 487-501.
7. G. Andrews, R. Baxter, D. Bressoud, W. Burge, P. Forrester, and G. Viennot, *Partitions with prescribed hook differences*, Eur. J. Comb. **8** (1987), 341-350.
8. D. Bressoud, *A generalization of the Rogers-Ramanujan identities for all moduli*, J. Comb. Th. A **27** (1979), 64-68.
9. ———, *The Borwein conjecture and partitions with prescribed hook differences*, Electron. J. Combin. **3** (1996).
10. ———, private communication (1999).
11. D. Bressoud, M. E. H. Ismail, and D. Stanton, *Change of base in Bailey pairs*, Ramanujan J. **4** (2000), 435-453.
12. G. Gasper and M. Rahman, *Basic Hypergeometric Series*, Cambridge University Press, Cambridge, 1990.
13. M. E. H. Ismail, D. Kim, and D. Stanton, *Lattice paths and positive trigonometric sums*, Const. Approx. **15** (1999), 69-81.
14. L. Slater, *Further identities of the Rogers-Ramanujan type*, Proc. Lon. Math. Soc. (2) **54** (1952), 147–167.
15. S. Warnaar, *The generalized Borwein conjecture. I. The Burge transform*, preprint (2000).
16. ———, *Supernomial coefficients, Bailey's lemma and Rogers-Ramanujan-type identities. A survey of results and open problems*, Sem. Lothar. Combin. **42** (1999).

17. _____, *50 years of Bailey's lemma*, preprint (2000).

SCHOOL OF MATHEMATICS, UNIVERSITY OF MINNESOTA, MINNEAPOLIS, MINNESOTA 55455
E-mail address: stanton@math.umn.edu

Multiple Polylogarithms: A Brief Survey

Douglas Bowman and David M. Bradley

ABSTRACT. We survey various results and conjectures concerning multiple polylogarithms and the multiple zeta function. Among the results, we announce our resolution of several conjectures on multiple zeta values. We also provide a new integral representation for the general multiple polylogarithm, and develop a q-analogue of the shuffle product.

1. Introduction

In recent years, nested harmonic sums have attracted increasing attention in both the mathematics and physics communities. The sums occur within the context of knot theory and quantum field theory, yet their rich structure offers much to fascinate theoreticians in such diverse areas as algebra, number theory, and combinatorics. Multiple polylogarithms generalize the aforementioned nested sums, as well as the Riemann zeta function and the classical polylogarithm, while still retaining many interesting properties. Their study has led to many beautiful yet unproven conjectures, including evaluations at arbitrary depth discovered with the use of recently developed integer relations-finding algorithms and high precision numerical computations in the thousands of digits.

Multiple polylogarithms [13, 46, 48] are multiply nested sums of the form

$$(1.1) \qquad \operatorname{Li}_{s_1,\ldots,s_k}(z_1,\ldots,z_k) := \sum_{n_1 > \cdots > n_k > 0} \prod_{j=1}^{k} n_j^{-s_j} z_j^{n_j},$$

where $s_1, \ldots, s_k$ and $z_1, \ldots, z_k$ are complex numbers suitably restricted so that the sum (1.1) converges. Instances of multiple polylogarithms have occurred in several disparate fields, such as combinatorics (analysis of quad-trees [41, 59] and of lattice reduction algorithms [37]), knot theory [24, 26, 25, 60, 61, 77], perturbative quantum field theory [7, 19, 20, 22] and mirror symmetry [55]. There is also quite

---

1991 *Mathematics Subject Classification*. Primary 33E20; Secondary 11G55, 11M99, 40B05.

Key words and phrases. Euler sums, multiple zeta values, polylogarithms, multiple harmonic series, quantum field theory, knot theory, Riemann zeta function.

The first author was supported in part by NSF Grant DMS-9705782.

The second author was partially supported by the University of Maine summer faculty research fund.

sophisticated work relating polylogarithms and their generalizations to arithmetic and algebraic geometry, and to algebraic K-theory [8, 27, 28, 46, 47, 48, 79, 80, 81].

Figuring prominently are the nested sums (1.1) in which each $z_j = 1$. These latter are now commonly referred to as multiple zeta values [12, 13, 17, 68, 82] and are denoted by

$$(1.2) \qquad \zeta(s_1,\ldots,s_k) := \sum_{n_1>\cdots>n_k>0} \prod_{j=1}^{k} n_j^{-s_j}.$$

The study of such sums goes back to Euler [39], who showed that

$$(1.3) \qquad 2\zeta(m,1) = m\zeta(m+1) - \sum_{j=1}^{m-2} \zeta(m-j)\zeta(j+1), \qquad 2 \leq m \in \mathbf{Z}.$$

It can be shown that the sum (1.2) is absolutely convergent in the region

$$\{(s_1,\ldots,s_k) \in \mathbf{C}^k : \sum_{j=1}^{r} \Re(s_j) > r \text{ for } 1 \leq r \leq k\}.$$

(The condition given in Proposition 1 of [83] is insufficient to guarantee absolute convergence.)

Define the *depth* of the multiple polylogarithm (1.1) to be the number k of nested summations. A good deal of work on multiple polylogarithms, and more specifically multiple zeta values, has been motivated by the problem of determining which sums can be expressed (say polynomially with rational coefficients) in terms of other sums of lesser depth. Settling this question in complete generality is currently beyond the reach of number theory. For example, proving the irrationality of expressions such as $\zeta(5,3)/\zeta(5)\zeta(3)$ appears to be impossible with current techniques. Nevertheless, considerable progress has been made with regard to proving specific classes of reductions, even at arbitrary depth. The first nontrivial success at arbitrary depth was the settling [12, 13] of Zagier's conjecture [82]

$$(1.4) \qquad \zeta(\underbrace{3,1,3,1,\ldots,3,1}_{2n}) = 4^{-n}\zeta(\underbrace{4,4,\ldots,4}_{n}) = \frac{2\pi^{4n}}{(4n+2)!}, \qquad 0 \leq n \in \mathbf{Z},$$

in which the $2n$ and n beneath the underbraces in (1.4) denote the depth of the respective multiple zeta values. Subsequent work (see eg. [16, 17, 18]) has largely focused on developing a suitable framework for dealing with ultimately periodic argument strings in general, and additionally sums of multiple zeta values whose set of argument strings is fixed by the action of certain subgroups of the group of permutations.

It is instructive to trace the development of the subject and see for oneself how ad hoc techniques and considerations have in many cases evolved into more systematic methods of ongoing interest. In this connection, one might begin by citing the partial fractions technique of Euler [39] and Nielsen [67], subsequently employed by many others eg. [10, 15, 52, 56, 64, 66, 76], and which Ohno recently parlayed in his exceedingly clever proof of the cyclic sum formula [57, 69]. Techniques based on elementary integration formulæ and identities for special functions tailored to specific examples eg. [9, 10, 38, 66] have evolved [11, 13] into quite general, sophisticated, and powerful analytic methods [16, 18]. The naïve approach of deriving elementary series transformation identities and solving the resulting systems of linear equations [15, 72], used to prove reducibility results of depth three or less, has been largely superceded (eg. by methods based on contour integration [40]) and supplanted by considerations of the shuffle and stuffle [13] multiplications, and relatedly the harmonic algebra and the algebra of quasi-symmetric functions [53, 54, 57].

Computational issues—both numeric and symbolic—have also come into play. Relations satisfied by multiple polylogarithms, and multiple zeta values in particular, can be exploited by symbolic computer algebra systems to prove reductions of small weight [65]. (Here the *weight* of the multiple zeta value (1.2) is simply the sum of the arguments $s_1 + \cdots + s_k$.) Interest in high-precision, rapid computation of multiple zeta values [35, 36] (see also [13, §7.2]) has been stimulated by the ability to numerically hunt for or rule out identities (to a high degree of probability) with the aid of recently developed integer relations-finding algorithms [42, 50, 62].

In addition to the as yet unsolved problem of classifying all possible relationships between multiple zeta values at positive integer arguments, one can also consider (1.2) as a function of the complex numbers $s_1, \ldots, s_k$ and consider questions regarding analytic continuation, trivial zeros, and values at the non-positive integers. The analytic continuation of (1.2) in the case $k = 2$ was established by Atkinson [5] via the Poisson summation formula, and later by Apostol and Vu [3], who used the Euler-Maclaurin summation formula. Subsequently, Arakawa and Kaneko [4] proved that if $s_2, \ldots, s_k$ are fixed positive integers, then (1.2) can be meromorphically continued as a function of s_1 to the whole complex s_1-plane. The analytic continuation of (1.2) as a function defined on $\mathbf{C}^k$ was established by Akiyama, Egami and Tanigawa [1] using the Euler-Maclaurin summation formula. An independent approach due to Zhao [83] uses properties of Gelfand and Shilov's generalized functions [44]. Zhao also attempts a discussion of trivial zeros for $k \leq 3$. To our knowledge, no-one has yet determined the trivial zeros of (1.2) for general k.

The issue of values of (1.2) at the non-positive integers is subtle, since for $k > 1$ the result will in general depend on the order in which the respective limits are taken. Thus, for example, if n is a non-negative integer, $s(k,j)$ and $S(k,j)$ denote the Stirling numbers of the first and second kind, respectively, and B_j denotes the jth Bernoulli number, then [2]

$$\lim_{s_1 \to -n} \lim_{s_2 \to 0} \cdots \lim_{s_k \to 0} \zeta(s_1, \ldots, s_k) = \frac{(-1)^{n+1}}{n+1} \sum_{j=1}^{n+1} \frac{(-1)^{k+j} j! S(n+1, j)}{k+j},$$

whereas

$$\lim_{s_k \to 0} \cdots \lim_{s_2 \to 0} \lim_{s_1 \to -n} \zeta(s_1, \ldots, s_k) = (-1)^k \delta_{n,0} - \frac{1}{(k-1)!} \sum_{j=1}^{k} \frac{s(k,j) B_{n+j}}{n+j},$$

where $\delta_{n,0} = 0$ if $n > 0$ and $\delta_{0,0} = 1$. In particular (cf. also [83])

$$\lim_{s_1 \to 0} \lim_{s_2 \to 0} \zeta(s_1, s_2) = \frac{1}{3}, \quad \text{but} \quad \lim_{s_2 \to 0} \lim_{s_1 \to 0} \zeta(s_1, s_2) = \frac{5}{12}.$$

We are unaware of any systematic treatment in the case of *arbitrary* non-positive integer arguments.

1.1. Notation. Let $\sigma_1,\ldots,\sigma_k \in \{-1,1\}$. We will have occasion to discuss the particular sums of the form

$$(1.5) \qquad \operatorname{Li}_{s_1,\ldots,s_k}(\sigma_1 x, \sigma_2, \ldots, \sigma_k) = \sum_{n_1 > \cdots > n_k > 0} x^{n_1} \prod_{j=1}^{k} n_j^{-s_j} \sigma_j^{n_j},$$

in which $0 \leq x \leq 1$ is real and $s_1,\ldots,s_k$ are positive integers with $x = s_1 = \sigma_1 = 1$ excluded for convergence. Accepted practice dictates that (1.5) may be abbreviated by $\zeta_x(s_1,\ldots,s_k)$ with a bar placed over s_j if and only if $\sigma_j = -1$. When $x = 1$, these are called Euler sums. Thus a multiple zeta value is an Euler sum with no alternations. We adopt the convention that $\zeta_x() := 1$ when no arguments are present ($k = 0$). We also drop the subscript x when $x = 1$ since $\zeta_1(s_1,\ldots,s_k)$ agrees with (1.2) when each s_j is bar-free. For example,

$$\zeta(\overline{2},1) = \sum_{n=1}^{\infty} \frac{(-1)^n}{n^2} \sum_{m=1}^{n-1} \frac{1}{m}.$$

It will be convenient to abbreviate strings of repeated arguments by using exponentiation to denote concatenation powers. Then the first two members of (1.4) may be written $\zeta(\{3,1\}^n) = 4^{-n}\zeta(\{4\}^n)$.

Finally, as customary the Gaussian hypergeometric function and the logarithmic derivative of the Euler gamma function are denoted by

$$F(a,b;c;x) = \sum_{n=0}^{\infty} x^n \prod_{j=0}^{n-1} \frac{(a+j)(b+j)}{(1+j)(c+j)} \quad \text{and} \quad \psi = \frac{\Gamma'}{\Gamma},$$

respectively. We also abbreviate the set of the first k positive integers $\{1,2,\ldots,k\}$ by N_k.

2. Stuffles

For the sake of brevity and simplicity, we shall restrict the discussion in this section to multiple zeta values. For a discussion of the more general polylogarithmic case, see [13].

As Hoffman [57] observed, one can view multiple zeta values as values of a homomorphism on a commutative **Q**-algebra in two ways; the **Q**-algebra multiplications have been referred to elsewhere [13] as "shuffle" and "stuffle." It is conjectured that all relations between multiple zeta values are a consequence of the collision of the two multiplications, provided one admits the divergent sums (1.2) with $s_1 = 1$ (suitably renormalized) into the model. However, there seems little hope of proving this conjecture in the near future, and at present a wide variety of analytic, algebraic, and combinatorial techniques are used to prove identities for multiple zeta values.

Stuffle relations, or more simply stuffles—see §3.1 and §5 below for a discussion of shuffles—arise when one multiplies two nested series of the form (1.2) and expands the product distributively. Thus if u and v are (ordered) lists of positive integers, then

$$\zeta(u)\zeta(v) = \sum_{w \in u*v} \zeta(w),$$

where $u * v$ is the multiset defined by the recursion

(2.1) $\quad su * tv = s(u * tv) \cup t(su * v) \cup (s+t)(s * t), \quad 1 \leq s, t \in \mathbf{Z}.$

In (2.1) it is to be understood that if M is a multiset of lists and a is an integer, then aM denotes the multiset of lists obtained by placing a at the front of each list in M. For example $(s,t) * u = \{(s,t,u), (s,t+u), (s,u,t), (s+u,t), (u,s,t)\}$ and correspondingly $\zeta(s,t)\zeta(u) = \zeta(s,t,u) + \zeta(s,t+u) + \zeta(s,u,t) + \zeta(s+u,t) + \zeta(u,s,t)$.

Let $f(|u|, |v|)$ denote the number of lists in $u * v$. The recursive decomposition (2.1) shows that the generating function

$$F(x,y) := \sum_{m=0}^{\infty} \sum_{n=0}^{\infty} f(m,n) x^m y^n$$

satisfies the functional equation $F(x,y) = 1 + xF(x,y) + yF(x,y) + xyF(x,y)$. It follows that $F(x,y) = (1 - x - y - xy)^{-1}$ and hence that

(2.2) $\quad f(m,n) = \sum_{k=0}^{m} \binom{m}{k} \binom{n+k}{m} = \sum_{k=0}^{\min(m,n)} \binom{n}{k} \binom{m}{k} 2^k.$

One can also give a direct, combinatorial proof of (2.2) by considering how the indices interlace in the product of two nested series of the form (1.2).

There are interesting connections between stuffles, polyominoes, and codes which we briefly indicate. To begin, note that a stuffle counted by $f(m,n)$ can be viewed as a pair (ϕ, ψ) of order-preserving injections

$$\phi : N_m \to N_r, \qquad \psi : N_n \to N_r$$

where r is chosen so that $\max(m,n) \leq r \leq m+n$ and the union of the images of ϕ and ψ is all of N_r. One can associate to such a pair a sequence of integers $b_1, \ldots, b_m$ by defining $a_1 = \phi(1) - 1$ and $a_j = \phi(j) - \phi(j-1) - 1$ for $2 \leq j \leq m$ and then letting

$$b_j = \begin{cases} -a_j & \text{if } \phi(j) \text{ is in the image of } \psi, \\ a_j & \text{otherwise} \end{cases}$$

for each $j \in N_m$. Since ϕ is order-preserving, $a_j \geq 0$ for each $j \in N_m$ and $\sum_{j=1}^{m} |b_j| = \phi(m) - m \leq n$. Conversely, given a sequence of integers $b_1, \ldots, b_m$ satisfying $\sum_{j=1}^{m} |b_j| \leq n$, the pair (ϕ, ψ) is uniquely determined. Let $p = |\{j : b_j < 0\}|$. We have $r = m+n-p$, $\phi(1) = |b_1|+1$ and $\phi(j) = \phi(j-1) + |b_j| + 1$ for $2 \leq j \leq m$. Put $\psi(j) = \phi(j)$ if $b_j < 0$. The remaining values of ψ are determined by the requirement that it be an order-preserving injective map of N_n to N_r. Thus, there is a one-to-one correspondence between the stuffles counted by $f(n,m)$ and the sets of integer lattice points whose cardinalities satisfy

(2.3) $\quad \left| \left\{ (b_1, \ldots, b_m) \in \mathbf{Z}^m : \sum_{j=1}^{m} |b_j| \leq n \right\} \right| = \left| \left\{ (b_1, \ldots, b_n) \in \mathbf{Z}^n : \sum_{j=1}^{n} |b_j| \leq m \right\} \right|,$

the identity (2.3) holding in view of the obvious symmetry $f(m,n) = f(n,m)$.

Define an n-dimensional polyomino formed by adding m coats to a single-celled polyomino, where a coat consists of just enough cells to cover each previously exposed $(n-1)$-dimensional face. There is clearly a bijection between such polyominos and the second set of lattice points (2.3). The relationship is explored in greater detail in [**45**].

3. Integral Representations

3.1. The Drinfeld Integral.
There is also a representation for multiple zeta values in terms of an "iterated" (Drinfeld) integral due to Kontsevich [**82**]. For real $0 \le x \le 1$ and positive integers $s_1, \ldots, s_k$ with $x = s_1 = 1$ excluded for convergence, we have

$$(3.1) \qquad \zeta_x(s_1, \ldots, s_k) = \int \prod_{j=1}^{k} \left(\prod_{r=1}^{s_j - 1} \frac{dt_r^{(j)}}{t_r^{(j)}} \right) \frac{dt_{s_j}^{(j)}}{1 - t_{s_j}^{(j)}},$$

where the integral is over the simplex

$$x > t_1^{(1)} > \cdots > t_{s_1}^{(1)} > \cdots > t_1^{(k)} > \cdots > t_{s_k}^{(k)} > 0,$$

and is abbreviated by

$$(3.2) \qquad \int_0^x \prod_{j=1}^{k} a^{s_j - 1} b, \qquad a = dt/t, \quad b = dt/(1-t).$$

Making the simultaneous change of variable $t \mapsto 1 - t$ at each level of integration and then reversing the order of integration makes transparent the duality identity for multiple zeta values:

$$(3.3) \qquad \zeta(s_1 + 2, \{1\}^{r_1}, \ldots, s_k + 2, \{1\}^{r_k}) = \zeta(r_k + 2, \{1\}^{s_k}, \ldots, r_1 + 2, \{1\}^{s_1}),$$

first conjectured in [**52**] and proved in [**82**].

A related integral representation enabled Ohno [**68**] to prove the following beautiful generalization of (3.3). Let

$$S(p_1, \ldots, p_n; m) := \sum_{c_1 + \cdots + c_n = m} \zeta(p_1 + c_1, \ldots, p_n + c_n),$$

where the sum is over all non-negative integers $c_1, \ldots, c_n$ which sum to m. As in (3.3) define the dual argument lists

$$p := (s_1 + 2, \{1\}^{r_1}, \ldots, s_k + 2, \{1\}^{r_k})$$

and

$$p' := (r_k + 2, \{1\}^{s_k}, \ldots, r_1 + 2, \{1\}^{s_1}).$$

Then [**68**] $S(p; m) = S(p'; m)$. When $m = 0$, Ohno's result reduces to (3.3). Another interesting specialization is obtained by taking $p = (k+1)$ and $m = n - k - 1$; one then deduces Granville's theorem [**49**], originally conjectured independently by Courtney Moen [**52**] and Michael Schmidt [**64**]:

$$\sum_{s_1 + \cdots + s_k = n} \zeta(s_1, \ldots, s_k) = \zeta(n),$$

where the sum is over all positive integers $s_1, \ldots, s_k$ which sum to n and $s_1 > 1$.

The iterated integral representation is also responsible for a second multiplication rule satisfied by multiple zeta values. Suppose that $x, y \in \mathbf{R}$ and $f_j : [y, x] \to \mathbf{R}$ are integrable functions for $j = 1, 2, \ldots, n$. It is customary to make the abbreviation

$$(3.4) \qquad \int_y^x \prod_{j=1}^{n} \alpha_j := \int_{x > t_1 > t_2 > \cdots > t_n > y} \prod_{j=1}^{n} f_j(t_j) \, dt_j, \qquad \alpha_j := f_j(t_j) \, dt_j,$$

with the convention that (3.4) is equal to 1 if $n = 0$ regardless of the values of x and y. There is an alternative definition of iterated integrals which explains their name. For $j = 1, 2, \ldots, n$ again define the 1-forms α_j by $\alpha_j := f_j(t_j)\, dt_j$. Then put

$$(3.5) \qquad \int_y^x \alpha_1 \alpha_2 \cdots \alpha_n := \begin{cases} \int_y^x f_1(t_1) \int_y^{t_1} \alpha_2 \cdots \alpha_n\, dt_1 & \text{if } n > 0 \\ 1 & \text{if } n = 0. \end{cases}$$

Expanding out this second definition, it is easy to see that it coincides with the definiton as an integral over a simplex. Both definitions occur frequently in the literature.

Clearly the product of two iterated integrals of the form (3.4) consists of a sum of iterated integrals involving all possible interlacings of the variables. Therefore, if we denote the set of all $(m+n)!/m!\,n!$ permutations σ of the indices N_{m+n} satisfying $\sigma^{-1}(j) < \sigma^{-1}(k)$ for all $1 \leq j < k \leq m$ and $m+1 \leq j < k \leq m+n$ by $\mathrm{Shuff}(m,n)$, then we have the self-evident formula

$$(3.6) \qquad \left(\int_y^x \prod_{j=1}^m \alpha_j\right)\left(\int_y^x \prod_{j=m+1}^{m+n} \alpha_j\right) = \sum_{\sigma \in \mathrm{Shuff}(m,n)} \int_y^x \prod_{j=1}^{m+n} \alpha_{\sigma(j)},$$

and so define the shuffle product $\sqcup\!\sqcup$ by

$$(3.7) \qquad \left(\prod_{j=1}^m \alpha_j\right) \sqcup\!\sqcup \left(\prod_{j=m+1}^{m+n} \alpha_j\right) := \sum_{\sigma \in \mathrm{Shuff}(m,n)} \prod_{j=1}^{m+n} \alpha_{\sigma(j)}.$$

Thus, the sum is over all non-commutative products (counting multiplicity) of length $m+n$ in which the relative orders of the factors in the products $\alpha_1 \alpha_2 \cdots \alpha_m$ and $\alpha_{m+1} \alpha_{n+2} \cdots \alpha_{m+n}$ are preserved. The term "shuffle" is used because such permutations arise in riffle shuffling a deck of $m + n$ cards cut into one pile of m cards and a second pile of n cards.

The study of shuffles and iterated integrals was pioneered by Chen [30, 31] and subsequently formalized by Ree [74]. As with the case of stuffles, one can view an element of $\mathrm{Shuff}(m,n)$ as a pair of order-preserving injections (ϕ, ψ) where now $\phi : N_m \to N_{m+n}$ and $\psi : N_n \to N_{m+n}$ have disjoint images. One can then define a vector $(a_1, \ldots, a_m)$ of non-negative integers by $a_1 = \phi(1) - 1$ and $a_j = \phi(j) - \phi(j-1) - 1$ for $2 \leq j \leq m$. Since ϕ is order-preserving, $a_j \geq 0$ for each $j \in N_m$ and $\sum_{j=1}^m a_j = \phi(m) - m \leq n$. Conversely, if we have such a vector of non-negative integers, then $\phi(1) = a_1 + 1$ and $\phi(j) = \phi(j-1) + a_j + 1$ for $2 \leq j \leq m$ defines an order-preserving injection $\phi : N_m \to N_{m+n}$, and hence a shuffle. Thus, there is a one-to-one correspondence between $\mathrm{Shuff}(m,n)$ and the sets of non-negative integer lattice points whose cardinalities satisfy

$$\left|\left\{(a_1, \ldots, a_m) \in \mathbf{Z}_{\geq 0}^m : \sum_{j=1}^m a_j \leq n\right\}\right| = \left|\left\{(a_1, \ldots, a_n) \in \mathbf{Z}_{\geq 0}^n : \sum_{j=1}^n a_j \leq m\right\}\right|,$$

the latter identity holding in light of the fact that $\mathrm{Shuff}(m,n)$ is clearly symmetric in m and n. A deeper study of the algebra and combinatorics of shuffles leads to an alternative proof of (1.4) and generalizations thereof; see §5.

3.2. A New Integral Representation. In light of the usefulness of the various integral representations, it may be of interest to give here a new integral representation for (1.1). The new representation appears to embody properties of both the Drinfeld and partition integrals of [**13**], and therefore may be useful in proving certain results for multiple polylogarithms which have thus far withstood attacks based on traditional methods. The derivation employs MacMahon's Omega operator, which discards terms with non-positive exponents from formal Laurent series in $\lambda_1,\ldots,\lambda_k$. Thus, in view of (1.1), if $0 \le x_1,\ldots,x_k \le 1$, we may write

$$\mathrm{Li}_{s_1,\ldots,s_k}(x_1,\ldots,x_k)$$
$$= \Omega \prod_{j=1}^{k} \sum_{n_j>0} n_j^{-s_j} \left(x_j \lambda_j \lambda_{j-1}^{-1}\right)^{n_j}, \qquad \lambda_0 := 1$$
$$= \Omega \prod_{j=1}^{k} \mathrm{Li}_{s_j}\left(x_j \lambda_j \lambda_{j-1}^{-1}\right)$$
$$= \Omega \prod_{j=1}^{k} \int_{1>u_1^{(j)}>\cdots>u_{s_j}^{(j)}>0} \left(\prod_{r=1}^{s_j-1} \frac{du_r^{(j)}}{u_r^{(j)}} \right) \frac{x_j \lambda_j \lambda_{j-1}^{-1} \, du_{s_j}^{(j)}}{1 - x_j \lambda_j \lambda_{j-1}^{-1} u_{s_j}^{(j)}}$$
$$= \Omega \prod_{j=1}^{k} \int_{1>u_1^{(j)}>\cdots>u_{s_j}^{(j)}>0} \left(\prod_{r=1}^{s_j-1} \frac{du_r^{(j)}}{u_r^{(j)}} \right) \sum_{m_j=1}^{\infty} \left(x_j \lambda_j \lambda_{j-1}^{-1}\right)^{m_j} \left(u_{s_j}^{(j)}\right)^{m_j-1} du_{s_j}^{(j)}$$
$$= \int_{\Delta(\vec{s})} \left\{ \prod_{j=1}^{k} \left(\prod_{r=1}^{s_j-1} \frac{du_r^{(j)}}{u_r^{(j)}} \right) \right\} \sum_{m_1>\cdots>m_k>0} \prod_{j=1}^{k} \left(x_j u_{s_j}^{(j)}\right)^{m_j} \frac{du_{s_j}^{(j)}}{u_{s_j}^{(j)}},$$

where $\Delta(\vec{s})$ denotes the set of all integration variables satisfying

$$1 > u_1^{(j)} > u_2^{(j)} > \cdots > u_{s_j}^{(j)} > 0$$

for $j=1,2,\ldots,k$. Since $0 \le y_j < 1$ for each $j=1,2,\ldots,k$ implies

$$\sum_{m_1>\cdots>m_k>0} \prod_{j=1}^{k} y_j^{m_j} = \sum_{n_1=1}^{\infty} \cdots \sum_{n_k=1}^{\infty} y_1^{n_1+\cdots+n_k} y_2^{n_2+\cdots+n_k} \cdots y_k^{n_k}$$
$$= \frac{y_1}{1-y_1} \cdot \frac{y_1 y_2}{1-y_1 y_2} \cdots \frac{y_1 y_2 \cdots y_k}{1-y_1 y_2 \cdots y_k},$$

it follows that

(3.8) $$\mathrm{Li}_{s_1,\ldots,s_k}(x_1,\ldots,x_k) = \int_{\Delta(\vec{s})} \prod_{j=1}^{k} \left\{ \tau\left(\prod_{m=1}^{j} x_m u_{s_m}^{(m)} \right) \prod_{r=1}^{s_j} \frac{du_r^{(j)}}{u_r^{(j)}} \right\},$$

where $\tau(x) := x/(1-x)$.

4. Generating Functions

In many cases, generating functions provide the best means of stating reductions involving one or more parameters. A specific example of this which also illustrates how knowledge of the subject has progressed is given first. We then outline a systematic approach for tackling multiple zeta values with periodic argument lists, followed by additional examples to illustrate the richness of the theory.

4.1. Two-Parameter Symmetry. In connection with Euler's result (1.3), Markett [**64**] derived

$$\zeta(s,1,1) = \frac{1}{6}s(s+1)\zeta(s+2) - \frac{1}{2}(s-1)\zeta(2)\zeta(s) - \frac{s}{4}\sum_{n=0}^{s-4}\zeta(s-n-1)\zeta(n+3)$$

$$(4.1) \qquad + \frac{1}{6}\sum_{n=0}^{s-4}\zeta(s-n-2)\sum_{m=0}^{n}\zeta(n-m+2)\zeta(m+2), \qquad 3 \le s \in \mathbf{Z},$$

via elementary but intricate series manipulations and partial fraction identities. An equivalent formula is proved in [**10**] using elementary facts about the dilogarithm, the polygamma function and the higher derivatives of the Euler beta function. For larger values of n, the representation of $\zeta(s, \{1\}^n)$ in terms of values of the Riemann zeta function becomes increasingly complicated. Nevertheless, there is an elegant generating function formulation which we restate here.

THEOREM 4.1 ([**11**]). *The bivariate formal power series identity*

$$(4.2) \qquad \sum_{m=0}^{\infty}\sum_{n=0}^{\infty} x^{m+1}y^{n+1}\zeta(m+2,\{1\}^n)$$

$$= 1 - \exp\left\{\sum_{k=2}^{\infty}\frac{1}{k}\left(x^k + y^k - (x+y)^k\right)\zeta(k)\right\}$$

holds.

COROLLARY 4.2. *Let n and s be non-negative integers with $s \ge 2$. Then $\zeta(s, \{1\}^n)$ lies in the polynomial ring $\mathbf{Q}[\zeta(2), \zeta(3), \ldots, \zeta(s+n)]$.*

By comparing coefficients of $x^{s-1}y^{n+1}$ on both sides of (4.2), one sees that in fact, $\zeta(s, \{1\}^n)$ is a rational linear combination of products of Riemann zeta values such that the sum of the arguments in each product is equal to $s + n$. Moreover, Euler's result (1.3) is an immediate consequence of comparing coefficients of $x^{s-1}y^2$. Similarly, Markett's formula (4.1) can be obtained most easily by comparing coefficients of $x^{s-1}y^3$. Finally, as the right hand side of (4.2) is evidently symmetric in x and y, the left hand side must also be. Thus Theorem 4.1 implies the special case $\zeta(m+2, \{1\}^n) = \zeta(n+2, \{1\}^m)$ of the duality formula (3.3). It would be interesting to find a generating function formulation of duality at full strength.

4.2. Periodic Argument Lists. Results such as (1.4) and (4.2) suggest that one might profit from a more systematic study of multiple zeta values whose argument lists form an ultimately periodic sequence. This is indeed the case; such a study forms the basis of some of our current work in progress [**18**].

4.2.1. *Period One.* The case of all identical arguments is quite well understood. Nevertheless, there are a few items of interest worth recording here, in particular a connection to the problem of determining the number of unordered factorizations of an integer.

For $\Re(s) > 1$, equation (1.2) implies

$$(4.3) \qquad \sum_{k=0}^{\infty} t^{ks}\zeta(\{s\}^k) = \prod_{j=1}^{\infty}\left(1 + \frac{t^s}{j^s}\right).$$

If in (4.3) we take s to be an even integer, say $s = 2n$ where n is a positive integer, then we may rewrite (4.3) in the form

$$(4.4) \qquad \sum_{k=0}^{\infty} (-1)^k t^{2kn} \zeta(\{2n\}^k) = \prod_{j=0}^{n-1} \operatorname{sinc}(\pi t \rho^j),$$

where $\rho = e^{\pi i/n}$ and $\operatorname{sinc} x = \sin x / x$ for $x \neq 0$; $\operatorname{sinc} 0 := 1$. The identity (4.4) is one of many possible generalizations of Euler's formula for $\zeta(2n)$, and moreover shows that $\zeta(\{2n\}^k)$ is a rational multiple of π^{2kn}.

Differentiating both sides of (4.3) and equating coefficients yields the recurrence

$$(4.5) \qquad k\zeta(\{s\}^k) = \sum_{j=1}^{k} (-1)^{j+1} \zeta(js) \zeta(\{s\}^{k-j}), \qquad 0 \le k \in \mathbf{Z}, \quad \Re(s) > 1,$$

which is really just a special case of Newton's formula

$$k e_k = \sum_{j=1}^{k} (-1)^{j+1} p_j e_{k-j}, \qquad 0 \le k \in \mathbf{Z},$$

relating the elementary symmetric functions and power sum symmetric functions

$$e_k = \sum_{j_1 > \cdots > j_k > 0} \prod_{m=1}^{k} x_{j_m}, \qquad p_k := \sum_{j>0} x_j^k.$$

Substituting $1/j^s$ for each indeterminate x_j yields $e_k = \zeta(\{s\}^k)$ and $p_k = \zeta(ks)$.

From (4.5) it follows that if k is a positive integer and $\Re(s) > 1$, then $\zeta(\{s\}^k)$ lies in the polynomial ring $\mathbf{Q}[\zeta(s), \zeta(2s), \ldots, \zeta(ks)]$. In fact, there is an explicit formula for $\zeta(\{s\}^k)$ in terms of a sum over partitions of k.

DEFINITION 4.3. Let r be a non-negative integer and let $\alpha = (\alpha_1, \alpha_2, \ldots)$ be a non-negative integer partition of r. Let $m_j = \#\{i : \alpha_i = j\}$ be the number of parts of size j, and put $c_\alpha = \prod_{j \ge 1} m_j! (-j)^{m_j}$. Furthermore, abbreviate $r = \sum_{j \ge 1} \alpha_j$ by $|\alpha|$ and $\prod_{j \ge 1} p_{\alpha_j}$ by p_α.

In view of the generic relationship [63]

$$\sum_{k=0}^{\infty} e_k t^k = \exp\left\{-\sum_{r=1}^{\infty} \frac{(-t)^r p_r}{r}\right\} = \sum_\alpha (-t)^{|\alpha|} c_\alpha^{-1} p_\alpha,$$

for $\Re(s) > 1$ we therefore have

$$\sum_{k=0}^{\infty} t^k \zeta(\{s\}^k) = \exp\left\{-\sum_{r=1}^{\infty} \frac{(-1)^r \zeta(rs) t^r}{r}\right\} = \sum_\alpha (-t)^{|\alpha|} c_\alpha^{-1} \prod_{\alpha_j > 0} \zeta(\alpha_j s),$$

i.e.

$$(4.6) \qquad \zeta(\{s\}^k) = (-1)^k \sum_{|\alpha|=k} c_\alpha^{-1} \prod_{\alpha_j > 0} \zeta(\alpha_j s).$$

We note the following connection with factorisatio numerorum [51]. (See also [29, 70, 78].) Let α be as in Definition 4.3. Define the unrestricted divisor function associated with the partition α by

$$d_\alpha(m) = \sum_{\prod_{j \ge 1} d_j^{\alpha_j} = m} 1.$$

For example $d_{1,1}$ is the ordinary divisor function, and $d_2(m) = 1$ if m is a perfect square and zero otherwise.

PROPOSITION 4.4. *Let $\tau_k(m)$ denote the number of unordered factorizations of m into k distinct factors. Then*

$$\tau_k(m) = (-1)^k \sum_{|\alpha|=k} c_\alpha^{-1} d_\alpha(m).$$

PROOF. Observe that for $\Re(s) > 1$,

$$\zeta(\{s\}^k) = \sum_{n_1 > \cdots > n_k > 0} \prod_{j=1}^k n_j^{-s} = \sum_{m=1}^\infty \tau_k(m)\, m^{-s}.$$

Now compare coefficients of m^{-s} in (4.6). □

EXAMPLE 4.5. Since $\zeta(\{s\}^2) = \frac{1}{2}\zeta^2(s) - \frac{1}{2}\zeta(2s)$, we get $\tau_2(m) = \frac{1}{2}d_{1,1}(m) - \frac{1}{2}d_2(m)$. In particular $\tau_2(12) = 3$.

4.2.2. *Period Two and Beyond.* In contrast with the situation in which all arguments are identical, much remains to be explored in the case of argument strings of period two and higher. In [**13**] and [**16**] differential equations were found to be a useful technique for analyzing the generating functions for period 2. We summarize here some results from [**13**] and [**16**] to indicate the richness and complexity of the resulting formulæ arising from the solution of the associated fourth order differential equation.

DEFINITION 4.6. For $0 \leq x \leq 1$ and $z \in \mathbf{C}$, let

$$\begin{aligned}
Y_1(x,z) &:= F(z,-z;1;x), \\
Y_2(x,z) &:= (1-x)F(1+z, 1-z; 2; 1-x), \\
G(z) &:= \tfrac{1}{4}\{\psi(1+iz) + \psi(1-iz) - \psi(1+z) - \psi(1-z)\}.
\end{aligned}$$

THEOREM 4.7 ([**13**]). *Let Y_1 be as in Definition 4.6. Then for $0 \leq x \leq 1$ and $|z| < 1$,*

$$(4.7) \qquad \sum_{n=0}^\infty (-1)^n z^{4n} 4^n \zeta_x(\{3,1\}^n) = Y_1(x,z) Y_1(x,iz).$$

THEOREM 4.8 ([**16**]). *Let Y_1, Y_2 and G be as in Definition 4.6. Then for $0 \leq x \leq 1$ and $|z| < 1$,*

$$(4.8) \quad \sum_{n=0}^\infty (-1)^n z^{4n+2} 4^n \zeta_x(3, \{1,3\}^n) = G(z) Y_1(x,z) Y_1(x,iz)$$

$$- \frac{Y_1(x,iz) Y_2(x,z)}{4 Y_1(1,z)} + \frac{Y_1(x,z) Y_2(x,iz)}{4 Y_1(1,iz)}.$$

Note that (4.7) proves (1.4). Similarly (4.8) proves

$$\begin{aligned}
\zeta(3, \{1,3\}^n) &= 4^{-n} \sum_{k=0}^n \zeta(4k+3)\zeta(\{4\}^{n-k}) \\
&= \sum_{k=0}^n \frac{2\pi^{4k}}{(4k+2)!} \left(-\frac{1}{4}\right)^{n-k} \zeta(4n-4k+3),
\end{aligned}$$

which escaped the extensive numerical and symbolic searches carried out in the preparation of [11, 12, 13]. Differentiation of (4.8) followed by a delicate analysis of the asymptotic behaviour of the requisite hypergeometric functions at their singular points proves [16] the reduction

$$\zeta(2,\{1,3\}^n)$$
$$= 4^{-n}\sum_{k=0}^{n}(-1)^k\zeta(\{4\}^{n-k})\left\{(4k+1)\zeta(4k+2) - 4\sum_{j=1}^{k}\zeta(4j-1)\zeta(4k-4j+3)\right\}$$

conjectured in [11, 13].

The proof of Theorem 4.8 hinges on showing that both sides of (4.7) and (4.8) are annihilated by the same fourth order differential operator. In [13], computer algebra was used to establish this for (4.7). At the time, a conceptual proof was unavailable. Subsequently the present authors (see [16]) found a conceptual proof of the following more general result, which is perhaps best understood in the context of work going back to Orr [71] and Clausen [34] on differential equations satisfied by a product of hypergeometric series. The result is shown in [17] to be closely related to the combinatorial "shuffle" approach outlined in §5, and may be stated as follows.

LEMMA 4.9. *Let K be a differential field of characteristic not equal to 2 and let D be a derivation on K. For each $k \in K$, define a derivation $D_k := kD$. Let t be a constant, and suppose that for some $f, g, u, v \in K$ the differential equations $(D_f D_g + t)u = 0$ and $(D_f D_g - t)v = 0$ hold. Then uv is annihilated by the differential operator $(D_f^2 D_g^2 + 4t^2)$.*

In particular, taking $f(x) = 1 - x$, $g(x) = x$ and $t = z^2$, given that Y_1 and Y_2 satisfy $(D_f D_g + z^2)y = 0$, Lemma 4.9 shows that each of the three linearly independent functions $Y_1(x,z)Y_1(x,iz)$, $Y_1(x,iz)Y_2(x,z)$, and $Y_1(x,z)Y_2(x,iz)$ are annihilated by the operator $D_f^2 D_g^2 + 4z^4$. That $L(x,z)$ and $S(x,z)$ are annihilated by the same operator follows easily from the integral representation (3.1), whence Theorem 4.8 is proved.

Since $D_f^2 D_g^2 + 4z^4$ is a fourth order differential operator, one might legitimately ask in what context the fourth linearly independent solution $Y_2(x,z)Y_2(x,iz)$ arises. It turns out that due to the double logarithmic singularity arising from the product of the underlying hypergeometric functions at $x = 1$, it is easier to ascribe a meaning to this solution in the case of alternating sums (1.5). Recalling the generating function

$$(4.9) \quad A(z) := \sum_{n=0}^{\infty} z^n \zeta(\{\overline{1}\}^n) = \prod_{j=1}^{\infty}\left(1 + \frac{(-1)^j z}{j}\right) = \frac{\Gamma(1/2)}{\Gamma(1+z/2)\Gamma(1/2 - z/2)}$$

from [11], we have

THEOREM 4.10 ([16]). *Let $0 \leq x \leq 1$, and $|t| < \infty$. Put $z = (1+i)t/2$, $s = (1+x)/2$, and let $U(s,z) = Y_1(s,z) - zY_2(s,z)$, where Y_1 and Y_2 are as in Definition 4.6. Then,*

$$(4.10) \quad \sum_{n=0}^{\infty}\left[t^{2n}\zeta_x(\{\overline{1},1\}^n) + t^{2n+1}\zeta_x(\overline{1},\{1,\overline{1}\}^n)\right] = \frac{U(s,-z)U(s,iz)}{A(-z)A(iz)}.$$

Theorem 4.10 is a bivariate generalization of the conjecture [**11**, equation (14)] in the case $x = 1$, and may be viewed as an analytic extension of the purely combinatorial identity (5.4) below.

In recent work [**18**], the authors have greatly extended the differential equation approach. The authors have obtained results on more general generating functions which include not only multiple zeta values, but polylogarithmic and hyperlogarithmic [**58**] values in general. In fact, from the point of view of iterated integrals, arbitrary forms may occur in the iterated integrals studied. The differential equations are still present. The authors have classified various bases for the solutions of the differential equations, given matrices for change of basis, and found the explicit representations of the monodromy matrices of the associated differential equations. These results actually stand out with greater distinction in a more general setting. Taking arbitrary 1-forms on a manifold M, an explicit homomorphism is obtained from $\pi_1(M, x_0)$ into $\mathrm{GL}_n(\mathbf{C})$. This gives rise to a transport between the manifold M and its principle bundle constructed from the representation into $\mathrm{GL}_n(\mathbf{C})$. Finally these results can be cast yet more generally in the setting of differentiable spaces. Our homomorphism is similar to the celebrated homomorphism of K. T. Chen [**30, 31, 32, 33**] in that it is built out of a generating function of iterated integrals. The essential difference is that Chen's homomorphism maps into a formal Lie group, while our homomorphism maps into $\mathrm{GL}_n(\mathbf{C})$. Will our homomorphism give different information than Chen's? We are currently investigating the geometric implications of our work in this area.

5. Shuffles and Cyclic Insertion

As in [**65**] (cf. also [**12, 74**]) let A be a finite set and let A^* denote the free monoid generated by A. We regard A as an alphabet, and the elements of A^* as words formed by concatenating any finite number of letters (repetitions permitted) from the alphabet A. By linearly extending the concatenation product to the set $\mathbf{Q}\langle A \rangle$ of rational linear combinations of elements of A^*, we obtain a non-commutative polynomial ring with the elements of A being indeterminates and with multiplicative identity 1 denoting the empty word.

The shuffle product (3.7) is alternatively defined first on words by the recursion

$$(5.1) \quad \begin{cases} \forall w \in A^*, & 1 \shuffle w = w \shuffle 1 = w, \\ \forall a, b \in A, \quad \forall u, v \in A^*, & au \shuffle bv = a(u \shuffle bv) + b(au \shuffle v), \end{cases}$$

and then extended linearly to $\mathbf{Q}\langle A \rangle$. One checks that the shuffle product so defined is associative and commutative, and thus $\mathbf{Q}\langle A \rangle$ equipped with the shuffle product becomes a commutative $\mathbf{Q}$-algebra, denoted $\mathrm{Sh}_{\mathbf{Q}}[A]$. Radford [**73**] has shown that $\mathrm{Sh}_{\mathbf{Q}}[A]$ is isomorphic to the polynomial algebra $\mathbf{Q}[L]$ obtained by adjoining the transcendence basis L of Lyndon words to the field $\mathbf{Q}$ of rational numbers.

The recursive definition (5.1) has its analytical motivation in the formula for integration by parts—equivalently, the product rule for differentiation. Thus, if we put $a = f(t)\,dt$, $b = g(t)\,dt$ and

$$F(x) := \int_y^x (au \shuffle bv) = \left(\int_y^x f(t) \int_y^t u\,dt \right)\left(\int_y^x g(t) \int_y^t v\,dt \right),$$

then writing $F(x) = \int_y^x F'(s)\,ds$ and applying the product rule for differentiation yields

$$\begin{aligned}F(x) &= \int_y^x \left(f(s)\int_y^s u\right)\left(\int_y^s g(t)\int_y^t v\,dt\right)ds \\ &\quad + \int_y^x g(s)\left(\int_y^s f(t)\int_y^t u\,dt\right)\int_y^s v\,ds \\ &= \int_y^x \left[a(u \shuffle bv) + b(au \shuffle v)\right].\end{aligned}$$

Alternatively, by viewing F as a function of y, we see that the recursion (5.1) could equally well have been stated as

(5.2) $\quad\begin{cases}\forall w \in A^*, & 1 \shuffle w = w \shuffle 1 = w, \\ \forall a,b \in A, \ \forall u,v \in A^*, & ua \shuffle vb = (u \shuffle vb)a + (ua \shuffle v)b.\end{cases}$

Of course, both definitions are equivalent to (3.7).

The combinatorial proof [12] of Zagier's conjecture (1.4) hinged on expressing the sum of the words comprising the shuffle product of $(ab)^n$ with $(ab)^m$ as a linear combination of basis subsums. In [17] a more comprehensive study of the shuffle algebra $\text{Sh}_\mathbf{Q}[a,b]$ is undertaken, and as a consequence correspondingly deeper results for multiple zeta values are obtained. To highlight the most interesting of these results, we first recall the following

DEFINITION 5.1 ([12]). For integers $m \geq n \geq 0$ let $S_{m,n}$ denote the set of words occurring in the shuffle product $(ab)^n \shuffle (ab)^{m-n}$ in which the subword a^2 appears exactly n times, and let $T_{m,n}$ be the sum of the $m!/(2n)!(m-2n)!$ distinct words in $S_{m,n}$. For all other integer pairs (m,n) it is convenient to define $T_{m,n} := 0$.

One then has

THEOREM 5.2 ([17]). *Let x and y be commuting indeterminates, and let m be a non-negative integer. In the commutative polynomial ring $(\text{Sh}_\mathbf{Q}[a,b])[x,y]$ we have the shuffle convolution formula*

(5.3) $$\sum_{k=0}^m x^k y^{m-k}\left[(ab)^k \shuffle (ab)^{m-k}\right] = \sum_{n=0}^{\lfloor m/2 \rfloor} (4xy)^n (x+y)^{m-2n} T_{m,n}.$$

A special case of Theorem 5.2 implies the intriguing shuffle factorization due to Broadhurst, and which in turn implies (1.4):

(5.4) $$A\left(\frac{z}{1-i}\right) \shuffle A\left(\frac{z}{1+i}\right) = M(z) \in (\text{Sh}_\mathbf{Q}[a,b])[[z]], \quad i^2 = -1,$$

where

$$A(z) := \sum_{n=0}^\infty (z^2 ab)^n (1+za) \quad \text{and} \quad M(z) := \sum_{n=0}^\infty (z^4 a^2 b^2)^n (1 + za + z^2 a^2 + z^3 a^2 b).$$

The experts will recognize (4.9) as the analytic version of $A(z)$ above, in which $a = -dt/(1+t)$ and $b = dt/(1-t)$. Similarly for $M(z)$ and the left hand size of (4.10) when $x = 1$.

In addition, Theorem 5.2 plays a key role in a remarkable combinatorial generalization of (1.4) which we proceed to describe. Let $S_{m,n}$ be as in Definition 5.1.

Note that each word in $S_{m,n}$ has a unique representation

$$(5.5) \qquad (ab)^{m_0} \prod_{k=1}^{n} (a^2 b)(ab)^{m_{2k-1}} b(ab)^{m_{2k}},$$

in which $m_0, m_1, \ldots, m_{2n}$ are non-negative integers with sum $m - 2n$. Conversely, every ordered $(2n+1)$-tuple $(m_0, m_1, \ldots, m_{2n})$ of non-negative integers with sum $m - 2n$ gives rise to a unique word in $S_{m,n}$ via (5.5). Thus, a bijective correspondence φ is established between the set $S_{m,n}$ and the set $C_{2n+1}(m-2n)$ of ordered non-negative integer compositions of $m - 2n$ with $2n + 1$ parts. In view of the relationship (3.1) expressing multiple zeta values as iterated integrals, it therefore makes sense to define

$$Z(\vec{s}) := \int_0^1 \varphi(\vec{s}), \qquad \vec{s} \in C_{2n+1}(m-2n),$$

where as in (3.2), we now identify the abstract letters a and b with the differential 1-forms dt/t and $dt/(1-t)$, respectively. Thus, if $\vec{s} = (m_0, m_1, \ldots, m_{2n})$, then

$$\begin{aligned} Z(\vec{s}) &= \int_0^1 (ab)^{m_0} \prod_{k=1}^{n} (a^2 b)(ab)^{m_{2k-1}} b(ab)^{m_{2k}} \\ &= \zeta(\{2\}^{m_0}, 3, \{2\}^{m_1}, 1, \{2\}^{m_2}, 3, \{2\}^{m_3}, 1, \ldots, 3, \{2\}^{m_{2n-1}}, 1, \{2\}^{m_{2n}}), \end{aligned}$$

in which the argument string consisting of m_j consecutive twos is inserted after the jth element of the string $\{3, 1\}^n$ for each $j = 0, 1, 2, \ldots, 2n$. It turns out [17] that

$$(5.6) \qquad \sum_{\vec{s} \in C_{2n+1}(m-2n)} Z(\vec{s}) = \frac{2\pi^{2m}}{(2m+2)!} \binom{m+1}{2n+1},$$

for all non-negative integers m and n with $m \geq 2n$. The proof uses Theorem 5.2 at essentially full strength combined with some tricky generatingfunctionology. Observe that equation (1.4) is the special case of (5.6) in which $m = 2n$, since $Z(\{0\}^{2n+1}) = \zeta(\{3,1\}^n)$.

A more compelling formulation of (5.6) can be given as follows. Again, let $\vec{s} = (m_0, m_1, \ldots, m_{2n})$ and put

$$\mathcal{C}(\vec{s}) := Z(\vec{s}) + \sum_{j=1}^{2n} Z(m_j, m_{j+1}, \ldots, m_{2n}, m_0, \ldots, m_{j-1}).$$

In other words, sum over all cyclic permutations of the argument list $\vec{s}$. Then [17]

$$(5.7) \qquad \sum_{\vec{s} \in C_{2n+1}(m-2n)} \mathcal{C}(\vec{s}) = Z(m) \times |C_{2n+1}(m-2n)| = \frac{\pi^{2m}}{(2m+1)!} \binom{m}{2n}$$

is an equivalent formulation of (5.6). Here, we have used

$$Z(m) = \zeta(\{2\}^m) = \frac{\pi^{2m}}{(2m+1)!}, \qquad 0 \leq m \in \mathbf{Z},$$

which follows from (4.4). The cyclic insertion conjecture [13] can be restated as the assertion that $\mathcal{C}(\vec{s}) = Z(m)$ for all $\vec{s} \in C_{2n+1}(m-2n)$ and integers $m \geq 2n \geq 0$. Thus, (5.7) reduces the problem to that of establishing the invariance of $\mathcal{C}(\vec{s})$ on $C_{2n+1}(m-2n)$. It is likely that this remaining step can be accomplished using only the shuffle property of multiple zeta values in conjunction with (4.4).

6. Dimension Conjectures

Broadhurst [21] has conjectures concerning the size of various bases (graded by weight and depth) for expressing multiple zeta values in terms of either irreducible multiple zeta values, or irreducible Euler sums, and also for expressing Euler sums in terms of irreducible Euler sums. The adjunction of additional differential forms appears to simplify the problem at each stage. Thus, if $D(n,k)$ denotes the number of multiple zeta values of weight n and depth k in a minimal $\mathbf{Q}$-basis for reducing all multiple zeta values to a $\mathbf{Q}$-linear combination of products of basis multiple zeta values, it is conjectured that

$$\prod_{n\geq 3}\prod_{k\geq 1}(1-x^n y^k)^{D(n,k)} \stackrel{?}{=} 1 - \frac{x^3 y}{1-x^2} + \frac{x^{12} y^2 (1-y^2)}{(1-x^4)(1-x^6)}.$$

However, if we allow Euler sums into the basis, letting $M(n,k)$ denote the minimal number of Euler sums of weight n and depth k needed to reduce all multiple zeta values to basis Euler sums, then

$$\prod_{n\geq 3}\prod_{k\geq 1}(1-x^n y^k)^{E(n,k)} \stackrel{?}{=} 1 - \frac{x^3 y}{(1-x^2)(1-xy)}.$$

One can also consider the problem of reducing Euler sums in terms of basis Euler sums. Let $E(n,k)$ denote the minimal number of Euler sums of weight n and depth k required to reduce all Euler sums to basis Euler sums. It is conjectured that

$$\prod_{n\geq 3}\prod_{k\geq 1}(1-x^n y^k)^{E(n,k)} \stackrel{?}{=} 1 - \frac{x^3 y}{1-x^2}.$$

Adjoining forms associated with sixth roots of unity to the set of possible differential forms yields the multiple Clausen values [14], and here it is conjectured that the number $P(n,k)$ of irreducible multiple Clausen values of weight n and depth k is generated by

$$\prod_{n>1}\prod_{k>0}(1-x^n y^k)^{P(n,k)} \stackrel{?}{=} 1 - \frac{x^2 y}{1-x}.$$

7. q-Shuffles

Here we consider a q-analogue of the shuffle algebra discussed in §5. Let A be a set, not necessarily finite, and let $\eta : A \to A$ be bijective. Now form the free monoid generated by A and call it A^* as before. Extend the action of η to A^* in the obvious way so that η becomes an automorphism of A^*. Again regard A as an alphabet, and the elements of A^* as words formed by concatenating any finite number of letters (repetitions permitted) from the alphabet A. By linearly extending the concatenation product to the set $\mathbf{Q}\langle A\rangle$ of rational linear combinations of elements of A^*, we obtain a non-commutative polynomial ring with the elements of A being indeterminates and with multiplicative identity 1 denoting the empty word. It is clear that η now extends to an automorphism of $\mathbf{Q}\langle A\rangle$.

A q-shuffle algebra is defined to be the ordered pair $(\mathbf{Q}\langle A\rangle, \sqcup\!\sqcup_q)$ where $\sqcup\!\sqcup_q$ is a commutative and associative bilinear operator on $\mathbf{Q}\langle A\rangle$ satisfying the identity

(7.1) $\quad \begin{cases} \forall w \in A^*, & 1 \sqcup\!\sqcup_q w = w \sqcup\!\sqcup_q 1 = w, \\ \forall a, b \in A, \quad \forall u, v \in A^*, & au \sqcup\!\sqcup_q bv = a(u \sqcup\!\sqcup_q bv) + b(\eta(au) \sqcup\!\sqcup_q v). \end{cases}$

We denote a q-shuffle algebra over A by $\mathrm{Shq}_\mathbf{Q}[A]$. It will be observed that a q-shuffle algebra is a commutative $\mathbf{Q}$-algebra.

Our definition implies that there may be more than one way of writing the q-shuffle product of two words. For example, letting $a, b \in A$, it is easy to see that in $\mathrm{Shq}_\mathbf{Q}[A]$ one has $a \sqcup\!\sqcup_q b = ab + b\eta a = ba + a\eta b$. As the length of the words being multiplied increases the number of different expressions also grows.

The motivation for our definition of $\mathrm{Shq}_\mathbf{Q}[A]$ is not difficult to see. As the shuffle algebra is motivated by the property (3.6) of iterated integrals, one wants a similar identity to hold for iterated Jackson q-integrals [43]. Recall the definition of a Jackson q-integral. For $x > 0$, let $f : [0, x] \to \mathbf{R}$ be Riemann integrable. The Jackson q-integral of f on $[0, x]$ is defined by

(7.2) $$\int_0^x f(t)\, d_q t := \sum_{n \geq 0} f(xq^n)\, xq^n(1-q).$$

Because for any $0 < q < 1$ the sum on the right hand side of (7.2) is a Riemann sum for $\int_0^x f(t)\, dt$, it follows that the Jackson q-integral tends to the ordinary Riemann integral in the limit as q approaches 1.

One defines iterated Jackson q-integrals in exactly the same way that ordinary iterated integrals are defined by (3.5). To this end, for $j = 1, 2, \ldots, n$ let $f_j : [0, x] \to \mathbf{R}$ and $\omega_j := f_j(t_j)\, d_q t_j$. Then put

(7.3) $$\int_0^x \omega_1 \omega_2 \cdots \omega_n := \prod_{j=1}^n \int_0^{t_{j-1}} f_j(t_j)\, d_q t_j, \quad t_0 := x$$
$$= \begin{cases} \int_0^x f_1(t_1) \int_0^{t_1} \omega_2 \cdots \omega_n\, d_q t_1 & \text{if } n > 0 \\ 1 & \text{if } n = 0. \end{cases}$$

Here the fact that the 1-forms on the right hand side of (7.3) are q-difference 1-forms implies that the integral on the left hand side of (7.3) is a q-iterated integral and not an ordinary iterated integral.

Iterating the definition of the q-iterated integral, one finds that

(7.4) $$\int_0^x \omega_1 \cdots \omega_k = \sum_{n_1, \ldots, n_k \geq 0} f_1(xq^{n_k}) f_2(xq^{n_k + n_{k-1}}) \cdots f_k(xq^{n_k + \cdots + n_1})$$
$$\times q^{kn_k + (k-1)n_{k-1} + \cdots + 2n_2 + n_1}(1-q)^k x^k,$$

but this is not a very convenient expression. To simplify (7.4) it helps to make the following change of indices:

$$l_i := \sum_{j=k-i+1}^k n_j, \quad 1 \leq i \leq k.$$

Then (7.4) reduces to

$$\int_0^x \omega_1 \cdots \omega_k = (1-q)^k \sum_{0 \le l_1 \le l_2 \le \cdots \le l_k} xq^{l_1} f(xq^{l_1}) xq^{l_2} f(xq^{l_2}) \cdots xq^{l_k} f(xq^{l_k}). \tag{7.5}$$

This in turn motivates the definition of q-difference 1-forms by the equation

$$\omega_i = \omega_i(t) = f_i(t)\, d_q t := f_i(t) t(1-q). \tag{7.6}$$

Notice that with this definition, the q-integral reduces to a summation operator on these 1-forms:

$$\int_0^x \omega = \sum_{j \ge 0} \omega(t_0 q^j),$$

which agrees with the original definition by virtue of $t_0 = x$.

Now fix $0 < q < 1$ and define the Rogers [75] q-difference operator η acting on continuous functions $f : [0, \infty] \to \mathbf{R}$ by the equation $(\eta f)(x) := f(xq)$. Also define the q-derivative in the usual way:

$$(D_q f)(x) := \frac{f(x) - f(xq)}{x(1-q)}.$$

The relevant fact about the q-derivative is the following well-known property:

$$D_q \int_0^x f(t)\, d_q t = f(x). \tag{7.7}$$

We are ready to give the motivation for the q-shuffle product. The alphabet A now consists of q-difference 1-forms, and the automorphism η is Rogers' q-difference operator. The action of η is extended to forms by the equation $\eta \omega = \omega(tq) = f(tq)tq(1-q)$. This of course defines a new form $\omega' = \eta \omega$ by $\omega' = g(t)t(1-q)$, where $g(t) = qf(tq)$. The alphabet A needs to be infinite to account for all the forms $\eta^j \omega$ for $j \in \mathbf{Z}$. The action of η is now extended to $\mathbf{Q}\langle A \rangle$ in the obvious way. It clearly forms an automorphism of this algebra. We wish to define the q-shuffle product so that for $u, v \in A^*$ the following equation is true:

$$\int_0^x u \shuffle_q v = \left(\int_0^x u\right)\left(\int_0^x v\right). \tag{7.8}$$

To accomplish this, one applies the q-analogue of the argument given in §5 for deriving the recursive definition of the shuffle product. Take $a, b \in A$ and $u, v \in A^*$. Put $a = f(t)\, d_q t$, $b = g(t)\, d_q t$ and

$$F(x) := \int_0^x (au \shuffle_q bv) = \left(\int_0^x f(t) \int_0^t u\, d_q t\right)\left(\int_0^x g(t) \int_0^t v\, d_q t\right).$$

Writing $F(x) = \int_0^x F'(s)\,d_q s$ where $F'(s) = (D_q F)(s)$, and applying the q-product rule for q-differentiation $((D_q fg) = (D_q f)g + (\eta f)(D_q g))$ yields

$$\begin{aligned}
F(x) &= \int_0^x \left(f(s)\int_0^s u\right)\left(\int_0^s g(t)\int_0^t v\,d_q t\right) d_q s \\
&\quad + \int_0^x g(s)\left(\int_0^{sq} f(t)\int_0^t u\,d_q t\right)\int_0^s v\,d_q s \\
&= \int_0^x [a(u \sqcup\!\sqcup_q bv) + b(\eta(au) \sqcup\!\sqcup v)],
\end{aligned}$$

where the first equality follows from (7.7) and the product rule for D_q. Hence the inductive definition of the q-shuffle product results. Notice that commutativity and associativity of $\sqcup\!\sqcup_q$ follow immediately from (7.8).

We will conclude with a few examples of the q-shuffle product illustrating how several equivalent sums can arise from this product. Taking $\omega_1 \sqcup\!\sqcup_q \omega_2\omega_3$ using the inductive definition gives (among several possibilities):

$$\begin{aligned}
\omega_1 \sqcup\!\sqcup_q \omega_2\omega_3 &= \omega_1\omega_2\omega_3 + \omega_2(\eta\omega_1)\omega_3 + \omega_2\omega_3(\eta^2\omega_1) \\
&= \omega_1\omega_2\omega_3 + \omega_2(\eta\omega_1\omega_3) + \omega_2\omega_3(\eta\omega_1).
\end{aligned}$$

Writing $\omega_i = \omega_i(x)$, these equations translate into the easily verifiable generic series identities:

$$\begin{aligned}
&\sum_{0\leq l_1}\omega_1(xq^{l_1})\sum_{0\leq l_2\leq l_3}\omega_2(xq^{l_2})\omega_3(xq^{l_3}) \\
&= \sum_{0\leq l_1\leq l_2\leq l_3}\omega_1(xq^{l_1})\omega_2(xq^{l_2})\omega_3(xq^{l_3}) + \sum_{0\leq l_2\leq l_1\leq l_3}\omega_2(xq^{l_2})\omega_1(xq^{l_1+1})\omega_3(xq^{l_3}) \\
&\quad + \sum_{0\leq l_2\leq l_3\leq l_1}\omega_2(xq^{l_2})\omega_3(xq^{l_3})\omega_1(xq^{l_1+2}) \\
&= \sum_{0\leq l_1\leq l_2\leq l_3}\omega_1(xq^{l_1})\omega_2(xq^{l_2})\omega_3(xq^{l_3}) + \sum_{0\leq l_2\leq l_1\leq l_3}\omega_2(xq^{l_2})\omega_1(xq^{l_1+1})\omega_3(xq^{l_3+1}) \\
&\quad + \sum_{0\leq l_2\leq l_3\leq l_1}\omega_2(xq^{l_2})\omega_3(xq^{l_3})\omega_1(xq^{l_1+1}).
\end{aligned}$$

Taking the q-shuffle product as acting on the non-commutative polynomials in forms, it follows that all these different expressions for the q-shuffle product tend to the ordinary shuffle product in the limit as q approaches 1. Further results about q-shuffle algebras and their combinatorics will be given elsewhere.

References

[1] S. Akiyama, S. Egami, and Y. Tanigawa, *An analytic continuation of multiple zeta functions and their values at non-positive integers*, preprint.
[2] S. Akiyama and Y. Tanigawa, *Multiple zeta values at non-positive integers*, preprint.
[3] T. M. Apostol and T. H. Vu, *Dirichlet series related to the Riemann zeta function*, J. Number Theory, **19** (1984), 85–102.
[4] T. Arakawa and M. Kaneko, *Multiple zeta values, poly-Bernoulli numbers, and related zeta functions*, Nagoya Math. J., **153** (1999), 189–209.
[5] F. V. Atkinson, *The mean-value of the Riemann zeta function*, Acta Math., **81** (1949), 353–376.
[6] David H. Bailey, Jonathan M. Borwein and Roland Girgensohn, *Experimental evaluation of Euler sums*, Experiment. Math., **3** (1994), 17–30.

[7] D. T. Barfoot and David J. Broadhurst, $Z_2 \times S_6$ symmetry of the two-loop diagram, Zeit. für Physik C **41** (1988), 81–85.

[8] A. A. Beilinson, A. B. Goncharov, V. V. Schechtman and A. N. Varchenko, *Aomoto dilogarithms, mixed Hodge structures and motivic cohomology of pairs of triangles on the plane*, The Grothendieck Festschrift, Vol. I, Progr. Math. **86**, Birkhäuser, Boston, (1990), 135–171.

[9] David Borwein and Jonathan M. Borwein, *On an intriguing integral and some series related to $\zeta(4)$*, Proc. Amer. Math. Soc., **123** (1995), 1191–1198.

[10] David Borwein, Jonathan M. Borwein and Roland Girgensohn, *Explicit evaluation of Euler sums*, Proc. Edinburgh Math. Soc., **38** (1995), 227–294.

[11] Jonathan M. Borwein, David M. Bradley and David J. Broadhurst, *Evaluations of k-fold Euler/Zagier sums: a compendium of results for arbitrary k*, Electronic J. Combin., **4** (1997), no. 2, #R5. Wilf Festschrift.

[12] Jonathan M. Borwein, David M. Bradley, David J. Broadhurst and Petr Lisoněk, *Combinatorial aspects of multiple zeta values*, Electronic J. Combinatorics, **5** (1998), no. 1, #R38.

[13] _____ *Special values of multiple polylogarithms*, Trans. Amer. Math. Soc., **355** (2001), no. 3, 907–941.

[14] Jonathan M. Borwein, David J. Broadhurst and Joel Kamnitzer, *Central binomial sums, multiple Clausen values, and zeta values*, Experiment. Math. **10** (2001), no. 1, 25–34.

[15] Jonathan M. Borwein and Roland Girgensohn, *Evaluation of triple Euler sums*, Electronic J. Combin., **3** (1996) #R23.

[16] Douglas Bowman and David M. Bradley, *Resolution of some open problems concerning multiple zeta evaluations of arbitrary depth*, submitted October, 1999.

[17] _____ *The algebra and combinatorics of shuffles and multiple zeta values*, to appear in J. Combinatorial Theory, Series A.

[18] _____ *On multiple polylogarithms possessing ultimately periodic argument lists*, in preparation.

[19] David J. Broadhurst, *Evaluation of a class of Feynman diagrams for all numbers of loops and dimensions*, Phys. Lett. B **164** (1985), 356–360.

[20] _____, *Exploiting the 1,440-fold symmetry of the master two-loop diagram*, Zeit. Phys. C, **32** (1986), 249–253.

[21] _____ *Conjectured Enumeration of irreducible multiple zeta values, from knots and Feynman diagrams*, preprint.

[22] _____ *On the enumeration of irreducible k-fold Euler sums and their roles in knot theory and field theory*, to appear in J. Math. Phys.

[23] _____ *Massive 3-loop Feynman diagrams reducible to SC^* primitives of algebras of the sixth root of unity*, Eur. Phys. J. C Part. Fields **8** (1999), no. 2, 313–333.

[24] David J. Broadhurst, John A. Gracey and Dirk Kreimer, *Beyond the triangle and uniqueness relations; non-zeta terms at large N from positive knots*, Zeit. Phys. C, **75** (1997), no. 3, 559–574.

[25] David J. Broadhurst and Dirk Kreimer, *Association of multiple zeta values with positive knots via Feynman diagrams up to 9 loops*, Phys. Lett. B, **393** (1997) no. 3-4, 403–412.

[26] David J. Broadhurst and Dirk Kreimer, *Knots and numbers in ϕ^4 theory to 7 loops and beyond*, Int. J. Mod. Phys., **C6** (1995), no. 4, 519–524.

[27] Jerzy Browkin, *Conjectures on the dilogarithm*, K-Theory, **3** (1989), no. 1, 29–56.

[28] _____ *K-theory, cyclotomic equations and Clausen's function*, in Structural Properties of Polylogarithms, edited by Leonard Lewin, Amer. Math. Soc. Mathematical Surveys and Monographs **37**, Providence, RI, 1991, 233–273.

[29] E. R. Canfield, Paul Erdös and Carl Pomerance, *On a problem of Oppenheim concerning "factorisatio numerorum,"* J. Number Theory, **17** (1983), 1–28.

[30] Kuo-Tsai Chen, *Iterated integrals and exponential homomorphisms*, Proc. London Math. Soc., (3) **4** (1954), 502–512.

[31] _____ *Integration of paths, geometric invariants and a generalized Baker-Hausdorff formula*, Ann. of Math., **65** (1957), No. 1, 163–178.

[32] _____ *Integration of paths– a faithful representation of paths by noncommutative formal power series*, Trans. Amer. Math. Soc., **98** (1958), No. 2, 395–407.

[33] _____ *Differential Forms and Homotopy Groups*, J. Differential Geometry, **6**, (1971), 231–246.

[34] T. Clausen, *Ueber die Fälle, wenn die reihe von der form ... ein quadrat von der form ... hat*, J. reine angew. Math., **3** (1828), 89–91.

[35] Richard E. Crandall, *Fast evaluation of multiple zeta sums*, Math. Comp., **67** (1998), no. 223, 1163–1172.

[36] Richard E. Crandall and Joe P. Buhler, *On the evaluation of Euler sums*, Experiment. Math., **3** (1994), no. 4, 275–285.

[37] Hervé Daudé, Philippe Flajolet and Brigitte Vallée, *An average-case analysis of the Gaussian algorithm for lattice reduction*, Combin. Probab. Comput., **6** (1997), no. 4, 397–433.

[38] P. J. De Doelder, *On some series containing $\psi(x) - \psi(y)$ and $(\psi(x) - \psi(y))^2$ for certain values of x and y*, J. Comput. Appl. Math. **37** (1991), 124–141.

[39] Leonhard Euler, *Meditationes circa singulare serierum genus*, Novi Comm. Acad. Sci. Petropol., **20** (1775), 140–186, Reprinted in "Opera Omnia", ser. I, **15**, B. G. Teubner, Berlin, 1927, pp. 217–267.

[40] Philippe Flajolet and Bruno Salvy, *Euler sums and contour integral representations*, Experiment. Math. **7** (1998), no. 1, 15–35.

[41] Philippe Flajolet, Gilbert Labelle, Louise Laforest and Bruno Salvy, *Hypergeometrics and the cost structure of quadtrees*, Random Structures and Algorithms, **7** (1995), no. 2, 117–144.

[42] H. R. P. Ferguson and R. W. Forcade, *Generalization of the Euclidean algorithm for real numbers to all dimensions higher than two*, Bull. Amer. Math. Soc., **1** (1979), 912–914.

[43] George Gasper and Mizan Rahman, *Basic Hypergeometric Series*, Cambridge University Press, Cambridge, 1990.

[44] I. M. Gelfand and G. E. Shilov, *Generalized Functions* vol. I, Academic Press, New York, 1964.

[45] Solomon W. Golomb and Lloyd R. Welch, *Perfect codes in the Lee metric and the packing of polyominoes*, SIAM J. Appl. Math., **18** (1970), no. 2, 302–317.

[46] Alexander B. Goncharov, *Polylogarithms in arithmetic and geometry*, Proceedings of the International Congress of Mathematicians, **1, 2** (Zürich, 1994), 374–387, Birkhäuser, Basel, 1995.

[47] _____ *The double logarithm and Manin's complex for modular curves*, Math. Res. Lett., **4** (1997), no. 5, 617–636.

[48] _____ *Multiple polylogarithms, cyclotomy and modular complexes*, Math. Res. Lett., **5** (1998), no. 4, 497–516.

[49] Andrew Granville, *A decomposition of Riemann's zeta-function*, in Analytic Number Theory, London Mathematical Society Lecture Notes Series 247, Cambridge University Press, Y. Motohashi ed., (1997), 95–101.

[50] J. Hastad, B. Just, J. C. Lagarias and C. P. Schnorr, *Polynomial time algorithms for finding integer relations among real numbers*, SIAM J. Comp., **18** (1988), 859–881.

[51] Doug Hensley, *The distribution of the number of factors in a factorization*, J. Number Theory, **26** (1987), 179–191.

[52] Michael E. Hoffman, *Multiple harmonic series*, Pacific J. Math., **152** (1992), no. 2, 275–290.

[53] _____ *The algebra of multiple harmonic series*, J. Algebra, **194** (1997), 477–495.

[54] _____ *Quasi-shuffle products*, J. Algebraic Combin., **11** (2000), 49–68.

[55] _____ *Periods of mirrors and multiple zeta values*, to appear in Proc. Amer. Math. Soc.

[56] Michael E. Hoffman and Courtney Moen, *Sums of triple harmonic series*, J. Number Theory, **60** (1996), 329–331.

[57] Michael E. Hoffman and Yasuo Ohno, *Relations of multiple zeta values and their algebraic expression*, submitted.

[58] J. A. Lappo-Danilevsky, *Mémoires sur la théorie des systémes des équations différentielles linéaires*, Chelsea, 1953.

[59] Gilbert Labelle and Louise Laforest, *Combinatorial variations on multidimensional Quadtrees* J. Combinatorial Theory, Series A, **69** (1995), no. 1, 1–16.

[60] Tu Quoc Thang Le and Jun Murakami, *Kontsevich's integral for the Homfly polynomial and relations between values of multiple zeta functions*, Topology and its Applications, **62** (1995), no. 2, 193–206.

[61] _____ *Kontsevich's integral for the Kauffman polynomial*, Nagoya Math. J., **142** (1996), 39–65.

[62] A. K. Lenstra, H. W. Lenstra and L. Lovasz, *Factoring polynomials with rational coefficients*, Math. Annalen, **261** (1982), 515–534.

[63] I. G. MacDonald, *Symmetric functions and Hall polynomials*, Clarendon Press, Oxford, 1995.
[64] Clemens Markett, *Triple sums and the Riemann zeta function*, J. Number Theory, **48** (1994), 113–132.
[65] Hoang Ngoc Minh and Michel Petitot, *Polylogarithms and the Riemann ζ function*, Discrete Math., **217** (2000), no. 1–3, 273–292.
[66] L. J. Mordell, *On the evaluation of some multiple series*, J. London Math Soc. **33** (1958), no. 2, 368–371.
[67] Niels Nielsen, *Die Gammafunktion*, Chelsea, New York, 1965, pp. 47–51.
[68] Yasuo Ohno, *A generalization of the duality and sum formulas on the multiple zeta values*, J. Number Theory, **74** (1999), 39–43.
[69] _____ *A proof of the cyclic sum conjecture for multiple zeta values*, preprint.
[70] A. Oppenheim, *On an arithmetic function*, J. London Math. Soc., **1** (1926), 205–211.
[71] W. Orr, *Theorems relating to the product of two hypergeometric series*, Trans. Camb. Phil. Soc., **17** (1899), 1–15.
[72] R. Sita Ramachandra Rao and M. V. Subbarao, *Transformation formulae for multiple series*, Pacific J. Math., **113** (1984), no. 2, 471–479.
[73] David E. Radford, *A natural ring basis for the shuffle algebra and an application to group schemes*, J. Algebra, **58** (1979), 432–454.
[74] Rimhak Ree, *Lie elements and an algebra associated with shuffles*, Ann. of Math., **62** (1958), No. 2, 210–220.
[75] L. J. Rogers, *On the expansion of some infinite products*, Proc. Lond. Math. Soc., **24**, (1893), 337–352.
[76] M. V. Subbarao and R. Sitaramachandrarao, *On some infinite series of L. J. Mordell and their analogues*, Pacific J. Math., **119** (1985), no. 1, 245–255.
[77] Takashi Takamuki, *The Kontsevich invariant and relations of multiple zeta values*, Kobe J. Math. **16** (1999), no. 1, 27–43.
[78] Richard Warlimont, *Factorisatio numerorum with constraints*, J. Number Theory, **45** (1993), 186–199.
[79] Zdzislaw Wojtkowiak, *The basic structure of polylogarithmic functional equations*, in Structural Properties of Polylogarithms, edited by Leonard Lewin, Amer. Math. Soc. Mathematical Surveys and Monographs **37**, Providence, RI, 1991, 205–231.
[80] _____ *Functional equations of iterated integrals with regular singularities*, Nagoya Math. J., **142** (1996), 145–159.
[81] _____ *Mixed Hodge structures and iterated integrals I*, June, 1999. [K-theory preprint #351, http://www.math.uiuc.edu/K-theory]
[82] Don Zagier, *Values of zeta functions and their applications*, First European Congress of Mathematics, Vol. II, Birkhäuser, Boston, 1994, 497–512.
[83] Jianqiang Zhao, *Analytic continuation of multiple zeta functions*, Proc. Amer. Math. Soc., **128** (2000), 1275–1283.

UNIVERSITY OF ILLINOIS AT URBANA-CHAMPAIGN, DEPARTMENT OF MATHEMATICS, 273 ALTGELD HALL, 1409 W. GREEN ST., URBANA, IL 61801 U.S.A.
E-mail address: `bowman@math.uiuc.edu`

DEPARTMENT OF MATHEMATICS AND STATISTICS, UNIVERSITY OF MAINE, 5752 NEVILLE HALL, ORONO, ME 04469-5752 U.S.A.
E-mail address: `dbradley@e-math.ams.org`

Contemporary Mathematics
Volume **291**, 2001

Swinnerton-Dyer Type Congruences for certain Eisenstein series

Matthew Boylan

ABSTRACT. We consider a normalized Eisenstein series of weight k on a congruence subgroup of type $\Gamma_0(N)$ with Nebentypus character χ which vanishes at all cusps of $\Gamma_0(N)$ inequivalent to the cusp at infinity. We determine conditions on N, k, χ, and an ideal $\mathfrak{a}$ in certain number fields, under which their Fourier series are congruent to 1 (mod $\mathfrak{a}$).

1. Introduction and Statement of Results

If $k \geq 4$ is an even integer, then it is well-known [**K, pg.111**] that the normalized Eisenstein series given by

$$E_k(z) = 1 - \frac{2k}{B_k} \sum_{n \geq 1} \sigma_{k-1}(n) q^n$$

is a modular form of weight k with respect to $SL(2, \mathbb{Z})$, where B_k is the kth Bernoulli number, $q := e^{2\pi i z}$, and $\sigma_k(n)$ is the function which sums the kth powers of the positive divisors of n. Swinnerton-Dyer [**Sw-D**] showed that $E_k(z)$ satisfies the following congruence property:

THEOREM (SWINNERTON-DYER). *If $\ell \geq 5$ is a prime, then $E_k(z) \equiv 1 \pmod{\ell}$ if and only if $k \equiv 0 \pmod{\ell - 1}$.*

This follows from the Von Staudt-Claussen Theorem regarding the divisibility of the denominators of Bernoulli numbers.

Here we generalize Swinnerton-Dyer's result to certain Eisenstein series in spaces of modular forms of weight k on a congruence subgroup of type $\Gamma_0(N)$ with Nebentypus character χ. These spaces are denoted by $M_k(\Gamma_0(N), \chi)$. For background on integer weight modular forms, see [**K**]. Following the methods in Sections 1-3 of Chapter VII of Schoeneberg's book, *Elliptic Modular Functions*, we develop the Fourier expansion at infinity of a normalized Eisenstein series $E_{N,k,\chi}(z) \in M_k(\Gamma_0(N), \chi)$ which vanishes at all cusps of $\Gamma_0(N)$ inequivalent to

---

1991 *Mathematics Subject Classification.* Primary 11F33.
Key words and phrases. Congruences for Eisenstein series.

© 2001 American Mathematical Society

the cusp at infinity, and we state conditions on N, k, and χ guaranteeing the existence of these series. (Schoeneberg does this for Eisenstein series without character on an arbitrary subgroup of level N). Using these expansions, we obtain conditions on N, k, and χ, and an ideal $\mathfrak{a}$ in certain number fields, under which $E_{N,k,\chi}(z) \equiv 1$ (mod $\mathfrak{a}$).

Theorem 1.1 lists formulas for $E_{N,k,\chi}(z)$ when they exist.

THEOREM 1.1. *Suppose χ is a Dirichlet character with modulus N and conductor f, and $\tau_m(d, \chi) := \sum_{h=1}^{m-1} \chi(h)\zeta_m^{dh}$, where $\zeta_m := e^{\frac{2\pi i}{m}}$. Suppose also that if χ is nontrivial and $N = 1$ or 2, then $k \geq 4$ is an even integer satisfying $\chi(-1) = (-1)^k$, and if χ is nontrivial and $N > 2$, then $k \geq 3$ is an integer satisfying $\chi(-1) = (-1)^k$. Then the series $E_{N,k,\chi}(z)$ given by the following formulas are normalized modular forms in $M_k(\Gamma_0(N), \chi)$ which vanish at all cusps of $\Gamma_0(N)$ inequivalent to the cusp at infinity.*

1. *If χ is trivial and $N = 1$, then for an even integer $k \geq 4$,*

$$(1) \qquad E_{N,k,\chi}(z) = E_k(z) = 1 - \frac{2k}{B_k} \sum_{n \geq 1} \sigma_{k-1}(n) q^n.$$

If χ is trivial and $N > 1$, then for an even integer $k \geq 4$,

$$(2) \qquad E_{N,k,\chi}(z) = 1 - \frac{2k\phi(N)}{N^k B_k \prod_{p|N}\left(1 - \frac{1}{p^k}\right)} \sum_{n \geq 1} \left(\sum_{\substack{d|n \\ d>0}} d^{k-1} \frac{\mu(N/\gcd(d,N))}{\phi(N/\gcd(d,N))} \right) q^n,$$

where ϕ denotes Euler's phi function, and μ denotes the Möbius function.

2. *If χ is nontrivial, then*

$$(3) \qquad E_{N,k,\chi}(z) = 1 - \frac{k}{(\frac{N}{f})^k \tau_f(1, \overline{\chi}) B_{k,\chi}} \sum_{n \geq 1} \left(\sum_{\substack{d|n \\ d>0}} d^{k-1}(\tau_N(d, \overline{\chi}) + (-1)^k \tau_N(-d, \overline{\chi})) \right) q^n,$$

where $B_{k,\chi}$ is the generalized Bernoulli number associated to χ.

3. *If χ is nontrivial and primitive, then*

$$(4) \qquad E_{N,k,\chi}(z) = 1 - \frac{2k}{B_{k,\chi}} \sum_{n \geq 1} \left(\sum_{\substack{d|n \\ d>0}} \chi(d) d^{k-1} \right) q^n.$$

In what follows, let $K_\chi = \mathbb{Q}(\chi)$ denote the extension of $\mathbb{Q}$ obtained by adjoining the values of χ, let O_{K_χ} denote the ring of integers of K_χ, and denote by $O_{K_{\chi,N}}$ the ring of integers of $K_{\chi,N} = \mathbb{Q}(\chi, \zeta_N)$. We also define $\mathrm{ord}_m(n)$ to be the power of m dividing n if m and n are integers. If $\alpha = \frac{a}{b} \in \mathbb{Q}$, then $\mathrm{ord}_m(\alpha) := \mathrm{ord}_m(a) - \mathrm{ord}_m(b)$. Theorems 1.2 and 1.3 generalize Swinnerton-Dyer's Theorem to the series (2) and (4) listed in Theorem 1.1.

THEOREM 1.2. *Suppose that ℓ is an odd prime, χ is the trivial Dirichlet character modulo N, and $k \geq 4$ is an even integer. Then the following are true:*
1. $E_{N,k,\chi}(z) \equiv 1 \pmod{\ell}$ *if and only if $k \equiv 0 \pmod{\ell-1}$ and $N = \ell^t$ for some nonnegative integer t.*
2. $E_{N,k,\chi}(z) \equiv 1 \pmod{2}$ *if and only if $N = 2^a p^b$, where p is an odd prime satisfying $\mathrm{ord}_2(p^k - 1) = 1 + \mathrm{ord}_2(k)$ and a and b are nonnegative integers.*

REMARK. *Theorems 1.2.1 and 1.2.2 contain Swinnerton-Dyer's Theorem as a special case, the case where $N = 1$.*

THEOREM 1.3. *Suppose that ℓ is an odd rational prime, $\mathfrak{a}$ is an ideal in $O_{K_\chi, N}$ with the property that $\mathfrak{a} \nmid (2)$, and χ is a nontrivial primitive Dirichlet character. Then for an integer $k \geq 3$ satisfying $\chi(-1) = (-1)^k$, we have:*

1. *If N has at least two distinct prime divisors, then*
$$E_{N,k,\chi}(z) \not\equiv 1 \pmod{\mathfrak{a}}.$$

2. *If $N = 4$ and $\mathfrak{a} \nmid (4)$, then*
$$E_{4,k,\chi}(z) \not\equiv 1 \pmod{\mathfrak{a}}.$$
If $N = 2^t$ for some integer $t \geq 3$, then
$$E_{2^t,k,\chi}(z) \not\equiv 1 \pmod{\mathfrak{a}}.$$

3. *If $N = \ell$, then*
$$E_{\ell,k,\chi}(z) \not\equiv 1 \pmod{\mathfrak{a}}$$
unless there is a primitive root g of $\mathbb{Z}/\ell\mathbb{Z}$ satisfying
$$\mathfrak{p} = \gcd(\ell, 1 - \chi(g)g^k) \neq (1),$$
where $\mathfrak{p}$ is an ideal in O_{K_χ}. In this case
$$E_{\ell,k,\chi}(z) \equiv 1 \pmod{\mathfrak{p}^{\mathrm{ord}_\mathfrak{p}(\ell)}}.$$

4. *If $N = \ell$ and $\chi = \left(\frac{\bullet}{\ell}\right)$, the Legendre symbol, then*
$$E_{\ell,k,\left(\frac{\bullet}{\ell}\right)} \equiv 1 \pmod{\ell}$$
if and only if $k \equiv \frac{\ell-1}{2} \pmod{\ell - 1}$.

5. *If $N = \ell^t$ for some integer $t \geq 2$, and if $\gcd(\ell, 1 - \chi(g)g^k) = (1)$ for every primitive root g of $\mathbb{Z}/\ell\mathbb{Z}$, then*
$$E_{\ell^t,k,\chi}(z) \not\equiv 1 \pmod{\mathfrak{a}}.$$

Note that if $\mathfrak{a}$ is an ideal in $O_{K_\chi,N}$ and if j is a positive integer, then $E^j_{N,k,\chi}(z) \equiv 1 \pmod{\mathfrak{a}}$ whenever $E_{N,k,\chi}(z) \equiv 1 \pmod{\mathfrak{a}}$, where $E^j_{N,k,\chi}(z) \in M_{jk}(\Gamma_0(N), \chi)$.

2. Background on Schoeneberg's Eisenstein Series

Before proceeding with the proof of Theorems 1.1-1.3, we describe the basic properties of Schoeneberg's primitive and reduced Eisenstein series, the building blocks for the Eisenstein series $E_{N,k,\chi}(z)$ that we construct. We keep the notation from Schoeneberg's book in what follows.

If $f : \mathbb{H} \mapsto \hat{\mathbb{C}}$, where $\mathbb{H}$ is the upper half plane and $\hat{\mathbb{C}} = \mathbb{C} \cup \{\infty\}$, and if $S = \begin{bmatrix} a & b \\ c & d \end{bmatrix} \in SL(2,\mathbb{Z})$, then $f(z)|_k S := (cz+d)^{-k} f(Sz)$. Suppose that $N \geq 1$ and $k \geq 3$ are integers, and $\mathbf{m} = \begin{bmatrix} m_1 \\ m_2 \end{bmatrix}$ and $\mathbf{a} = \begin{bmatrix} a_1 \\ a_2 \end{bmatrix}$ are pairs of integers. Schoeneberg defines the inhomogenous Eisenstein series of weight k and level N as follows [**Sc, pg.155, (2)**]:

$$G_{N,k,\mathbf{a}}(z) = \sum_{\substack{m_1 \equiv a_1 \pmod{N} \\ m_2 \equiv a_2 \pmod{N} \\ \mathbf{m} \neq \mathbf{0}}} (m_1 z + m_2)^{-k},$$

(In his notation, Schoeneberg refers to modular forms having *dimension* $-k < 0$, rather than having *weight* $k > 0$, which means the same. We prefer to use the term *weight*.) If $\gcd(a_1, a_2, N) = 1$, then $G_{N,k,\mathbf{a}}(z)$ is called a *primitive* Eisenstein series.

The relevant facts about primitive Eisenstein series are these:

1. [**Sc, pg.155, Thm.1**] For all $\mathbf{a}$, $G_{N,k,\mathbf{a}}(z) \in M_k(\Gamma(N))$.
2. [**Sc, pg.155, (3)**] For all $\mathbf{a}$,

(5) $$G_{N,k,-\mathbf{a}}(z) = (-1)^k G_{N,k,\mathbf{a}}(z).$$

3. [**Sc, pg.155, (3)**] If $\mathbf{a} \equiv \mathbf{a}_1 \pmod{N}$, then

(6) $$G_{N,k,\mathbf{a}}(z) = G_{N,k,\mathbf{a}_1}(z).$$

4. [**Sc, pg.156, (4)**] If $A \in SL(2,\mathbb{Z})$, then

(7) $$G_{N,k,\mathbf{a}}(z)|_k A = G_{N,k,A'\mathbf{a}}(z),$$

where A' is the transpose of A.

5. [**Sc, pg.157, (5)**] For integers a and b, define

(8) $$\delta\left(\frac{a}{b}\right) = \begin{cases} 1 & \text{if } b \mid a, \\ 0 & \text{if } b \nmid a. \end{cases}$$

Then

$$G_{N,k,\mathbf{a}}(z) = \sum_{\nu \geq 0} \alpha_\nu(N,k,\mathbf{a}) e^{\frac{2\pi i \nu z}{N}},$$

where

(9) $$\alpha_0(N,k,\mathbf{a}) = \delta\left(\frac{a_1}{N}\right) \sum_{\substack{m_2 \equiv a_2 \pmod{N} \\ m_2 \neq 0}} m_2^{-k},$$

and for $\nu \geq 1$,

$$\alpha_\nu(N, k, \mathbf{a}) = \frac{(-2\pi i)^k}{N^k(k-1)!} \sum_{\substack{m \mid \nu \\ \frac{\nu}{m} \equiv a_1 \pmod{N}}} m^{k-1} \mathrm{sgn}(m) \zeta_N^{a_2 m}. \tag{10}$$

One may also define *reduced* Eisenstein series. If $\gcd(a_1, a_2, N) = 1$, then these may be written [**Sc, pg.158, (7)**]:

$$G^*_{N,k,\mathbf{a}}(z) = \sum_{\substack{\mathbf{m} \equiv \mathbf{a} \pmod{N} \\ \gcd(m_1, m_2) = 1}} (m_1 z + m_2)^{-k}.$$

Facts 1-4 concerning primitive Eisenstein series also hold for reduced Eisenstein series. The reduced Eisenstein series are expressible as a linear combination of primitive Eisenstein series [**Sc, pg.159, (9)**]:

$$G^*_{N,k,\mathbf{a}}(z) = \sum_{t \pmod{N}} \left(\sum_{\substack{dt \equiv 1 \pmod{N} \\ d > 0}} \frac{\mu(d)}{d^k} \right) G_{N,k,t\mathbf{a}}(z). \tag{11}$$

We now proceed with the proof of Theorem 1.1

3. The Proof of Theorem 1.1

We note that $\Gamma_0(N) = \bigcup_{\nu=1}^{\mu_1} \Gamma(N) A_\nu$, where the coset representatives A_ν lie in the set

$$\left\{ \begin{bmatrix} \alpha_\nu & \beta_\nu \\ \gamma_\nu & \delta_\nu \end{bmatrix} \in SL(2, \mathbb{Z}) \right\}, \tag{12}$$

where $\alpha_\nu \in (\mathbb{Z}/N\mathbb{Z})^*$, $\beta_\nu \in \mathbb{Z}/N\mathbb{Z}$, $\gamma_\nu \equiv 0 \pmod{N}$, $\delta_\nu \equiv \alpha_\nu^{-1} \pmod{N}$, and $\mu_1 = [\Gamma_0(N) : \Gamma(N)]$. We suppose that χ is a Dirichlet character with modulus N and conductor f. As our goal is to construct modular forms for $M_k(\Gamma_0(N), \chi)$, we impose the condition that $k \geq 3$ is an integer with the property that

$$\chi(-1) = (-1)^k. \tag{13}$$

If $A = \begin{bmatrix} a & b \\ c & d \end{bmatrix} \in \Gamma_0(N)$, then the transformation law satisfied by modular forms $f(z) \in M_k(\Gamma_0(N), \chi)$ is given by:

$$f(z)|_k A = \chi(d) f(z). \tag{14}$$

An application of (14) using $A = -I \in \Gamma_0(N)$ shows that the spaces $M_k(\Gamma_0(N), \chi)$ contain only the modular form which is identically zero when (13) does not hold. We claim that

$$G^*_{\Gamma_0(N), k, \chi, \begin{bmatrix} 0 \\ 1 \end{bmatrix}}(z) = \sum_{\nu=1}^{\mu_1} \overline{\chi}(\delta_\nu) G^*_{N, k, \begin{bmatrix} 0 \\ 1 \end{bmatrix}}(z)|_k A_\nu$$

is a modular form in $M_k(\Gamma_0(N), \chi)$ with the property that it vanishes at all cusps of $\Gamma_0(N)$ inequivalent to the cusp at infinity.

Observing that $G^*_{\Gamma_0(N),k,\chi,\begin{bmatrix}0\\1\end{bmatrix}}(z)$ is a linear combination of reduced, and hence, by (11), primitive Eisenstein series, we impose the additional condition that $k \geq 4$ is an even integer when $N = 1$ or 2. When $N = 1$ or 2 and $k \geq 3$ is an odd integer, it follows by (6) and (5) that

$$G_{N,k,\mathbf{a}}(z) = G_{N,k,-\mathbf{a}}(z)$$
$$= -G_{N,k,\mathbf{a}}(z).$$

This shows that $G_{N,k,\mathbf{a}}(z) = 0$ in this case.

To verify that $G^*_{\Gamma_0(N),k,\chi,\begin{bmatrix}0\\1\end{bmatrix}}(z) \in M_k(\Gamma_0(N),\chi)$, we only need to show that it satisfies (14) since it clearly satisfies the remaining defining properties of a modular form in $M_k(\Gamma_0(N),\chi)$. If $A = \begin{bmatrix} a & b \\ c & d \end{bmatrix} \in \Gamma_0(N)$, then $A_\nu A = G_\nu A_{\nu'}$ for some $G_\nu \in \Gamma(N)$, and for some ν' uniquely determined by ν which runs through $\{1,...,\mu_1\}$ as ν does. Moreover, $A_{\nu'} = \begin{bmatrix} \alpha_{\nu'} & \beta_{\nu'} \\ \gamma_{\nu'} & \delta_{\nu'} \end{bmatrix}$, with $\delta_\nu \equiv d^{-1}\delta_{\nu'} \pmod{N}$. Therefore,

$$G^*_{\Gamma_0(N),k,\chi,\begin{bmatrix}0\\1\end{bmatrix}}(z)|_k A = \sum_{\nu=1}^{\mu_1} \overline{\chi}(\delta_\nu) G^*_{N,k,\begin{bmatrix}0\\1\end{bmatrix}}(z)|_k A_\nu A$$

$$= \sum_{\nu'=1}^{\mu_1} \overline{\chi}(d^{-1}\delta_{\nu'}) G^*_{N,k,\begin{bmatrix}0\\1\end{bmatrix}}(z)|_k G_\nu A_{\nu'}$$

$$= \chi(d) \sum_{\nu'=1}^{\mu_1} \overline{\chi}(\delta_{\nu'}) G^*_{N,k,\begin{bmatrix}0\\1\end{bmatrix}}(z)|_k A_{\nu'}$$

$$= \chi(d) G^*_{\Gamma_0(N),k,\chi,\begin{bmatrix}0\\1\end{bmatrix}}(z),$$

so $G^*_{\Gamma_0(N),k,\chi,\begin{bmatrix}0\\1\end{bmatrix}}(z) \in M_k(\Gamma_0(N),\chi)$.

Next, we calculate the value of $G^*_{\Gamma_0(N),k,\chi,\begin{bmatrix}0\\1\end{bmatrix}}(z)$ at an arbitrary cusp $\frac{-d}{c}$. To do this, we form $A = \begin{bmatrix} a & b \\ c & d \end{bmatrix} \in SL(2,\mathbb{Z})$, and consider $G^*_{\Gamma_0(N),k,\chi,\begin{bmatrix}0\\1\end{bmatrix}}(z)|_k A^{-1} = \sum_{n\geq 0} r(n) q^{\frac{n}{N}}$. The value of $G^*_{\Gamma_0(N),k,\chi,\begin{bmatrix}0\\1\end{bmatrix}}(z)$ at $\frac{-d}{c}$ is $r(0)$. The first step in the calculation is to simplify $G^*_{\Gamma_0(N),k,\chi,\begin{bmatrix}0\\1\end{bmatrix}}(z)|_k A^{-1}$ using (7) twice and (12):

$$G^*_{\Gamma_0(N),k,\chi,\begin{bmatrix}0\\1\end{bmatrix}}(z)|_k A^{-1} = \sum_{\nu=1}^{\mu_1} \overline{\chi}(\delta_\nu) G^*_{N,k,\begin{bmatrix}0\\1\end{bmatrix}}(z)|_k A_\nu A^{-1}$$

$$= \sum_{\nu=1}^{\mu_1} \overline{\chi}(\delta_\nu) G^*_{N,k,\begin{bmatrix}\gamma_\nu\\\delta_\nu\end{bmatrix}}(z)|_k A^{-1}$$

$$= N \sum_{h\in(\mathbb{Z}/N\mathbb{Z})^*} \overline{\chi}(h) G^*_{N,k,\begin{bmatrix}0\\h\end{bmatrix}}(z)|_k A^{-1}$$

(15)
$$= N \sum_{h\in(\mathbb{Z}/N\mathbb{Z})^*} \overline{\chi}(h) G^*_{N,k,\begin{bmatrix}-ch\\ah\end{bmatrix}}(z).$$

In view of (15), we simplify $G^*_{N,k,\left[\begin{smallmatrix}-ch\\ah\end{smallmatrix}\right]}(z)$ using (11):

$$G^*_{N,k,\left[\begin{smallmatrix}-ch\\ah\end{smallmatrix}\right]}(z) = \sum_{\nu \geq 0} \alpha^*_\nu\left(N,k,\left[\begin{smallmatrix}-ch\\ah\end{smallmatrix}\right]\right) e^{\frac{2\pi i\nu z}{N}}$$

(16)
$$= \sum_{t\in(\mathbb{Z}/N\mathbb{Z})^*}\left(\sum_{\substack{dt\equiv 1\ (\mathrm{mod}\ N)\\ d>0}} \frac{\mu(d)}{d^k}\right) G_{N,k,\left[\begin{smallmatrix}-tch\\tah\end{smallmatrix}\right]}(z).$$

We follow Schoeneberg's computation of $\alpha^*_0\left(N,k,\left[\begin{smallmatrix}-ch\\ah\end{smallmatrix}\right]\right)$ and note that the first equality is obtained by applying (9) and (16) [**Sc, pg.160**]:

$$\alpha^*_0\left(N,k,\left[\begin{smallmatrix}-ch\\ah\end{smallmatrix}\right]\right) = \sum_{t\in(\mathbb{Z}/N\mathbb{Z})^*}\left(\sum_{\substack{dt\equiv 1\ (\mathrm{mod}\ N)\\ d>0}}\frac{\mu(d)}{d^k}\right)\delta\left(\frac{-tch}{N}\right)\sum_{\substack{m\equiv tah\ (\mathrm{mod}\ N)\\ m\neq 0}} m^{-k}$$

$$= \delta\left(\frac{-ch}{N}\right)\sum_{d>0}\sum_{\substack{md\equiv ah\ (\mathrm{mod}\ N)\\ m\neq 0}} \frac{\mu(d)}{(dm)^k}$$

(17)
$$= \delta\left(\frac{-ch}{N}\right)\sum_{\substack{m\equiv ah\ (\mathrm{mod}\ N)\\ m\neq 0}} m^{-k}\sum_{\substack{d|m\\ d>0}}\mu(d)$$

Observe that since

(18)
$$\sum_{\substack{d|m\\ d>0}}\mu(d) = \begin{cases} 1 & \text{if } m=1, \text{or} -1, \\ 0 & \text{if } m\neq 1, \text{or} -1, \end{cases}$$

it follows that $\alpha^*_0\left(N,k,\left[\begin{smallmatrix}-ch\\ah\end{smallmatrix}\right]\right) \neq 0$ if and only if $c \equiv 0 \pmod{N}$ and $d \equiv h \pmod{N}$, i.e., if and only if $\frac{-d}{c}$ is $\Gamma(N)$-equivalent to the cusp at infinity. Continuing our calculation, we now have

(19)
$$r(0) = N\sum_{h\in(\mathbb{Z}/N\mathbb{Z})^*}\overline{\chi}(h)\alpha^*_0\left(N,k,\left[\begin{smallmatrix}-ch\\ah\end{smallmatrix}\right]\right)$$

by (15). Therefore, $r(0) = 0$ at all cusps of $\Gamma(N)$ inequivalent to the cusp at infinity, and hence, at all cusps of $\Gamma_0(N)$ inequivalent to the cusp at infinity since $\Gamma(N) \subset \Gamma_0(N)$.

It remains to show that the value of $G^*_{\Gamma_0(N),k,\chi,\left[\begin{smallmatrix}0\\1\end{smallmatrix}\right]}(z)$ at the cusp at infinity is nonzero. Combining the previous facts given by (17), (19), (18), and (13), we calculate:

$$r(0) = N\sum_{h\in(\mathbb{Z}/N\mathbb{Z})^*}\overline{\chi}(h)\sum_{\substack{m\equiv h\ (\mathrm{mod}\ N)\\ m\neq 0}}m^{-k}\sum_{\substack{d|m\\ d>0}}\mu(d)$$

$$= N(\overline{\chi}(1)(1)^k + \overline{\chi}(-1)(-1)^k)$$

(20)
$$= 2N.$$

This proves that $G^*_{\Gamma_0(N),k,\chi,\begin{bmatrix}0\\1\end{bmatrix}}(z)$ satisfies the cusp conditions stated in Theorem 1.1.

We now develop the normalized Fourier expansion of $G^*_{\Gamma_0(N),k,\chi,\begin{bmatrix}0\\1\end{bmatrix}}(z)$ at infinity. Recall that k is even when $N = 1$ or 2. Letting $s_t := \sum_{\substack{dt \equiv 1 \pmod{N} \\ d>0}} \frac{\mu(d)}{d^k}$ for $t \in (\mathbb{Z}/N\mathbb{Z})^*$, we simplify $G^*_{\Gamma_0(N),k,\chi,\begin{bmatrix}0\\1\end{bmatrix}}(z)$ using (15) and (11):

$$G^*_{\Gamma_0(N),k,\chi,\begin{bmatrix}0\\1\end{bmatrix}}(z) = N \sum_{h \in (\mathbb{Z}/N\mathbb{Z})^*} \overline{\chi}(h) G^*_{N,k,\begin{bmatrix}0\\h\end{bmatrix}}(z)$$

$$= N \sum_{h \in (\mathbb{Z}/N\mathbb{Z})^*} \overline{\chi}(h) \sum_{t \in (\mathbb{Z}/N\mathbb{Z})^*} s_t G_{N,k,\begin{bmatrix}0\\th\end{bmatrix}}(z).$$

Letting $c(N) := N \sum_{t \in (\mathbb{Z}/N\mathbb{Z})^*} s_t \chi(t)$, a constant dependent only on N, and making the change of variable $j = th$, we obtain

$$(21) \qquad G^*_{\Gamma_0(N),k,\chi,\begin{bmatrix}0\\1\end{bmatrix}}(z) = c(N) \sum_{j \in (\mathbb{Z}/N\mathbb{Z})^*} \overline{\chi}(j) G_{N,k,\begin{bmatrix}0\\j\end{bmatrix}}(z).$$

Substituting (9) and (10) in (21), we have

$$(22) \quad G^*_{\Gamma_0(N),k,\chi,\begin{bmatrix}0\\1\end{bmatrix}}(z) = c(N) \sum_{j \in (\mathbb{Z}/N\mathbb{Z})^*} \overline{\chi}(j) \times$$

$$\left(\sum_{\substack{m \equiv j \pmod{N} \\ m \neq 0}} m^{-k} + \frac{(-2\pi i)^k}{N^k(k-1)!} \sum_{n \geq 1} \left(\sum_{\substack{d \mid n \\ \frac{n}{d} \equiv 0 \pmod{N}}} d^{k-1} \operatorname{sgn}(d) \zeta_N^{dj} \right) q^{\frac{n}{N}} \right).$$

Using (22) and (13), observe that the constant term $r(0)$ may also be expressed as

$$r(0) = c(N) \sum_{j \in (\mathbb{Z}/N\mathbb{Z})^*} \overline{\chi}(j) \sum_{\substack{m \equiv j \pmod{N} \\ m \neq 0}} m^{-k}$$

$$= c(N) \sum_{m \geq 1} \overline{\chi}(m) m^{-k} (1 + \overline{\chi}(-1)(-1)^k)$$

$$(23) \qquad = 2c(N) L(k, \overline{\chi}),$$

where $L(k,\chi)$ is the Dirichlet L-function associated to χ. Substituting (23) in (22) gives us

$$G^*_{\Gamma_0(N),k,\chi,\begin{bmatrix}0\\1\end{bmatrix}}(z) = c(N) \times$$

$$\left(2L(k,\overline{\chi}) + \frac{(-2\pi i)^k}{N^k(k-1)!} \sum_{n\geq 1} \left(\sum_{j\in(\mathbb{Z}/N\mathbb{Z})^*} \overline{\chi}(j) \sum_{\substack{d|n \\ \frac{n}{d}\equiv 0 \pmod{N}}} d^{k-1} sgn(d)\zeta_N^{dj}\right) q^{\frac{n}{N}}\right).$$

It is clear from (20) and (23) that $c(N) \neq 0$. Therefore, we can define

$$E_{N,k,\chi}(z) := (2L(k,\overline{\chi})c(N))^{-1} G^*_{\Gamma_0(N),k,\chi,\begin{bmatrix}0\\1\end{bmatrix}}(z) = 1 + \frac{(-2\pi i)^k}{2N^k(k-1)!L(k,\overline{\chi})} \times$$

$$\sum_{n\geq 1} \left(\sum_{j\in(\mathbb{Z}/N\mathbb{Z})^*} \overline{\chi}(j) \sum_{\substack{d|n \\ \frac{n}{d}\equiv 0 \pmod{N}}} d^{k-1} sgn(d)\zeta_N^{dj}\right) q^{\frac{n}{N}} \in M_k(\Gamma_0(N),\chi).$$

Noting that

$$\sum_{j\in(\mathbb{Z}/N\mathbb{Z})^*} \overline{\chi}(j) \sum_{d|n} d^{k-1} \mathrm{sgn}(d)\zeta_N^{dj} = \sum_{j\in(\mathbb{Z}/N\mathbb{Z})^*} \overline{\chi}(j) \sum_{\substack{d|n \\ d>0}} d^{k-1}(\zeta_N^{dj} + (-1)^k \zeta_N^{-dj})$$

$$= \sum_{\substack{d|n \\ d>0}} d^{k-1}(\tau_N(d,\overline{\chi}) + (-1)^k \tau_N(-d,\overline{\chi})),$$

and that the condition $\frac{n}{d} \equiv 0 \pmod{N}$ allows us to make the change of variable $n \to nN$, our formula becomes:

$$E_{N,k,\chi}(z) = 1 + \frac{(-2\pi i)^k}{2N^k(k-1)!L(k,\overline{\chi})} \sum_{n\geq 1} \left(\sum_{\substack{d|n \\ d>0}} d^{k-1}(\tau_N(d,\overline{\chi}) + (-1)^k \tau_N(-d,\overline{\chi}))\right) q^n.$$

We now simplify $E_{N,k,\chi}(z)$ in the three cases specified in Theorem 1.1 using certain well-known facts. If χ is the trivial character with modulus N, then $\overline{\chi} = \chi$ and k is even. We use the following facts to obtain formulas (1) and (2):

1. [**Ir-R, Thm.2, pg.231**] If k is a positive even integer, then

$$2\zeta(k) = \frac{(-1)^{\frac{k}{2}+1}(2\pi)^k B_k}{k!},$$

where $\zeta(s)$ is the Riemann zeta function.

2. [**Ir-R, pg.255**] If χ is trivial, then

$$L(k,\chi) = \begin{cases} \zeta(k) & \text{if } N=1, \\ \zeta(k)\prod_{p|N}\left(1-\frac{1}{p^k}\right) & \text{if } N>1. \end{cases}$$

3. [**A, pg.164**] If χ is trivial, then

$$\tau_N(d, \chi) = \tau_N(-d, \chi)$$
$$= \frac{\phi(N)\mu(N/\gcd(d,N))}{\phi(N/\gcd(d,N))}.$$

When χ is a nontrivial character with modulus N and conductor f, we use a different set of facts to produce the formula (3):

1. [**Ir-R, Prop.16.6.2**] If χ is nontrivial and k is a positive integer, then

$$L(1-k, \chi) = -\frac{B_{k,\chi}}{k}.$$

2. If k is a positive integer, then

$$\Gamma(k) = (k-1)!,$$

where $\Gamma(s)$ is the classical Γ-function.

3. If χ is nontrivial and k is a positive integer, and if we define

$$\delta_\chi := \begin{cases} 1 & \text{if } \chi(-1) = -1, \\ 0 & \text{if } \chi(-1) = 1, \end{cases}$$

then the functional equation for $L(k, \chi)$ is [**Iw, Ch.1, Sec.1.2**]:

$$L(k, \chi) = \frac{\tau_f(1, \chi)}{2i^{\delta_\chi}} \left(\frac{2\pi}{f}\right)^k \frac{L(1-k, \overline{\chi})}{\Gamma(k)\cos\left(\frac{\pi(k-\delta_\chi)}{2}\right)}.$$

If χ is a nontrivial primitive character with modulus N, we know the additional fact [**A, Thm.8.15**]:

$$\tau_N(d, \chi) = \chi(d)\tau_N(1, \chi),$$

which we substitute in (3) to obtain (4). This finishes the proof of Theorem 1.1.

4. The Proof of Theorem 1.2

The proof of Theorem 1.2 relies on some well-known facts about the ordinary Bernoulli numbers, B_k:

THEOREM 4.1 (VON STAUDT-CLAUSSEN) [**Ir-R, Thm.3, pg.233**]. *Suppose that ℓ is a prime and k is a positive even integer. If $\ell - 1 \nmid k$, then $\mathrm{ord}_\ell(B_k) \geq 0$, and if $\ell - 1 \mid k$, then $\mathrm{ord}_\ell(B_k) = -1$.*

THEOREM 4.2 [**Ir-R, Thm.15.2.4**]. *Suppose that ℓ is a prime and k is a positive even integer. If $\ell - 1 \nmid k$, then $\mathrm{ord}_\ell\left(\frac{B_k}{k}\right) \geq 0$.*

In the case where χ is the trivial character modulo 1 and ℓ is a rational prime, the desired result follows by applying Theorem 4.1 to formula (1). This is Swinnerton-Dyer's Theorem.

Therefore, we consider the cases in which $N > 1$. We let $E_{N,k,\chi}(z) = \sum_{n \geq 0} a(n)q^n$ and $c_N := \frac{N}{\prod_{p|N} p}$. To start, we simplify the coefficient $a(c_N)$:

$$a(c_N) = \frac{-2k\phi(N)}{N^k B_k \prod_{p|N}\left(1 - \frac{1}{p^k}\right)} \sum_{d | \frac{N}{\prod_{p|N} p}} d^{k-1} \frac{\mu(N/\gcd(d, N))}{\phi(N/\gcd(d, N))}.$$

Since

$$\mu(N/\gcd(d,N)) = \begin{cases} (-1)^{\omega(N)} & \text{if } d = N/\prod_{p|N} p, \\ 0 & \text{if } d < N/\prod_{p|N} p, \end{cases}$$

where $\omega(N)$ is the number of distinct prime divisors of N, it follows that

$$a(c_N) = \frac{(-1)^{\omega(N)+1} 2k}{B_k \prod_{p|N}(p^k - 1)}.$$

Note that

$$(24) \qquad \mathrm{ord}_\ell(a(c_N)) = \mathrm{ord}_\ell(2) + \mathrm{ord}_\ell(k) - \mathrm{ord}_\ell(B_k) - \sum_{p|N} \mathrm{ord}_\ell(p^k - 1)$$

for a given prime ℓ. Assuming for now that ℓ is odd, we analyze $a(c_N) \pmod{\ell}$ in several cases.

CASE 1. $\ell - 1 \nmid k$.

$\mathrm{ord}_\ell\left(\frac{B_k}{k}\right) \geq 0$ by Theorem 4.2, so $\mathrm{ord}_\ell(a(c_N)) \leq 0$ by (24), and hence, $E_{N,k,\chi}(z) \not\equiv 1 \pmod{\ell}$.

Cases 2a and 2b concern the situation where $\ell - 1 \mid k$. In this situation, Theorem 4.1 implies that $\mathrm{ord}_\ell(B_k) = -1$. We suppose that $\mathrm{ord}_\ell(k) = j$. Then

$$(25) \qquad k = \ell^j(\ell - 1)m = \phi(\ell^{j+1})m$$

for some positive integer m coprime to ℓ.

CASE 2A. $\ell - 1 \mid k$ and $\ell \nmid N$.

$\gcd(p, \ell) = 1$ for every prime $p \mid N$ since $\ell \nmid N$, so $p^k \equiv 1 \pmod{\ell^{j+1}}$ by (25). Using (24) we have

$$\mathrm{ord}_\ell(a(c_N)) \leq (j+1)(1 - \omega(N)).$$

$\omega(N) \geq 1$ since $N > 1$, so $\mathrm{ord}_\ell(a(c_N)) \leq 0$ in this case. Consequently, $E_{N,k,\chi}(z) \not\equiv 1 \pmod{\ell}$.

CASE 2B. $\ell - 1 \mid k$ and $\ell \mid N$.

Using (24) and (25) again, we have

$$\mathrm{ord}_\ell(a(c_N)) \leq (j+1)(2 - \omega(N)).$$

It follows that if N is not a positive power of ℓ, then $\mathrm{ord}_\ell(a(c_N)) \leq 0$, in which case $E_{N,k,\chi}(z) \not\equiv 1 \pmod{\ell}$.

Therefore, in the case where ℓ is an odd prime we know the following: $E_{N,k,\chi}(z) \not\equiv 1 \pmod{\ell}$ if $k \not\equiv 0 \pmod{\ell-1}$ or if $N \neq \ell^t$ for all positive integers t. We now prove the converse.

For an odd prime ℓ and a positive integer t, we simplify $E_{\ell^t,k,\chi}(z)$ by first observing that

$$\frac{\phi(\ell^t)\mu(\ell^t/\gcd(d,\ell^t))}{\phi(\ell^t/\gcd(d,\ell^t))} = \begin{cases} \ell^{t-1}(\ell-1) & \text{if } \mathrm{ord}_\ell(d) = t, \\ -\ell^{t-1} & \text{if } \mathrm{ord}_\ell(d) = t-1, \\ 0 & \text{if } \mathrm{ord}_\ell(d) \leq t-2. \end{cases}$$

(26)
$$= \ell^{t-1}\left(\delta\left(\frac{d}{\ell^t}\right)\ell - \delta\left(\frac{d}{\ell^{t-1}}\right)\right),$$

using the notation defined by (8). After substituting (26) in (2), we obtain

$$E_{\ell^t,k,\chi}(z) = 1 - \frac{2k\ell^{t-1}}{\ell^{tk}B_k\left(1-\frac{1}{\ell^k}\right)} \times$$

(27)
$$\left(\sum_{n\geq 1}\left(\sum_{\substack{d|n \\ d>0}} d^{k-1}\delta\left(\frac{d}{\ell^t}\right)\ell\right)q^n - \sum_{n\geq 1}\left(\sum_{\substack{d|n \\ d>0}} d^{k-1}\delta\left(\frac{d}{\ell^{t-1}}\right)\right)q^n\right).$$

Making the change of variables $d \to d\ell^t$ gives us

$$\sum_{n\geq 1}\left(\sum_{\substack{d|n \\ d>0}} d^{k-1}\delta\left(\frac{d}{\ell^t}\right)\ell\right)q^n = \ell^{t(k-1)+1}\sum_{n\geq 1}\left(\sum_{\substack{d|\frac{n}{\ell^t} \\ d>0}} d^{k-1}\right)q^n$$

(28)
$$= \ell^{t(k-1)+1}\sum_{n\geq 1}\sigma_{k-1}\left(\frac{n}{\ell^t}\right)q^n,$$

and similarly, making the change of variables $d \to d\ell^{t-1}$ gives us

(29)
$$\sum_{n\geq 1}\left(\sum_{\substack{d|n \\ d>0}} d^{k-1}\delta\left(\frac{d}{\ell^{t-1}}\right)\right)q^n = \ell^{(t-1)(k-1)}\sum_{n\geq 1}\sigma_k\left(\frac{n}{\ell^{t-1}}\right)q^n,$$

where $\sigma_k\left(\frac{a}{b}\right) = 0$ if a and b are integers with $b \neq 0$ but $\frac{a}{b} \notin \mathbb{Z}$. Substituting (28) and (29) in (27) yields

(30) $$E_{N,k,\chi}(z) = 1 - \frac{2k}{B_k(\ell^k-1)}\sum_{n\geq 1}\left(\ell^k\sigma_{k-1}\left(\frac{n}{\ell^t}\right) - \sigma_{k-1}\left(\frac{n}{\ell^{t-1}}\right)\right)q^n.$$

$E_{\ell^t,k,\chi}(z) \equiv 1 \pmod{\ell}$ since $\mathrm{ord}_\ell\left(\frac{2k}{B_k(\ell^k-1)}\right) \geq 1$ by Theorem 4.1. This proves Theorem 1.2.1.

Next, we assume $\ell = 2$. If $\mathrm{ord}_2(k) = j$, then

(31) $$k = 2^j m = \phi(2^{j+1})m$$

for some positive odd integer m. Moreover, Theorem 4.1 implies that $\mathrm{ord}_2(B_k) = -1$ for every positive even integer $k \geq 4$. We examine $\mathrm{ord}_2(a(c_N))$ in two cases.

CASE 1'. $2 \nmid N$.

$\mathrm{ord}_2(p^k - 1) \geq j + 1$ for every prime $p \mid N$ using (31), so

$$\mathrm{ord}_2(a(c_N)) \leq 1 + (j+1)(1 - \omega(N))$$

using (24). If $\omega(N) > 1$, it follows that $E_{N,k,\chi}(z) \not\equiv 1 \pmod{2}$. Furthermore, if $N = p^b$ for some odd prime p and positive integer b, and if $\mathrm{ord}_2(p^k - 1) > j + 1$, then $E_{p^b,k,\chi}(z) \not\equiv 1 \pmod{2}$.

CASE 2'. $2 \mid N$.

Using (31) and (24) as in case 1', we obtain

$$\mathrm{ord}_2(a(c_N)) \leq 1 + (j+1)(2 - \omega(N)).$$

If $\omega(N) > 2$, then $E_{N,k,\chi}(z) \not\equiv 1 \pmod{2}$. Also, if $N = 2^a p^b$ for some odd prime p and positive integers a and b, and if $\mathrm{ord}_2(p^k - 1) > j + 1$, then $E_{2^a p^b, k, \chi}(z) \not\equiv 1 \pmod{2}$.

Hence, $E_{N,k,\chi}(z) \not\equiv 1 \pmod{2}$ if N does not have the form $2^a p^b$, where p is an odd prime satisfying

$$(32) \qquad \mathrm{ord}_2(p^k - 1) = j + 1$$

and a and b are nonnegative integers. We therefore examine $E_{N,k,\chi}(z) \pmod{2}$ when N does have this form.

In the case where $N = 2^a$ for some positive integer a and in the case where $N = p^b$ where p is an odd prime satisfying (32) and b is a positive integer, the formulas for $E_{N,k,\chi}(z)$ are given by (30) with $\ell = 2$ and $\ell = p$, respectively. The reasoning used there can also be used to show that $E_{N,k,\chi}(z) \equiv 1 \pmod{2}$ in these cases.

In the case where $N = 2^a p^b$ for some odd prime p satisfying (32) and a and b positive integers, a formula for $E_{2^a p^b, k, \chi}(z)$ can be found by applying the same reasoning used to derive (30):

$$E_{2^a p^b, k, \chi}(z) = 1 - \frac{2k}{B_k(p^k - 1)(2^k - 1)} \sum_{n \geq 1} A(n) q^n,$$

where

$A(n) =$
$(2p)^k \sigma_{k-1}\left(\dfrac{n}{2^a p^b}\right) - p^k \sigma_{k-1}\left(\dfrac{n}{2^{a-1} p^b}\right) - 2^k \sigma_{k-1}\left(\dfrac{n}{2^a p^{b-1}}\right) + \sigma_{k-1}\left(\dfrac{n}{2^{a-1} p^{b-1}}\right).$

Now observe that $E_{2^a p^b, k, \chi}(z) \equiv 1 \pmod{2}$ since $\mathrm{ord}_2\left(\dfrac{2k}{B_k(p^k-1)(2^k-1)}\right) = 1$ using Theorem 4.1. This completes the proof of Theorem 1.2.

5. The Proof of Theorem 1.3

The proof of Theorem 1.3 follows by applying the Theorems of Carlitz (Theorems 5.1, 5.2) regarding the divisibility properties of the generalized Bernoulli numbers $B_{k,\chi}$ to the formula (4). We extend the definition of ord to rings of integers of number fields in the obvious way.

THEOREM 5.1 (CARLITZ) [C, Thm.1]. *Suppose that k is a positive integer, ℓ is a rational prime, and χ is a nontrivial primitive Dirichlet character with conductor N. Then*
$$\frac{B_{k,\chi}}{k} = \frac{\mathfrak{R}}{\mathfrak{D}},$$
where $\mathfrak{R}$ and $\mathfrak{D}$ are elements in O_{K_χ} with $\gcd(\mathfrak{R}, \mathfrak{D}) = 1$. If N has at least two distinct rational prime divisors, then $\mathfrak{D} = 1$. If $N = \ell^t$, then $\mathfrak{D}$ is a product of prime divisors of ℓ.

We denote the series in formula (4) by $E_{N,k,\chi}(z) = \sum_{n \geq 0} b(n) q^n$, and observe that
$$b(1) = \frac{-2k}{B_{k,\chi}}.$$
If N has at least two rational prime divisors, then there is an $\mathfrak{R} \in O_{K_\chi}$ for which $b(1) = \frac{-2}{\mathfrak{R}}$ by Theorem 5.1. It follows that if $\mathfrak{a}$ is an ideal in $O_{K_\chi, N}$ and $\mathfrak{a} \nmid (2)$, then $\mathrm{ord}_\mathfrak{a}(b(1)) \leq 0$, so that $E_{N,k,\chi}(z) \not\equiv 1 \pmod{\mathfrak{a}}$, proving Theorem 1.3.1. The proofs of Theorems 1.3.2-1.3.5 follow from Theorem 5.2.

THEOREM 5.2 (CARLITZ) [C, Thm.4]. *Suppose χ is a nontrivial primitive character with conductor N, and ℓ is an odd rational prime. Then the following are true.*

1. *If $N = \ell$, then $\frac{B_{k,\chi}}{k} \in \mathbb{Z}$ unless there is a primitive root g of $\mathbb{Z}/\ell\mathbb{Z}$ for which*

(33) $$\mathfrak{p} = \gcd(\ell, 1 - \chi(g)g^k) \neq (1),$$

where $\mathfrak{p}$ is an ideal in O_{K_χ}. In this case,
$$\ell B_{k,\chi} + 1 \equiv 0 \pmod{\mathfrak{p}^{1+\mathrm{ord}_\ell(k)}}.$$

2. *If $N = \ell^t$ for some integer $t \geq 2$, then $\frac{B_{k,\chi}}{k} \in \mathbb{Z}$ unless there is a primitive root g of $\mathbb{Z}/\ell\mathbb{Z}$ for which*
$$\mathfrak{p} = \gcd(\ell, 1 - \chi(g)g^k) \neq (1),$$
where $\mathfrak{p}$ is an ideal in O_{K_χ}. In this case,
$$(1 - \chi(1+\ell))\frac{B_{k,\chi}}{k} \equiv 1 \pmod{\mathfrak{p}}.$$

3. *If $N = 4$, then*
$$\frac{B_{k,\chi}}{k} \equiv \begin{cases} 1 \pmod{\frac{1}{2}} & \text{if } k \text{ odd,} \\ 0 \pmod{1} & \text{if } k \text{ even.} \end{cases}$$

4. *If $N = 2^t$, for some integer $t \geq 3$, then $\frac{B_{k,\chi}}{k} \in \mathbb{Z}$.*

Note that if (33) holds, and if we let $B_{k,\chi} = \frac{U_{k,\chi}}{V_{k,\chi}}$, where $U_{k,\chi}$ and $V_{k,\chi} \in O_{K_\chi}$ and $\gcd(U_{k,\chi}, V_{k,\chi}) = 1$, then

$$\text{ord}_{\mathfrak{p}}(B_{k,\chi}) = -\text{ord}_{\mathfrak{p}}(\ell) \leq -1. \qquad (34)$$

We now proceed with the proofs of Theorems 1.3.2-1.3.5. We assume in all cases that ℓ is an odd rational prime, $\mathfrak{a}$ is an ideal in $O_{K_\chi,N}$ with the property that $\mathfrak{a} \nmid (2)$, χ is a nontrivial primitive Dirichlet character, and $k \geq 3$ is an integer satisfying $\chi(-1) = (-1)^k$.

PROOF (OF THEOREM 1.3.2). If $N = 4$ and k is even, then χ is trivial. If $N = 4$ and k is odd, then $b(1) = \frac{-4}{s}$ for some nonzero integer s by Theorem 5.2.3. If $\mathfrak{a} \nmid (4)$, then $\text{ord}_{\mathfrak{a}}(b(1)) \leq 0$, so $E_{4,k,\chi}(z) \not\equiv 1 \pmod{\mathfrak{a}}$. If $N = 2^t$ for some integer $t \geq 3$, then $b(1) = \frac{-2}{j}$ for some nonzero integer j by Theorem 5.2.4. Hence, $\text{ord}_{\mathfrak{a}}(b(1)) \leq 0$, so $E_{2^t,k,\chi}(z) \not\equiv 1 \pmod{\mathfrak{a}}$.

PROOF (OF THEOREM 1.3.3). If $N = \ell$ and $\gcd(\ell, 1 - \chi(g)g^k) = (1)$ for every primitive root g of $\mathbb{Z}/\ell\mathbb{Z}$, then $b(1) = \frac{-2}{j}$ for some nonzero integer j by Theorem 5.2.1. Hence, $\text{ord}_{\mathfrak{a}}(b(1)) \leq 0$, so $E_{\ell,k,\chi}(z) \not\equiv 1 \pmod{\mathfrak{a}}$. If there is a primitive root g of $\mathbb{Z}/\ell\mathbb{Z}$ for which $\mathfrak{p} = \gcd(\ell, 1 - \chi(g)g^k) \neq (1)$, then $\text{ord}_{\mathfrak{p}}(B_{k,\chi}) = -\text{ord}_{\mathfrak{p}}(\ell) \leq -1$ by (34). It follows that $\text{ord}_{\mathfrak{p}}(b(n)) \geq 1$ for every $n \geq 1$ by formula (4), and thus, $E_{\ell,k,\chi}(z) \equiv 1 \pmod{\mathfrak{p}^{\text{ord}_{\mathfrak{p}}(\ell)}}$.

PROOF (OF THEOREM 1.3.4). Suppose $N = \ell$ and $\chi = \left(\frac{\cdot}{\ell}\right)$, the Legendre symbol. If we choose an arbitrary primitive root g of $\mathbb{Z}/\ell\mathbb{Z}$ and suppose that there is an $a \in \mathbb{Z}/\ell\mathbb{Z}$ for which $a^2 \equiv g \pmod{\ell}$, then $g^{\frac{\ell-1}{2}} \equiv a^{\ell-1} \equiv 1 \pmod{\ell}$ since $\gcd(a, \ell) = 1$. This contradicts the hypothesis that g is a primitive root of $\mathbb{Z}/\ell\mathbb{Z}$, so $\left(\frac{g}{\ell}\right) = -1$. Using this fact, observe that $\gcd(\ell, 1 - \left(\frac{g}{\ell}\right)g^k) = \gcd(\ell, 1 + g^k) \neq (1)$ if and only if $g^k \equiv -1 \pmod{\ell}$, i.e., if and only if $k \equiv \frac{\ell-1}{2} \pmod{\ell-1}$. In this case $\gcd(\ell, 1 - \left(\frac{g}{\ell}\right)g^k) = (\ell)$, so $\text{ord}_{\ell}\left(B_{k,\left(\frac{\cdot}{\ell}\right)}\right) \leq -1$ by (34). Using formula (4), it follows that $\text{ord}_{\ell}(b(n)) \geq 1$ for all $n \geq 1$ if and only if $k \equiv \frac{\ell-1}{2} \pmod{\ell-1}$.

PROOF (OF THEOREM 1.3.5). If $N = \ell^t$ for some integer $t \geq 2$, and if $\gcd(\ell, 1 - \chi(g)g^k) = (1)$ for every primitive root g of $\mathbb{Z}/\ell\mathbb{Z}$, then $b(1) = \frac{-2}{j}$ for some nonzero integer j by Theorem 5.2.2. Hence, $\text{ord}_{\mathfrak{a}}(b(1)) \leq 0$, so $E_{\ell^t,k,\chi}(z) \not\equiv 1 \pmod{\mathfrak{a}}$.

Q.E.D.

References

[A]   T. Apostol, *Introduction to Analytic Number Theory*, Springer-Verlag, 1976.

[C]   L. Carlitz, *Arithmetic Properties of Generalized Bernoulli Numbers*, J.Reine und Angew. Math. **202** (1959), 173-182.

[Ir-R]   K. Ireland and M. Rosen, *A Classical Introduction to Modern Number Theory*, Springer-Verlag, New York, 1990 GTM 84.

[Iw]   K. Iwasawa, *Lectures on p-Adic L-Functions*, Princeton Univ. Press, 1972.

[K]   N. Koblitz, *Introduction to Elliptic Curves and Modular Forms*, Springer-Verlag, New York, 1984.

[Sc]   B. Schoeneberg, *Elliptic Modular Functions*, Springer-Verlag New York Heidelberg Berlin, 1974.

[Sw-D] H.P.F. Swinnerton-Dyer, *On ℓ-adic representations and congruences for coefficients of modular forms*, Modular functions of one variable, Springer Lect. Notes in Math. **350** (1973), 1-55.

DEPARTMENT OF MATHEMATICS, UNIVERSITY OF WISCONSIN, MADISON, WISCONSIN 53706
E-mail address: `boylan@math.wisc.edu`

More Generating Functions for L-Function Values

Gwynneth G. H. Coogan

1. Introduction and Statement of Results

Let $Q(n)$ denote the number of partitions of the natural number n into distinct parts, and let $\eta(z) := \prod_{n=1}^{\infty}(1-q^n)$, where $q := e^{2\pi i z}$. Then we have the well-known generating function for $Q(n)$:

$$F(z) := \frac{\eta(48z)}{\eta(24z)} = \sum_{n=0}^{\infty} Q(n) q^{24n+1},$$

In his "lost" notebook, Ramanujan recorded, and Andrews proved [1], the following formula for the 'sums of tails' of the infinite product

$$\sum_{n=0}^{\infty} \left(F(z) - q(1+q^{24})(1+q^{48}) \cdots (1+q^{24n}) \right)$$

$$= F(z) \left(-\frac{1}{2} + \sum_{n=1}^{\infty} d(n) q^{24n} \right) + \frac{1}{2} M_1(q),$$

where $M_1(q)$ is the mock theta function given by:

(1.1) $\qquad q + \sum_{n=1}^{\infty} \frac{q^{12n^2+12n+1}}{(1-q^{24})(1-q^{48})\cdots(1-q^{24n})} = q + q^{25} - q^{49} + 2q^{73} - \cdots,$

and $d(n)$ is the usual divisor function.

Recently, Zagier proved a similar formula for the sums of tails of the η-function, namely

$$\sum_{n=0}^{\infty} \left(\eta(24z) - q(1-q^{24})(1-q^{48}) \cdots (1-q^{24n}) \right)$$

$$= \eta(24z) \left(-\frac{1}{2} + \sum_{n=1}^{\infty} d(n) q^{24n} \right) + \frac{1}{2} \sum_{n=1}^{\infty} \left(\frac{12}{n} \right) n q^{n^2}.$$

Notice that each of these formulas gives an extremely convenient way to analyze the error series or mock theta functions at roots of unity, places where mock theta functions are notoriously interesting. Zagier demonstrated a surprising result of this

---

1991 *Mathematics Subject Classification.* Primary 11B65, 11M35; Secondary 05A15.
The author thanks the National Science Foundation for their generous support.

© 2001 American Mathematical Society

type of analysis - a generating function for values of certain Dirichlet L - functions at negative integers:

$$(1.2) \quad -e^{-t/24} \sum_{n=0}^{\infty} (1-e^{-t})(1-e^{-2t}) \cdots (1-e^{-nt}) = \frac{1}{2} \sum_{n=0}^{\infty} L(\chi_{12}, -2n-1) \frac{(-t/24)^n}{n!}$$

where χ_{12} is the Kronecker character for $\mathbb{Q}(\sqrt{12})$.

The above results suggest that identities produced from the sums of tails of modular forms might provide a worthwhile way to produce or analyze mock theta functions. Indeed, Andrews, Jiménez-Urroz and Ono [**2**, Theorem 2] found the following general formula for sums of tails of q-series,

$$(1.3) \quad \sum_{n=0}^{\infty} \left(\frac{(a;q)_\infty (b;q)_\infty}{(q;q)_\infty (c;q)_\infty} - \frac{(a;q)_n (b;q)_n}{(q;q)_n (c;q)_n} \right)$$
$$= \frac{(a;q)_\infty (b;q)_\infty}{(q;q)_\infty (c;q)_\infty} \left(\sum_{n=1}^{\infty} \frac{q^n}{1-q^n} - \sum_{n=0}^{\infty} \frac{aq^n}{1-aq^n} - \sum_{n=1}^{\infty} \frac{(c/b;q)_n b^n}{(a;q)_n (1-q^n)} \right),$$

where we use the standard notation

$$(1.4) \quad (a;q)_\infty := \prod_{k=0}^{\infty} (1-aq^k)$$

and

$$(1.5) \quad (a;q)_n = \frac{(a;q)_\infty}{(aq^n;q)_\infty},$$

which is valid for all positive integers n. And, they also showed that when one considers the asymptotic behavior as $q \searrow \pm 1$ of several specializations of (1.3) these identities yield similar generating functions for values of L- functions at negative integers. In this paper we notice, as Zagier did for $\eta(24z)$, that interesting infinite families of generating functions for values of L-functions result when one considers the behavior of specializations of (1.3) and Theorem 1 [**2**], at more general roots of unity.

The first two theorems give us generating functions for $L(\phi, s)$, the generalized L-function defined by analytically continuing the Dirichlet series. Let,

$$L(\phi, s) := \sum_{n=1}^{\infty} \frac{\phi(n)}{n^s},$$

where $\phi(n)$ is a periodic function from $\mathbb{Z}$ to $\mathbb{C}$. In addition, we will need the following character. Let,

$$\chi_2(n) := \begin{cases} 1 & \text{if } n \equiv 1, 7 \pmod{8} \\ -1 & \text{if } n \equiv 3, 5 \pmod{8} \\ 0 & \text{otherwise} \end{cases}.$$

Throughout, define $\zeta_m := e^{\frac{2\pi i}{m}}$ where m is an odd integer.

THEOREM 1.1. *If m is an odd integer, let $\phi(n) = \zeta_m^{n^2}$, and let*

$$S(q) = \sum_{n=0}^{\infty} \frac{(1-q)(1-q^2) \cdots (1-q^n)}{(1+q)(1+q^2) \cdots (1+q^n)}.$$

Then,

$$\frac{1}{4}S(\zeta_m e^{-t}) = -\sum_{n=0}^{\infty}(-1)^n(4^{1+n}-1)L(\phi,-2n-1)\cdot\frac{t^n}{n!}$$

THEOREM 1.2. *If m is an odd integer, let $\psi(n) = \zeta_m^{\frac{n^2}{8}}\chi_2(n)$, and let*

$$T(q) = \sum_{n=1}^{\infty}\frac{-q^{1/8}(1-q^2)(1-q^4)\cdots(1-q^{2n})}{(1+q)(1+q^3)\cdots(1+q^{2n+1})}.$$

Then

$$2T(\zeta_m e^{-t}) = -\sum_{n=0}^{\infty}\left(\frac{-1}{8}\right)^n L(\psi,-2n-1)\cdot\frac{t^n}{n!}.$$

The next two theorems give us values of the Hecke L-function,

(1.6) $$L(\rho,s) = \sum_{n=1}^{\infty}\frac{a(n)}{n^s} := \sum_{\mathfrak{a}\subseteq\mathbb{Z}[\sqrt{6}]}\chi(\mathfrak{a})N\mathfrak{a}^{-s},$$

where χ is the order 2 character of conductor $4(3+\sqrt{6})$ on ideals in $\mathbb{Z}[\sqrt{6}]$ defined by

(1.7) $$\chi(\mathfrak{a}) := \begin{cases} i^{yx}\left(\frac{12}{x}\right) & \text{if } y \text{ is even,} \\ i^{yx+1}\left(\frac{12}{x}\right) & \text{if } y \text{ is odd,} \end{cases}$$

when $\mathfrak{a} = (x+y\sqrt{6})$. If $1 \leq r < 48m$ is an integer, then let $L_r(\rho_m,s)$ be the partial L-function defined by

(1.8) $$L_r(\rho_m,s) := \sum_{n\equiv r \pmod{48m}}\frac{a(n)\zeta_m^{n/24}}{n^s}.$$

By the orthogonality of the Dirichlet characters modulo $48m$ and the analytic continuation of the associated twists of $L(\rho,s)$, each $L_r(\rho_m,s)$ has analytic continuation to $\mathbb{C}$.

THEOREM 1.3. *If m is an odd integer, as a power series in t we have*

$$-2\zeta_m^{-1/24}e^{t/24}\sum_{n=0}^{\infty}(1-\zeta_m e^{-t})(1-\zeta_m^3 e^{-3t})\cdots(1-\zeta_m^{2n-1}e^{-(2n-1)t}) =$$

$$\sum_{n=0}^{\infty}(-1/24)^n\left(\sum_{q=0}^{m-1}L_{23+48q}(\rho_m,s)+L_{23+(2q+1)24}(\rho_m,s)\right)\cdot\frac{t^n}{n!}$$

THEOREM 1.4. *If m is an odd integer, as a power series in t we have*

$$-2\zeta_m^{1/24}e^{-t/24}\sum_{n=0}^{\infty}(1-\zeta_m e^{-t})(1+\zeta_m^3 e^{-3t})\cdots(1+(-1)^n\zeta_m^{2n-1}e^{-(2n-1)t}) =$$

$$\sum_{n=0}^{\infty}(-1/24)^n\left(\sum_{q=0}^{m-1}L_{1+48q}(\rho_m,s)+L_{1+(2q+1)24}(\rho_m,s)\right)\cdot\frac{t^n}{n!}$$

2. Proof of Theorems

We use the same method for each of the four proofs.

PROOF. *(of Theorem 1.2)* When one specializes (1.3) to the function

$$H(z) := \frac{\eta^2(16z)}{\eta(8z)} = \sum_{n=0}^{\infty} q^{(2n+1)^2} = \frac{q}{1-q^8} \prod_{i=1}^{\infty} \frac{(1-q^{16i})}{(1-q^{16i+8})},$$

one obtains:

$$\sum_{n=0}^{\infty} \left(H(z/8) - \frac{q^{1/8}}{1-q} \times \frac{(1-q^2)(1-q^4)\cdots(1-q^{2n})}{(1-q^3)(1-q^5)\cdots(1-q^{2n+1})} \right)$$

$$= H(z/8)\left(-\frac{1}{2} + \sum_{n=1}^{\infty} d_8(n) q^n \right) + \frac{1}{2} \sum_{n=0}^{\infty} (2n+1) q^{(2n+1)^2/8},$$

where $d_8(n) := \sum_{d|n} (-1)^d$. Notice that, as $q \to -1$, $H(z/8)$ vanishes to infinite order. In fact if we replace q by $\zeta_m q$, where $\zeta_m := e^{2\pi i/m}$, then as long as m is an odd integer $H(z/8)$ vanishes to infinite order as $q \to -1$. Define coefficients $c(n)$ by

$$-\zeta_m^{1/8} e^{-t/8} \sum_{n=0}^{\infty} \frac{(1-\zeta_m^2 e^{-2t})(1-\zeta_m^4 e^{-4t})\cdots(1-\zeta_m^{2n} e^{-2nt})}{(1+\zeta_m e^{-t})(1+\zeta_m^3 e^{-3t})\cdots(1-\zeta_m^{(2n+1)} e^{-(2n+1)t})} = \sum_{n=0}^{\infty} c(n) t^n.$$

Now we see that as $t \searrow 0$,

$$\sum_{n=0}^{\infty} c(n) t^n = \frac{1}{2} \sum_{n=0}^{\infty} (2n+1)(-1)^{n(n+1)/2} \zeta_m^{(2n+1)^2/8} e^{-t(2n+1)^2/8} = G(-\zeta_m e^{-t}).$$

Consider the Mellin transform of $G(-\zeta_m e^{-t})$:

$$\int_0^{\infty} G(-\zeta_m e^{-t}) t^{s-1} dt$$

$$= \frac{1}{2} \sum_{n=0}^{\infty} (2n+1)(-1)^{n(n+1)/2} \zeta_m^{(2n+1)^2/8} \int_0^{\infty} e^{-t(2n+1)^2/8} t^{s-1} dt.$$

Let $k = 2n+1$ in the sum and let $r = k^2 t/8$ in the integral to obtain:

$$(2.1) \quad \int_0^{\infty} G(-\zeta_m e^{-t}) t^{s-1} dt = \frac{8^s \Gamma(s)}{2} \cdot \sum_{k=0}^{\infty} \frac{\chi_2(k)}{k^{2s-1}} \zeta_m^{k^2/8} = \frac{8^s}{2} \cdot \Gamma(s) L(\psi, 2s-1),$$

where $\psi(n) = \chi_2(n) \zeta_m^{\frac{n^2}{8}}$.

On the other hand, by direct evaluation of the residues we get

$$\int_0^\infty G(-\zeta_m e^{-t})t^{s-1}dt = \int_0^1 G(-\zeta_m e^{-t})t^{s-1}dt + \int_1^\infty G(-\zeta_m e^{-t})t^{s-1}dt$$

$$= \int_0^1 \sum_{n=0}^\infty c(n)t^{n+s-1}dt + \int_1^\infty G(-\zeta_m e^{-t})t^{s-1}dt$$

$$= \int_0^1 \sum_{n=0}^{N-1} c(n)t^{n+s-1} + O(t^{N+s-1})dt + \int_1^\infty G(-\zeta_m e^{-t})t^{s-1}dt$$

$$= \sum_{n=0}^{N-1} \frac{c(n)}{s+n} + A(s),$$

where, $A(s)$ is analytic for $Re(s) > -N$. It is clear now that the residue at $s = -n$ is $c(n)$. By (2.1), this implies that

$$c(n) = Res_{s=-n}\left(\frac{8^s}{2}\Gamma(s)L(\psi, 2s-1)\right) = \frac{1}{2}\frac{(-1/8)^n}{n!} \cdot L(\psi, -2n-1).$$

This completes the proof of theorem (1.2). □

PROOF. *(of Theorem 1.1)* Theorem (1.1) concerns the function:

$$J(z) := \frac{\eta^2(z)}{\eta(2z)} = 1 + 2\sum_{n=0}^\infty (-1)^n q^{n^2} = \prod_{i=1}^\infty \frac{(1-q^i)}{(1+q^i)}.$$

When we specialize (1.3) to this function, we get the following identity.

$$\sum_{n=0}^\infty \left(J(z) - \frac{(1-q)(1-q^2)\cdots(1-q^n)}{(1+q)(1+q^2)\cdots(1+q^n)}\right)$$

$$= 2J(z)\left(\sum_{n=1}^\infty d_1(n)q^n\right) + 4\sum_{n=1}^\infty (-1)^n n q^{n^2},$$

where $d_1(n)$ is the number of odd divisors of n. The rest of the proof follows *mutatis mutandis*. □

Proof of Theorems 1.3 and 1.4

PROOF. We need the following two specializations of theorem 1 in [**2**]. First for the function:

$$K(z) := \frac{\eta(24z)}{\eta(48z)} = q^{-1}\prod_{i=1}^\infty (1 - q^{24(2i-1)}),$$

we have

$$\sum_{n=0}^\infty \left(K(z) - q^{-1}(1-q^{24})(1-q^{72})\cdots(1-q^{24(2n-1)})\right)$$

$$= K(z)\left(\sum_{n=1}^\infty d(n)q^{24n}\right) + M_2(q),$$

where $d(n) := \sum_{d|n} 1$, and $M_2(q)$ is the mock theta function by:

$$\sum_{n=1}^\infty \frac{(-1)^n q^{24n^2-1}}{(1-q^{24})(1-q^{72})\cdots(1-q^{24(2n-1)})} = -q^{23} - q^{47} - \cdots.$$

Next, for the function:
$$F(z) := \frac{\eta(48z)}{\eta(24z)} = q\prod_{i=1}^{\infty}(1+q^{24i}),$$
we have
$$\sum_{n=0}^{\infty}\left(F(z) - q(1+q^{24})(1+q^{48})\cdots(1+q^{24n})\right)$$
$$= F(z)\left(-\frac{1}{2} + \sum_{n=1}^{\infty}d(n)q^{24n}\right) + \frac{1}{2}M_1(q),$$
where $M_1(q)$ is the mock theta function given in 1.1. In [3] we see that the coefficients $a(n)$ defining $L(\rho,s)$ in (1.6) are defined by

(2.2) $$\sum_{n=1}^{\infty}a(n)q^n = M_1(q) + 2M_2(q).$$

The theorems follow from this fact, and the fact that the coefficients of $M_1(q)$ and $M_2(q)$ are supported on exponents for which $n \equiv 1 \pmod{24}$ and $n \equiv 23 \pmod{24}$ respectively. $\square$

References

[1] G.E. Andrews, *Ramanujan's "Lost" Notebook V: Euler's partition identity*, Advances in Mathematics **61** (1986), 156-164.
[2] G.E. Andrews, J. Jimenez-Urroz, K. Ono, *q-series identities and values of certain L-functions*, Duke Math. J., to appear.
[3] H. Cohen, *q-identities for Maass waveforms*, Inventiones Mathematicae **91** (1988), 409-422.
[4] D. Zagier, *Vassiliev invariants and a strange identity related to the Dedekind eta-function*, to appear, Topology.

DEPARTMENT OF MATHEMATICS, UNIVERSITY OF WISCONSIN, MADISON, WISCONSIN 53706
E-mail address: `gwynneth@math.wisc.edu`

Contemporary Mathematics
Volume **291**, 2001

On sums of an even number of squares, and an even number of triangular numbers: an elementary approach based on Ramanujan's $_1\psi_1$ summation formula

Shaun Cooper

ABSTRACT. An elementary and self contained method is given for determining formulas involving the number of representations of an integer as a sum of an even number of squares, and as an even number of triangular numbers. Our method uses Ramanujan's $_1\psi_1$ summation formula and Venkatachaliengar's elementary approach to elliptic functions.

1. Introduction

Let $r_{2s}(n)$ denote the number of representations of a non-negative integer n as a sum of $2s$ squares, and let $t_{2s}(n)$ denote the number of representations of n as a sum of $2s$ triangular numbers. Ramanujan [**16**, pp. 158–159] stated that

$$r_{2s}(n) = \delta_{2s}(n) + e_{2s}(n)$$

where $\delta_{2s}(n)$ is a divisor function and $e_{2s}(n)$ is a function of lower order than $\delta_{2s}(n)$. Ramanujan gave, without proofs, exact formulas for the generating functions of $e_{2s}(n)$ and $\delta_{2s}(n)$ for $1 \leq s \leq 12$, as well as the forms of these functions for general s. Mordell [**12**] utilised modular forms methods to give the first proof of Ramanujan's general results for $r_{2s}(n)$. An excellent account of the history of these results is given by Milne [**11**, pp. 4–5]. We add the remark that proofs of Ramanujan's formulas for $1 \leq s \leq 12$ were also given by Ramamani [**14**], who made extensive use of the Weierstrassian $\wp$ function. Ramamani also stated how her method could be used to obtain the analogous results for sums of triangular numbers, but she did not work out the details.

The purpose of this article is to give elementary and self contained proofs of Ramanujan's formulas for general s, as well as proofs of the corresponding results for triangular numbers. By "elementary" we mean that our approach, inspired by Venkatachaliengar's beautiful development [**17**] and the analysis in [**7**, pp. 112–117], is based entirely on Ramanujan's $_1\psi_1$ summation formula and straight forward manipulations of infinite series. Thus our methods are ones which Ramanujan

---

1991 *Mathematics Subject Classification.* Primary 11E25, 33E05; Secondary 05A15, 33D15.
Key words and phrases. Elliptic functions, sums of squares, sums of triangular numbers, Ramanujan's $_1\psi_1$ summation formula, Venkatachaliengar's elementary approach to elliptic functions.

© 2001 American Mathematical Society

could have used. "Self contained" means just that - all of the formulas we will require for series, products and their properties are stated and proved here. Our only assumption is that the reader is familiar with Ramanujan's $_1\psi_1$ summation formula.

The first half of this article is devoted to developing some basic properties of elliptic functions, following the approach of Venkatachaliengar. Since the details of Venkatachaliengar's work, in particular his Fundamental Multiplicative Identity and its consequences, are not widely known, we take this opportunity to reproduce some of his results here.

General results for sums of squares and sums of triangular numbers are developed in the second half of the article. From this, explicit, exact formulas for the generating functions of $e_{2s}(n)$ and $\delta_{2s}(n)$, as well as the analogous results for sums of triangular numbers, may be computed for any particular positive integral value of s. The explicit results for $1 \leq s \leq 25$ are given in Tables 2 and 3.

2. Elliptic functions

2.1. Venkatachaliengar's fundamental multiplicative identity. Let τ be a fixed complex number which satisfies $\operatorname{Im} \tau > 0$ and let $q = e^{i\pi\tau}$, so that $|q| < 1$. Ramanujan's $_1\psi_1$ summation formula is

$$\prod_{k=1}^{\infty} \frac{(1+zq^{2k-1})(1+q^{2k-1}/z)(1-q^{2k})(1-\alpha\beta q^{2k})}{(1+\alpha z q^{2k-1})(1+\beta q^{2k-1}/z)(1-\alpha q^{2k})(1-\beta q^{2k})}$$
$$= 1 + \left\{ \frac{1-\alpha}{1-\beta q^2} qz + \frac{1-\beta}{1-\alpha q^2} \frac{q}{z} \right\}$$
$$+ \left\{ \frac{(1-\alpha)(q^2-\alpha)}{(1-\beta q^2)(1-\beta q^4)} (qz)^2 + \frac{(1-\beta)(q^2-\beta)}{(1-\alpha q^2)(1-\alpha q^4)} \left(\frac{q}{z}\right)^2 \right\}$$
$$+ \left\{ \frac{(1-\alpha)(q^2-\alpha)(q^4-\alpha)}{(1-\beta q^2)(1-\beta q^4)(1-\beta q^6)} (qz)^3 \right.$$
$$\left. + \frac{(1-\beta)(q^2-\beta)(q^4-\beta)}{(1-\alpha q^2)(1-\alpha q^4)(1-\alpha q^6)} \left(\frac{q}{z}\right)^3 \right\} + \cdots \quad (2.1)$$

where $|\beta q| < |z| < 1/|\alpha q|$ and $|q| < 1$. This formula appears in Ramanujan's Second Notebook in Chapter 16, Entry 17 [15]. Hardy [8, pp. 222 - 223] described it as "a remarkable formula with many parameters". Proofs and more information about this formula can be found in [2], [3] and [5].

The Jordan-Kronecker function is defined as follows.

DEFINITION 2.1. [17, p. 37] Let

$$F(a,t) = \sum_{j=-\infty}^{\infty} \frac{t^j}{1-aq^{2j}}. \quad (2.2)$$

This series converges provided $|q^2| < |t| < 1$, and so long as $a \neq q^{2k}$, $k = 0, \pm 1, \pm 2, \ldots$. In Ramanujan's summation formula (2.1) take $\alpha = 1/a$, $\beta = a$, $z = -at/q$ and divide by $1 - a$ to obtain

$$F(a,t) = \frac{(at, q^2/at, q^2, q^2; q^2)_\infty}{(t, q^2/t, a, q^2/a; q^2)_\infty}. \quad (2.3)$$

This extends the definition of F to all values of a and t except for $a, t = q^{2k}$, $k = 0, \pm 1, \pm 2, \ldots$, where there are simple poles. The following properties are immediate from (2.3):

$$F(a,t) = F(t,a), \qquad (2.4)$$
$$F(a,t) = -F(1/a, 1/t), \qquad (2.5)$$
$$F(a,t) = tF(aq^2, t) = aF(a, tq^2). \qquad (2.6)$$

The twelve Jacobian elliptic functions correspond to the functions $F(A, Be^{i\theta})$, where $A = -1, q$ or $-q$, and $B = 1, -1, q$ or $-q$. The Weierstrass σ and $\wp$ functions are also related to the Jordan-Kronecker function F, and some of these connections will be given later.

Venkatachaliengar's development of elliptic functions is based on the following result.

THEOREM 2.2 (Fundamental multiplicative identity). [**17**, p. 37]

$$F(a,t)F(b,t) = t\frac{\partial}{\partial t}F(ab,t) + F(ab,t)(\rho_1(a) + \rho_1(b)), \qquad (2.7)$$

where the function ρ_1 is defined by

$$\rho_1(z) = \frac{1}{2} + \sum_{j}{}' \frac{z^j}{1 - q^{2j}}. \qquad (2.8)$$

The prime denotes that the summation is over all integers j excluding 0.

The series (2.8) defining ρ_1 converges in the annulus $|q^2| < |z| < 1$. Shortly we will obtain the analytic continuation of ρ_1, so the identity (2.7) will be valid for all values of a, b and t.

PROOF. For $|q^2| < |a|, |b| < 1$, we have

$$F(a,t)F(b,t) = \sum_{j=-\infty}^{\infty} \sum_{k=-\infty}^{\infty} \frac{a^j b^k}{(1 - tq^{2j})(1 - tq^{2k})}$$
$$= \sum_{j=-\infty}^{\infty} \frac{a^j b^j}{(1 - tq^{2j})^2} + \sum_{j \neq k} \frac{a^j b^k}{(1 - tq^{2j})(1 - tq^{2k})}. \qquad (2.9)$$

The first sum is

$$\sum_{j=-\infty}^{\infty} \frac{(ab)^j}{(1 - tq^{2j})^2} = \sum_{j=-\infty}^{\infty} \frac{\partial}{\partial t} \frac{(ab/q^2)^j}{(1 - tq^{2j})}$$
$$= \frac{\partial}{\partial t} F(ab/q^2, t)$$
$$= \frac{\partial}{\partial t}[tF(ab, t)]$$
$$= t\frac{\partial}{\partial t}F(ab, t) + F(ab, t) \qquad (2.10)$$

The penultimate step above follows from (2.6). The interchange of differentiation and summation is valid as all series converge absolutely and uniformly (in t) on compact sets which avoid the poles $t = q^{2k}$, $k = 0, \pm 1, \pm 2, \ldots$, provided $|q^4| < |ab| < |q^2|$. By analytic continuation, equation (2.10) continues to remain valid for $|q^4| < |ab| < 1$.

Using partial fractions, the second sum on the right hand side of (2.9) becomes

$$\sum_{j \neq k} \frac{a^j b^k}{(1-tq^{2j})(1-tq^{2k})}$$

$$= \sum_{j=-\infty}^{\infty} {\sum_{n}}' \frac{a^j b^{j+n}}{(1-tq^{2j})(1-tq^{2j+2n})}$$

$$= \sum_{j=-\infty}^{\infty} {\sum_{n}}' \frac{a^j b^{j+n}}{(1-tq^{2j})(1-q^{2n})} + \sum_{j=-\infty}^{\infty} {\sum_{n}}' \frac{a^j b^{j+n}}{(1-tq^{2j+2n})(1-q^{-2n})}$$

$$= \sum_{j=-\infty}^{\infty} \frac{a^j b^j}{(1-tq^{2j})} {\sum_{n}}' \frac{b^n}{(1-q^{2n})} + {\sum_{n}}' \frac{a^{-n}}{(1-q^{-2n})} \sum_{j=-\infty}^{\infty} \frac{a^{j+n} b^{j+n}}{(1-tq^{2j+2n})}$$

$$= F(ab,t) \left[{\sum_{n}}' \frac{a^n}{1-q^{2n}} + {\sum_{n}}' \frac{b^n}{1-q^{2n}} \right]$$

$$= F(ab,t) \left[\rho_1(a) + \rho_1(b) - 1 \right]. \tag{2.11}$$

All of the series in the derivation of (2.11) converge at least for $|q| < |a|, |b| < 1$ and $t \neq q^{2k}$, $k = 0, \pm 1, \pm 2, \ldots$, and so the series rearrangements above are valid. Now combine (2.9), (2.10) and (2.11). This gives (2.7) and proves the theorem. □

The analytic continuation of ρ_1 can be obtained as follows [**17**, p. 5].

$$\rho_1(z) = \frac{1}{2} + {\sum_{k}}' \frac{z^k}{1-q^{2k}}$$

$$= \frac{1}{2} + \sum_{k=1}^{\infty} \frac{z^k}{1-q^{2k}} + \sum_{k=1}^{\infty} \frac{z^{-k}}{1-q^{-2k}}$$

$$= \frac{1}{2} + \sum_{k=1}^{\infty} \frac{z^k(1-q^{2k}+q^{2k})}{1-q^{2k}} - \sum_{k=1}^{\infty} \frac{z^{-k} q^{2k}}{1-q^{2k}}$$

$$= \frac{1}{2} + \frac{z}{1-z} + \sum_{k=1}^{\infty} \frac{z^k q^{2k}}{1-q^{2k}} - \sum_{k=1}^{\infty} \frac{z^{-k} q^{2k}}{1-q^{2k}} \tag{2.12}$$

$$= \frac{1+z}{2(1-z)} + \sum_{k=1}^{\infty} \sum_{j=1}^{\infty} (z^k q^{2jk} - z^{-k} q^{2jk})$$

$$= \frac{1+z}{2(1-z)} + \sum_{j=1}^{\infty} \left(\frac{zq^{2j}}{1-zq^{2j}} - \frac{z^{-1} q^{2j}}{1-z^{-1} q^{2j}} \right). \tag{2.13}$$

This last series converges for all values of z except $z = q^{2k}$, $k = 0, \pm 1, \pm 2, \ldots$, where there are poles of order 1. Thus (2.13) gives the analytic continuation of the function ρ_1, and so the fundamental multiplicative identity (2.7) is valid for all values of a, b and t.

The function ρ_1 is related to the Weierstrass $\wp$ function in the following way. The Weierstrass $\wp$ function with periods 2π and $2\pi\tau$ is defined by

$$\wp(\theta) = \frac{1}{\theta^2} + {\sum_{j,k}}' \left[\frac{1}{(\theta - 2\pi k - 2\pi\tau j)^2} - \frac{1}{(2\pi k + 2\pi\tau j)^2} \right]. \tag{2.14}$$

The symbol $\sum_{j,k}'$ denotes a double sum over all integer values of j and k from $-\infty$ to ∞, excluding $(j,k) = (0,0)$. Using the results

$$\sum_{k=-\infty}^{\infty} \frac{1}{(\theta - 2\pi k)^2} = \frac{1}{4\sin^2 \frac{\theta}{2}}$$

and

$$\sum_{k=1}^{\infty} \frac{1}{k^2} = \frac{\pi^2}{6},$$

we have

$$\wp(\theta) = \frac{1}{\theta^2} + \sum_{k}' \left[\frac{1}{(\theta - 2\pi k)^2} - \frac{1}{(2\pi k)^2} \right]$$

$$+ \sum_{j}' \sum_{k=-\infty}^{\infty} \left[\frac{1}{(\theta - 2\pi k - 2\pi\tau j)^2} - \frac{1}{(2\pi k + 2\pi\tau j)^2} \right]$$

$$= \sum_{k} \frac{1}{(\theta - 2\pi k)^2} - \frac{2}{4\pi^2} \sum_{k=1}^{\infty} \frac{1}{k^2}$$

$$+ \sum_{j}' \sum_{k=-\infty}^{\infty} \left[\frac{1}{(\theta - 2\pi k - 2\pi\tau j)^2} - \frac{1}{(2\pi k + 2\pi\tau j)^2} \right]$$

$$= \frac{1}{4\sin^2 \frac{\theta}{2}} - \frac{1}{12} + \sum_{j}' \left[\frac{1}{4\sin^2(\frac{\theta}{2} - \pi\tau j)} - \frac{1}{4\sin^2 \pi\tau j} \right]$$

$$= -\frac{1}{12} - \frac{1}{2} \sum_{j=1}^{\infty} \frac{1}{\sin^2 \pi\tau j} + \frac{1}{4} \sum_{j=-\infty}^{\infty} \frac{1}{\sin^2(\frac{\theta}{2} + \pi\tau j)}.$$

Recall that $q = e^{i\pi\tau}$. Then

$$\wp(\theta) = -\frac{1}{12} + 2 \sum_{j=1}^{\infty} \frac{1}{(q^j - q^{-j})^2} - \sum_{j=-\infty}^{\infty} \frac{1}{(e^{i\theta/2} q^j - e^{-i\theta/2} q^{-j})^2}$$

$$= -\frac{1}{12} + 2 \sum_{j=1}^{\infty} \frac{q^{2j}}{(1 - q^{2j})^2} - \sum_{j=-\infty}^{\infty} \frac{e^{i\theta} q^{2j}}{(1 - e^{i\theta} q^{2j})^2}. \quad (2.15)$$

Continuing, we have

$$\wp(\theta) = -\frac{1}{12} + 2 \sum_{j=1}^{\infty} \frac{q^{2j}}{(1 - q^{2j})^2}$$

$$- \frac{e^{i\theta}}{(1 - e^{i\theta})^2} - \sum_{j=1}^{\infty} \left[\frac{e^{i\theta} q^{2j}}{(1 - e^{i\theta} q^{2j})^2} + \frac{e^{-i\theta} q^{2j}}{(1 - e^{-i\theta} q^{2j})^2} \right]$$

$$= -\frac{1}{12} + 2 \sum_{j=1}^{\infty} \frac{q^{2j}}{(1 - q^{2j})^2} + i \frac{d}{d\theta} \rho_1(e^{i\theta}).$$

Formula (2.13) was used to obtain the last line. Thus if we let

$$P = 1 - 24 \sum_{j=1}^{\infty} \frac{q^{2j}}{(1 - q^{2j})^2}, \quad (2.16)$$

then we have [**17**, p. 8]

$$\wp(\theta) = i\frac{d}{d\theta}\rho_1(e^{i\theta}) - \frac{P}{12}. \qquad (2.17)$$

2.2. Jacobian elliptic functions. We will now look at three special cases of the function $F(a,t)$, which we shall call f_1, f_2 and f_3. These functions will turn out to be the Jacobian elliptic functions cs, ns and ds, respectively, up to rescaling.

A number of properties of f_1, f_2 and f_3 (Fourier series, infinite product formulas, double periodicity, location of zeros and poles) will follow immediately from the definition of $F(a,t)$ and Ramanujan's $_1\psi_1$ summation formula. We will then use the fundamental multiplicative identity (2.7) to obtain some of the other properties of these functions, namely the connection with the $\wp$ function, elliptic analogues of the formula $\sin^2\theta + \cos^2\theta = 1$ and derivatives.

DEFINITION 2.3. [**17**, p. 111] Let

$$f_1(\theta) = \frac{1}{i}F(e^{i\pi}, e^{i\theta}), \qquad (2.18)$$

$$f_2(\theta) = \frac{e^{i\theta/2}}{i}F(e^{i\pi\tau}, e^{i\theta}), \qquad (2.19)$$

$$f_3(\theta) = \frac{e^{i\theta/2}}{i}F(e^{i\pi+i\pi\tau}, e^{i\theta}). \qquad (2.20)$$

The factors $1/i$ and $e^{i\theta/2}/i$ are included so that f_1, f_2 and f_3 will be real valued when θ is real.

2.3. Fourier series and infinite products. Fourier series expansions for the functions f_1, f_2 and f_3 follow directly from (2.2), the definition of F. For example

$$\begin{aligned}
f_1(\theta) &= \frac{1}{i}\sum_{j=-\infty}^{\infty}\frac{e^{ij\theta}}{1+q^{2j}} \qquad (2.21)\\
&= \frac{1}{i}\left[\frac{1}{2} + \sum_{j=1}^{\infty}\frac{e^{ij\theta}}{1+q^{2j}} + \sum_{j=1}^{\infty}\frac{e^{-ij\theta}}{1+q^{-2j}}\right]\\
&= \frac{1}{i}\left[\frac{1}{2} + \sum_{j=1}^{\infty}e^{ij\theta} - \sum_{j=1}^{\infty}\frac{q^{2j}e^{ij\theta}}{1+q^{2j}} + \sum_{j=1}^{\infty}\frac{q^{2j}e^{-ij\theta}}{1+q^{2j}}\right]\\
&= \frac{1}{i}\left[\frac{1}{2} + \frac{e^{i\theta}}{1-e^{i\theta}} - \sum_{j=1}^{\infty}\frac{q^{2j}}{1+q^{2j}}(e^{ij\theta} - e^{-ij\theta})\right]\\
&= \frac{1}{2}\cot\frac{\theta}{2} - 2\sum_{j=1}^{\infty}\frac{q^{2j}}{1+q^{2j}}\sin j\theta. \qquad (2.22)
\end{aligned}$$

Similarly,

$$f_2(\theta) = \frac{e^{i\theta/2}}{i} \sum_{j=-\infty}^{\infty} \frac{e^{ij\theta}}{1-q^{2j+1}} \tag{2.23}$$

$$= \frac{1}{2}\csc\frac{\theta}{2} + 2\sum_{j=1}^{\infty} \frac{q^{2j-1}}{1-q^{2j-1}} \sin(j-\frac{1}{2})\theta, \tag{2.24}$$

$$f_3(\theta) = \frac{e^{i\theta/2}}{i} \sum_{j=-\infty}^{\infty} \frac{e^{ij\theta}}{1+q^{2j+1}} \tag{2.25}$$

$$= \frac{1}{2}\csc\frac{\theta}{2} - 2\sum_{j=1}^{\infty} \frac{q^{2j-1}}{1+q^{2j-1}} \sin(j-\frac{1}{2})\theta. \tag{2.26}$$

The series (2.21), (2.23) and (2.25) converge in the region $0 < \operatorname{Im}\theta < \operatorname{Im}(2\pi\tau)$, while (2.22), (2.24) and (2.26) are valid in the region $-\operatorname{Im}(2\pi\tau) < \operatorname{Im}\theta < \operatorname{Im}(2\pi\tau)$.

Infinite product formulas follow from (2.3). We find that

$$f_1(\theta) = \frac{1}{i} \frac{(-e^{i\theta}, -q^2 e^{-i\theta}, q^2, q^2; q^2)_\infty}{(e^{i\theta}, q^2 e^{-i\theta}, -1, -q^2; q^2)_\infty} \tag{2.27}$$

$$= \frac{1}{2} \frac{(q^2;q^2)_\infty^2}{(-q^2;q^2)_\infty^2} \cot\frac{\theta}{2} \prod_{j=1}^{\infty} \frac{(1+2q^{2j}\cos\theta+q^{4j})}{(1-2q^{2j}\cos\theta+q^{4j})}, \tag{2.28}$$

$$f_2(\theta) = \frac{e^{i\theta/2}}{i} \frac{(qe^{i\theta}, qe^{-i\theta}, q^2, q^2; q^2)_\infty}{(e^{i\theta}, q^2 e^{-i\theta}, q, q; q^2)_\infty} \tag{2.29}$$

$$= \frac{1}{2} \frac{(q^2;q^2)_\infty^2}{(q;q^2)_\infty^2} \csc\frac{\theta}{2} \prod_{j=1}^{\infty} \frac{(1-2q^{2j-1}\cos\theta+q^{4j-2})}{(1-2q^{2j}\cos\theta+q^{4j})}, \tag{2.30}$$

$$f_3(\theta) = \frac{e^{i\theta/2}}{i} \frac{(-qe^{i\theta}, -qe^{-i\theta}, q^2, q^2; q^2)_\infty}{(e^{i\theta}, q^2 e^{-i\theta}, -q, -q; q^2)_\infty} \tag{2.31}$$

$$= \frac{1}{2} \frac{(q^2;q^2)_\infty^2}{(-q;q^2)_\infty^2} \csc\frac{\theta}{2} \prod_{j=1}^{\infty} \frac{(1+2q^{2j-1}\cos\theta+q^{4j-2})}{(1-2q^{2j}\cos\theta+q^{4j})}. \tag{2.32}$$

From the infinite product formulas we readily obtain the periodicity properties

$$f_1(\theta + 2\pi m + 2\pi\tau n) = (-1)^n f_1(\theta), \tag{2.33}$$
$$f_2(\theta + 2\pi m + 2\pi\tau n) = (-1)^m f_2(\theta), \tag{2.34}$$
$$f_3(\theta + 2\pi m + 2\pi\tau n) = (-1)^{m+n} f_3(\theta) \tag{2.35}$$

where m and n are integers. Thus f_1 is doubly periodic with periods 2π and $4\pi\tau$, f_2 is doubly periodic with periods 4π and $2\pi\tau$, while f_3 is doubly periodic with periods 4π and $2\pi + 2\pi\tau$.

Also from the infinite product expansions we see that f_1, f_2 and f_3 have zeros at $\theta = (2m+1)\pi + 2n\pi\tau$, $\theta = 2m\pi + (2n+1)\pi\tau$ and $\theta = (2m+1)\pi + (2n+1)\pi\tau$, respectively. The poles of f_1, f_2 and f_3 all occur at $\theta = 2m\pi + 2n\pi\tau$.

2.4. Connection with the Weierstrass $\wp$ function.

The Weierstrassian invariants e_1, e_2 and e_3 are defined by

$$e_1 = \wp(\pi), \qquad (2.36)$$
$$e_2 = \wp(\pi\tau), \qquad (2.37)$$
$$e_3 = \wp(\pi + \pi\tau). \qquad (2.38)$$

Let $b \to 1/a$ in the fundamental identity (2.7):

$$\lim_{b \to 1/a} F(a,t)F(b,t) = \lim_{b \to 1/a} t\frac{\partial}{\partial t}F(ab,t) + \lim_{b \to 1/a} F(ab,t)(\rho_1(a) + \rho_1(b)). \qquad (2.39)$$

The left hand side is just $F(a,t)F(1/a,t)$.
The first limit on the right hand side is

$$\lim_{b \to 1/a} t\frac{\partial}{\partial t}\sum_{j=-\infty}^{\infty} \frac{t^j}{1-abq^{2j}} = \lim_{b \to 1/a} {\sum_j}' \frac{jt^j}{1-abq^{2j}}$$
$$= {\sum_j}' \frac{jt^j}{1-q^{2j}} = t\frac{d}{dt}\rho_1(t).$$

From equation (2.13) it follows that $\rho_1(b) = -\rho_1(1/b)$. Using this and the infinite product formula (2.3) for the function F, the remaining limit on the right hand side of equation (2.39) becomes

$$\lim_{b \to 1/a} F(ab,t)(\rho_1(a) + \rho_1(b))$$
$$= \lim_{b \to 1/a}(1-ab)F(ab,t)\lim_{b \to 1/a}\frac{\rho_1(a) + \rho_1(b)}{1-ab}$$
$$= \lim_{b \to 1/a}(1-ab)\frac{(abt, q^2/abt, q^2, q^2; q^2)_\infty}{(t, q^2/t, ab, q^2/ab; q^2)_\infty}\lim_{b \to 1/a}\frac{\rho_1(a) - \rho_1(1/b)}{a - 1/b}\left(-\frac{1}{b}\right)$$
$$= (1)\rho_1'(a)(-a).$$

Thus [**17**, p. 112]

$$F(a,t)F(1/a,t) = t\frac{d}{dt}\rho_1(t) - a\frac{d}{da}\rho_1(a). \qquad (2.40)$$

On letting $a = e^{i\alpha}$, $t = e^{i\theta}$ and using equation (2.17), this becomes

$$F(e^{i\alpha}, e^{i\theta})F(e^{-i\alpha}, e^{i\theta}) = \wp(\alpha) - \wp(\theta). \qquad (2.41)$$

This formula can also be obtained by combining the two terms on the right hand side of (2.41) into a single series using (2.15), and then applying the ${}_6\psi_6$ summation formula.

Letting $\alpha = \pi$, $\alpha = \pi\tau$ and $\alpha = \pi + \pi\tau$ in (2.41), respectively, and simplifying, gives [**17**, p. 112]

$$f_1^2(\theta) = \wp(\theta) - e_1, \qquad (2.42)$$
$$f_2^2(\theta) = \wp(\theta) - e_2, \qquad (2.43)$$
$$f_3^2(\theta) = \wp(\theta) - e_3. \qquad (2.44)$$

Successively letting $\theta = \pi\tau$ in (2.42), $\theta = \pi + \pi\tau$ in (2.43) and $\theta = \pi$ in (2.44), and using the infinite products for f_1, f_2 and f_3, gives [**17**, p. 66]

$$e_1 - e_2 = \frac{1}{4}\frac{(-q;q^2)_\infty^4(q^2;q^2)_\infty^4}{(q;q^2)_\infty^4(-q^2;q^2)_\infty^4} \tag{2.45}$$

$$e_3 - e_2 = 4q\frac{(-q^2;q^2)_\infty^4(q^2;q^2)_\infty^4}{(-q;q^2)_\infty^4(q;q^2)_\infty^4} \tag{2.46}$$

$$e_1 - e_3 = \frac{1}{4}\frac{(q;q^2)_\infty^4(q^2;q^2)_\infty^4}{(-q^2;q^2)_\infty^4(-q;q^2)_\infty^4} \tag{2.47}$$

Note that since $\operatorname{Im}\tau > 0$ this implies that $e_1 \neq e_2 \neq e_3 \neq e_1$. Further, if τ is purely imaginary, then q is real, and so in this case we also have $e_1 > e_3 > e_2$.

If we define the modulus k and the complementary modulus k' by

$$k^2 = \frac{e_3 - e_2}{e_1 - e_2} = 16q\frac{(-q^2;q^2)_\infty^8}{(-q;q^2)_\infty^8}, \tag{2.48}$$

$$k'^2 = \frac{e_1 - e_3}{e_1 - e_2} = \frac{(q;q^2)_\infty^8}{(-q;q^2)_\infty^8}, \tag{2.49}$$

then clearly $k^2 + k'^2 = 1$, and hence we obtain Jacobi's formula [**17**, p. 62]

$$(q;q^2)_\infty^8 + 16q(-q^2;q^2)_\infty^8 = (-q;q^2)_\infty^8. \tag{2.50}$$

If the equations (2.42), (2.43) and (2.44) are combined two at a time to eliminate the $\wp(\theta)$ term, we obtain

$$f_2^2(\theta) - f_1^2(\theta) = e_1 - e_2, \tag{2.51}$$
$$f_2^2(\theta) - f_3^2(\theta) = e_3 - e_2, \tag{2.52}$$
$$f_3^2(\theta) - f_1^2(\theta) = e_1 - e_3. \tag{2.53}$$

These are the elliptic function analogues of the trigonometric identity $\sin^2\theta + \cos^2\theta = 1$. In fact, from (2.22)–(2.26) and (2.45)–(2.47), we have

$$\lim_{q\to 0} f_1(\theta) = \frac{1}{2}\cot\frac{\theta}{2}, \quad \lim_{q\to 0} f_2(\theta) = \lim_{q\to 0} f_3(\theta) = \frac{1}{2}\csc\frac{\theta}{2},$$

$$\lim_{q\to 0} e_1 = 1/6, \quad \lim_{q\to 0} e_2 = \lim_{q\to 0} e_3 = -1/12.$$

Therefore when $q = 0$, (2.51) and (2.53) reduce to

$$\frac{1}{4}\csc^2\frac{\theta}{2} - \frac{1}{4}\cot^2\frac{\theta}{2} = \frac{1}{4},$$

while (2.52) reduces to a tautology.

2.5. Derivatives. In the fundamental multiplicative identity (2.7), let $t = e^{i\theta}$ to get

$$F(a,e^{i\theta})F(b,e^{i\theta}) = \frac{1}{i}\frac{\partial}{\partial\theta}F(ab,e^{i\theta}) + F(ab,e^{i\theta})(\rho_1(a) + \rho_1(b)). \tag{2.54}$$

Now let $a = e^{i\pi}$ and $b = e^{i\pi\tau}$. From (2.13) we have $\rho_1(e^{i\pi}) = 0$, $\rho_1(e^{i\pi\tau}) = \frac{1}{2}$, hence

$$F(-1,e^{i\theta})F(q,e^{i\theta}) = \frac{1}{i}\frac{\partial}{\partial\theta}F(-q,e^{i\theta}) + \frac{1}{2}F(-q,e^{i\theta}). \tag{2.55}$$

The left hand side of this is

$$F(-1,e^{i\theta})F(q,e^{i\theta}) = if_1(\theta)ie^{-i\theta/2}f_2(\theta) = -e^{-i\theta/2}f_1(\theta)f_2(\theta).$$

The right hand side of (2.55) is

$$\frac{1}{i}\frac{\partial}{\partial \theta}\left(ie^{-i\theta/2}f_3(\theta)\right) + \frac{i}{2}e^{-i\theta/2}f_3(\theta)$$
$$= e^{-i\theta/2}f_3'(\theta) - \frac{i}{2}e^{-i\theta/2}f_3(\theta) + \frac{i}{2}e^{-i\theta/2}f_3(\theta)$$
$$= e^{-i\theta/2}f_3'(\theta).$$

Combining gives [17, p. 111]

$$f_3'(\theta) = -f_1(\theta)f_2(\theta). \tag{2.56}$$

Similarly, letting $a = e^{i\pi\tau}$, $b = e^{i\pi+i\pi\tau}$ and $a = e^{i\pi+i\pi\tau}$, $b = e^{i\pi}$ in (2.54) leads, respectively, to [17, p. 111]

$$f_1'(\theta) = -f_2(\theta)f_3(\theta), \tag{2.57}$$
$$f_2'(\theta) = -f_3(\theta)f_1(\theta). \tag{2.58}$$

3. Sums of squares and sums of triangular numbers

3.1. Introduction. Letting $\alpha, \beta \to 0$ in Ramanujan's sum (2.1) gives Jacobi's triple product identity:

$$\prod_{n=1}^{\infty}(1+zq^{2n-1})(1+q^{2n-1}/z)(1-q^{2n}) = \sum_{j=-\infty}^{\infty} q^{j^2}z^j. \tag{3.1}$$

Let us define two functions ϕ and ψ by

$$\phi(q) := \sum_{j=-\infty}^{\infty} q^{j^2}, \tag{3.2}$$

$$\psi(q) := \sum_{j=0}^{\infty} q^{j(j+1)/2}. \tag{3.3}$$

Thus ϕ and ψ are the generating functions for square numbers and triangular numbers, respectively.

Putting $z = 1$ in the Jacobi triple product identity (3.1) and using Euler's identity

$$\prod_{n=1}^{\infty}(1+q^{2n-1}) = \prod_{n=1}^{\infty}\frac{1}{(1+q^{2n})(1-q^{2n-1})}, \tag{3.4}$$

we obtain

$$\phi(q) = \sum_{j=-\infty}^{\infty} q^{j^2} = \prod_{n=1}^{\infty}(1+q^{2n-1})(1+q^{2n-1})(1-q^{2n})$$

$$= \prod_{n=1}^{\infty}\frac{(1+q^{2n-1})(1-q^{2n})}{(1-q^{2n-1})(1+q^{2n})} \tag{3.5}$$

$$= \prod_{n=1}^{\infty}\frac{1-(-q)^n}{1+(-q)^n}. \tag{3.6}$$

Similarly, letting $z = q$ in (3.1), and observing that

$$\sum_{j=-\infty}^{\infty} q^{j(j+1)} = 2\sum_{j=0}^{\infty} q^{j(j+1)},$$

we obtain

$$\psi(q^2) = \sum_{j=0}^{\infty} q^{j(j+1)} = \frac{1}{2} \prod_{n=1}^{\infty} (1+q^{2n})(1+q^{2(n-1)})(1-q^{2n})$$

$$= \prod_{n=1}^{\infty} (1+q^{2n})(1+q^{2n})(1-q^{2n})$$

$$= \prod_{n=1}^{\infty} \frac{(1+q^{2n})(1-q^{2n})}{(1+q^{2n-1})(1-q^{2n-1})} \tag{3.7}$$

$$= \prod_{n=1}^{\infty} \frac{1-q^{4n}}{1-q^{4n-2}}. \tag{3.8}$$

Furthermore, the modulus k and complementary modulus k' defined in (2.48) and (2.49) are related to the functions ϕ and ψ by

$$k = 4q^{1/2} \frac{\psi^2(q^2)}{\phi^2(q)}, \tag{3.9}$$

$$k' = \frac{\phi^2(-q)}{\phi^2(q)}. \tag{3.10}$$

We will denote the number of representations of an integer n as a sum of s squares by $r_s(n)$. That is, $r_s(n)$ is the number of solutions in integers of

$$x_1^2 + x_2^2 + \cdots + x_s^2 = n.$$

We pay attention to the sign and order of $x_1, x_2, \cdots, x_s$. For example,

$$1 = (\pm 1)^2 + 0^2 = 0^2 + (\pm 1)^2, \quad 13 = (\pm 3)^2 + (\pm 2)^2 = (\pm 2)^2 + (\pm 3)^2,$$

and so $r_2(1) = 4$ and $r_2(13) = 8$.

Similarly, we will denote the number of representations of an integer n as a sum of s triangular numbers by $t_s(n)$. That is, $t_s(n)$ is the number of solutions in *non-negative* integers of

$$\frac{x_1(x_1+1)}{2} + \frac{x_2(x_2+1)}{2} + \cdots + \frac{x_s(x_s+1)}{2} = n.$$

The condition that the x's be non-negative is used here without loss of generality, since allowing $x_1, x_2, \cdots, x_s$ to be any integers would merely multiply the number of such representations by 2^s.

As examples,

$$1 = \frac{(1)(2)}{2} + \frac{(0)(1)}{2} = \frac{(0)(1)}{2} + \frac{(1)(2)}{2},$$

$$6 = \frac{(3)(4)}{2} + \frac{(0)(1)}{2} = \frac{(0)(1)}{2} + \frac{(3)(4)}{2} = \frac{(2)(3)}{2} + \frac{(2)(3)}{2}$$

and so $t_2(1) = 2$ and $t_2(6) = 3$.

The $r_s(n)$ and $t_s(n)$ have the following generating functions[1]:

$$\sum_{j=0}^{\infty} r_s(j)q^j = \phi^s(q) = \left(\sum_{j=-\infty}^{\infty} q^{j^2}\right)^s,$$

$$\sum_{j=0}^{\infty} t_s(j)q^{2j} = \psi^s(q^2) = \left(\sum_{j=0}^{\infty} q^{j(j+1)}\right)^s.$$

3.2. Sums of two squares and two triangles. Taking $\alpha = \beta = -1$ and $z = 1$ in Ramanujan's sum (2.1) gives

$$\prod_{n=1}^{\infty} \frac{(1+q^{2n-1})^2(1-q^{2n})^2}{(1-q^{2n-1})^2(1+q^{2n})^2} = 1 + 4\left\{\frac{q}{1+q^2} + \frac{q^2}{1+q^4} + \frac{q^3}{1+q^6} + \cdots\right\}.$$

If we use equation (3.5) on the left hand side, then this becomes

$$\left(\sum_{j=-\infty}^{\infty} q^{j^2}\right)^2 = 1 + 4\left\{\frac{q}{1+q^2} + \frac{q^2}{1+q^4} + \frac{q^3}{1+q^6} + \cdots\right\}. \quad (3.11)$$

Equating the coefficients of q^n on both sides gives the following theorem of Jacobi:

THEOREM 3.1. *Let $d_1(n)$ be the number of divisors of n of the form $4m + 1$, and let $d_3(n)$ be the number of divisors of n of the form $4m + 3$. Then*

$$r_2(n) = 4\{d_1(n) - d_3(n)\}. \quad (3.12)$$

Similarly, if we take $\alpha = 1/q$, $\beta = q$ and $z = q$ in Ramanujan's sum (2.1) and use equation (3.7), then we obtain

$$\left(\sum_{j=0}^{\infty} q^{j(j+1)}\right)^2 = \prod_{n=1}^{\infty} \frac{(1+q^{2n})^2(1-q^{2n})^2}{(1+q^{2n-1})^2(1-q^{2n-1})^2}$$

$$= \frac{1}{1-q} - \frac{q}{1-q^3} + \frac{q^2}{1-q^5} - \frac{q^3}{1-q^7} + \cdots.$$

This implies the following theorem about sums of two triangles.

THEOREM 3.2.
$$t_2(n) = d_1(4n+1) - d_3(4n+1). \quad (3.13)$$

3.3. Higher derivatives. We begin by recording the values of $f_j^{(n)}$, $j = 1$, 2, 3, at the points π, $\pi\tau$ and $\pi + \pi\tau$. Differentiating equations (2.21) – (2.26) and using the results

$$\tan x = \sum_{j=1}^{\infty} \frac{(-1)^{j-1} 2^{2j}(2^{2j}-1)B_{2j}}{(2j)!} x^{2j-1},$$

$$\sec x = \sum_{j=0}^{\infty} \frac{(-1)^j E_{2j}}{(2j)!} x^{2j},$$

---

[1] Ramanujan [16, p. 157] defines $r_s(n)$ by $\phi^s(q) = 1 + 2\sum_{j=1}^{\infty} r_s(j)q^j$, which is different from the standard notation. We will not use Ramanujan's notation.

where $B_2 = 1/6$, $B_4 = -1/30$, $B_6 = 1/42$, $B_8 = -1/30$, $B_{10} = 5/66$, $\cdots$, are the Bernoulli numbers and $E_0 = 1$, $E_2 = -1$, $E_4 = 5$, $E_6 = -61$, $E_8 = 1385$, $E_{10} = -50521$, $\cdots$, are the Euler numbers, we get

$$f_1^{(2m-1)}(\pi) = \frac{(-1)^m(2^{2m}-1)B_{2m}}{2m} - \sum_{j=1}^{\infty} \frac{(-1)^j j^{2m-1} q^{2j}}{1+q^{2j}}, \quad (3.14)$$

$$f_2^{(2m)}(\pi) = \frac{(-1)^m}{2^{2m+1}}\left[E_{2m} - 4\sum_{j=1}^{\infty} \frac{(-1)^j(2j-1)^{2m} q^{2j-1}}{1-q^{2j-1}}\right], \quad (3.15)$$

$$f_3^{(2m)}(\pi) = \frac{(-1)^m}{2^{2m+1}}\left[E_{2m} + 4\sum_{j=1}^{\infty} \frac{(-1)^j(2j-1)^{2m} q^{2j-1}}{1+q^{2j-1}}\right], \quad (3.16)$$

$$f_1^{(2m)}(\pi\tau) = 2i(-1)^{m-1}\sum_{j=1}^{\infty} \frac{j^{2m} q^j}{1+q^{2j}}, \quad (3.17)$$

$$f_2^{(2m-1)}(\pi\tau) = \frac{(-1)^{m-1}}{2^{2m-2}}\sum_{j=1}^{\infty} \frac{(2j-1)^{2m-1} q^{j-1/2}}{1-q^{2j-1}}, \quad (3.18)$$

$$f_3^{(2m)}(\pi\tau) = \frac{i(-1)^{m-1}}{2^{2m-1}}\sum_{j=1}^{\infty} \frac{(2j-1)^{2m} q^{j-1/2}}{1+q^{2j-1}}, \quad (3.19)$$

$$f_1^{(2m)}(\pi+\pi\tau) = 2i(-1)^{m-1}\sum_{j=1}^{\infty} \frac{(-1)^j j^{2m} q^j}{1+q^{2j}}, \quad (3.20)$$

$$f_2^{(2m)}(\pi+\pi\tau) = \frac{(-1)^{m-1}}{2^{2m-1}}\sum_{j=1}^{\infty} \frac{(-1)^j(2j-1)^{2m} q^{j-1/2}}{1-q^{2j-1}}, \quad (3.21)$$

$$f_3^{(2m-1)}(\pi+\pi\tau) = \frac{i(-1)^{m-1}}{2^{2m-1}}\sum_{j=1}^{\infty} \frac{(-1)^j(2j-1)^{2m-1} q^{j-1/2}}{1-q^{2j-1}}. \quad (3.22)$$

Also,
$$f_1^{(2m)}(\pi) = f_2^{(2m-1)}(\pi) = f_3^{(2m-1)}(\pi) = 0,$$
$$f_1^{(2m-1)}(\pi\tau) = f_2^{(2m)}(\pi\tau) = f_3^{(2m-1)}(\pi\tau) = 0,$$
$$f_1^{(2m-1)}(\pi+\pi\tau) = f_2^{(2m-1)}(\pi+\pi\tau) = f_3^{(2m)}(\pi+\pi\tau) = 0.$$

Next, we calculate the values of $(f_j^2)^{(n)}$ at the points π, $\pi\tau$ and $\pi+\pi\tau$. By (2.42) – (2.44), f_1^2, f_2^2 and f_3^2 differ from each other only by additive constants, so it suffices to calculate the derivatives of f_1^2. Using (2.12), (2.17) and (2.42) we obtain

$$(f_1^2)^{(2m)}(\pi) = (-1)^m\left[\frac{(2^{2m+2}-1)}{2m+2}B_{2m+2} - 2\sum_{j=1}^{\infty} \frac{j^{2m+1} q^{2j}(-1)^j}{1-q^{2j}}\right], \quad (3.23)$$

$$(f_1^2)^{(2m)}(\pi\tau) = 2(-1)^{m+1}\sum_{j=1}^{\infty} \frac{j^{2m+1} q^j}{1-q^{2j}}, \quad (3.24)$$

$$(f_1^2)^{(2m)}(\pi+\pi\tau) = 2(-1)^{m+1}\sum_{j=1}^{\infty} \frac{j^{2m+1}(-1)^j q^j}{1-q^{2j}}. \quad (3.25)$$

Also,
$$(f_1^2)^{(2m-1)}(\pi) = (f_1^2)^{(2m-1)}(\pi\tau) = (f_1^2)^{(2m-1)}(\pi+\pi\tau) = 0.$$

Now we shall use (2.56) – (2.58) to compute these derivatives in another way. We have

$$\begin{aligned}
f_1' &= -f_2 f_3, \\
f_1'' &= f_1 f_2^2 + f_1 f_3^2, \\
f_1''' &= -4f_1^2 f_2 f_3 - f_2 f_3^3 - f_2^3 f_3 \\
f_1'''' &= 4f_1^3 f_2^2 + 4f_1^3 f_3^2 + f_1 f_2^4 + f_1 f_3^4 + 14 f_1 f_2^2 f_3^2, \\
&\vdots \quad \vdots
\end{aligned}$$

Similar results can be written down for derivatives of f_2 and f_3 by permuting the subscripts and using symmetry. By induction we obtain

$$f_1^{(n)} = \sum c(n; \lambda_1, \lambda_2, \lambda_3) f_1^{\lambda_1} f_2^{\lambda_2} f_3^{\lambda_3}, \tag{3.26}$$

$$f_2^{(n)} = \sum c(n; \lambda_1, \lambda_2, \lambda_3) f_1^{\lambda_2} f_2^{\lambda_1} f_3^{\lambda_3}, \tag{3.27}$$

$$f_3^{(n)} = \sum c(n; \lambda_1, \lambda_2, \lambda_3) f_1^{\lambda_3} f_2^{\lambda_2} f_3^{\lambda_1}. \tag{3.28}$$

In each case the sum is over all $\lambda_1, \lambda_2, \lambda_3$ satisfying $\lambda_1 + \lambda_2 + \lambda_3 = n+1$, $0 \leq \lambda_1, \lambda_2, \lambda_3 \leq n$, and $\lambda_1 + 1, \lambda_2, \lambda_3$ and n all have the same parity. In addition, the coefficients $c(n; \lambda_1, \lambda_2, \lambda_3)$ are all integers, and

$$c(n; \lambda_1, \lambda_2, \lambda_3) = c(n; \lambda_1, \lambda_3, \lambda_2), \tag{3.29}$$

$$(-1)^n \sum c(n; \lambda_1, \lambda_2, \lambda_3) = n! \tag{3.30}$$

We will see later in (3.44) that

$$\sum_{j=1}^{m} c(2m; 2m+1-2j, 2j, 0) = |E_{2m}|. \tag{3.31}$$

In (3.30) the sum is over the same range of values as for (3.26) – (3.28).

Similarly,

$$(f_1^2)^{(n)} = \sum c(n; \lambda_1, \lambda_2, \lambda_3) f_1^{\lambda_1} f_2^{\lambda_2} f_3^{\lambda_3} \tag{3.32}$$

where this time the sum is over all $\lambda_1, \lambda_2, \lambda_3$ satisfying $\lambda_1 + \lambda_2 + \lambda_3 = n+2$, $0 \leq \lambda_1, \lambda_2, \lambda_3 \leq n$, and $\lambda_1, \lambda_2, \lambda_3$ and n all have the same parity. In addition, the coefficients $c(n; \lambda_1, \lambda_2, \lambda_3)$ are integers,

$$c(n; \lambda_1, \lambda_2, \lambda_3) = c(n; \lambda_i, \lambda_j, \lambda_k) \tag{3.33}$$

for any permutation (i, j, k) of $(1, 2, 3)$, and

$$(-1)^n \sum c(n; \lambda_1, \lambda_2, \lambda_3) = (n+1)! \tag{3.34}$$

We will see later (3.49) that

$$\sum_{j=1}^{m-1} c(2m-2; 2m-2j, 2j, 0) = \frac{2^{2m}(2^{2m}-1)|B_{2m}|}{2m}. \tag{3.35}$$

In (3.34) the sum is over the same range of values as for (3.32).

TABLE 1. Values of f_1, f_2, f_3.

θ	π	$\pi\tau$	$\pi + \pi\tau$
$f_1(\theta)$	0	$\frac{1}{2i}\phi^2(q)$	$\frac{1}{2i}\phi^2(-q)$
$f_2(\theta)$	$\frac{1}{2}\phi^2(q)$	0	$\frac{1}{2}\left\{4q^{1/2}\psi^2(q^2)\right\}$
$f_3(\theta)$	$\frac{1}{2}\phi^2(-q)$	$\frac{1}{2i}\left\{4q^{1/2}\psi^2(q^2)\right\}$	0

We also record the values of $f_j(\pi)$, $f_j(\pi\tau)$ and $f_j(\pi + \pi\tau)$, $j = 1, 2, 3$. For example, using (2.27) and then (3.5) gives

$$f_1(\pi\tau) = \frac{1}{2i} \frac{(-q, -q, q^2, q^2; q^2)_\infty}{(q, q, -q^2, -q^2; q^2)_\infty} = \frac{-i}{2}\phi^2(q).$$

The results are summarised in Table 1.

3.4. Sums of $4m+2$ squares. Formulas for sums of $4m+2$ squares, $m \geq 1$, can be obtained by evaluating

$$2^{2m+1}\left[f_2^{(2m)}(\pi) + i(-1)^m f_1^{(2m)}(\pi\tau)\right] \tag{3.36}$$

in two different ways.

First, using (3.15) and (3.17) gives

$$2^{2m+1}\left[f_2^{(2m)}(\pi) + i(-1)^m f_1^{(2m)}(\pi\tau)\right]$$
$$= |E_{2m}| - 4(-1)^m \sum_{j=1}^\infty \frac{(-1)^j (2j-1)^{2m} q^{2j-1}}{1 - q^{2j-1}} + 2^{2m+2} \sum_{j=1}^\infty \frac{j^{2m} q^j}{1 + q^{2j}}. \tag{3.37}$$

Now use (3.26) and (3.27) to re-evaluate (3.36). We start with

$$i(-1)^m f_1^{(2m)}(\pi\tau) = i(-1)^m \sum c(2m; \lambda_1, \lambda_2, \lambda_3)(f_1(\pi\tau))^{\lambda_1}(f_2(\pi\tau))^{\lambda_2}(f_3(\pi\tau))^{\lambda_3}.$$

Since $f_2(\pi\tau) = 0$, the only nonzero terms in this sum will occur when $\lambda_2 = 0$. Consequently

$$i(-1)^m f_1^{(2m)}(\pi\tau)$$
$$= i(-1)^m \sum_{\mu=1}^m c(2m; 2m+1-2\mu, 0, 2\mu)(f_1(\pi\tau))^{2m+1-2\mu}(f_3(\pi\tau))^{2\mu}$$
$$= \frac{1}{2^{2m+1}} \sum_{\mu=1}^m c(2m; 2m+1-2\mu, 0, 2\mu)(\phi^2(q))^{2m+1-2\mu}(4q^{1/2}\psi^2(q^2))^{2\mu}$$
$$= \frac{\phi^{4m+2}(q)}{2^{2m+1}} \sum_{\mu=1}^m c(2m; 2m+1-2\mu, 0, 2\mu)\left(\frac{4q^{1/2}\psi^2(q^2)}{\phi^2(q)}\right)^{2\mu}$$
$$= \frac{\phi^{4m+2}(q)}{2^{2m+1}} \sum_{\mu=1}^m c(2m; 2m+1-2\mu, 0, 2\mu)k^{2\mu}. \tag{3.38}$$

Similarly,
$$f_2^{(2m)}(\pi) = \frac{\phi^{4m+2}(q)}{2^{2m+1}} \sum_{\mu=1}^{m} c(2m; 2m+1-2\mu, 0, 2\mu) k'^{2\mu}. \tag{3.39}$$

Adding (3.38) and (3.39) and multiplying by 2^{2m+1} gives
$$2^{2m+1} \left[f_2^{(2m)}(\pi) + i(-1)^m f_1^{(2m)}(\pi\tau) \right]$$
$$= \phi^{4m+2}(q) \sum_{\mu=1}^{m} c(2m; 2m+1-2\mu, 0, 2\mu) \left(k^{2\mu} + k'^{2\mu} \right). \tag{3.40}$$

Next observe that
$$\begin{aligned}
k^2 + k'^2 &= 1, \\
k^4 + k'^4 &= k^2(1-k'^2) + k'^2(1-k^2) = 1 - 2(kk')^2, \\
k^{2m} + k'^{2m} &= k^{2m-2}(1-k'^2) + k'^{2m-2}(1-k^2) \\
&= k^{2m-2} + k'^{2m-2} - (kk')^2(k^{2m-4} + k'^{2m-4}), \quad m = 3, 4, \cdots,
\end{aligned} \tag{3.41}$$

and so by induction $k^{2m} + k'^{2m}$ is a polynomial in $(kk')^2$ of degree $\lfloor m/2 \rfloor$, with integer coefficients and constant term 1. That is,
$$k^{2m} + k'^{2m} = 1 + \sum_{j=1}^{\lfloor m/2 \rfloor} a_j (kk')^{2j} \tag{3.42}$$

for some integers a_j. Using this in (3.40) gives
$$2^{2m+1} \left[f_2^{(2m)}(\pi) + i(-1)^m f_1^{(2m)}(\pi\tau) \right] = \phi^{4m+2}(q) \left\{ d_0 - \sum_{j=1}^{\lfloor m/2 \rfloor} d_j (kk')^{2j} \right\}, \tag{3.43}$$

for some integers $d_0, d_1, \cdots, d_{\lfloor m/2 \rfloor}$, where
$$d_0 = \sum_{\mu=1}^{m} c(2m; 2m+1-2\mu, 0, 2\mu).$$

Comparing the coefficients of q^0 in (3.37) and (3.43) gives
$$d_0 = \sum_{\mu=1}^{m} c(2m; 2m+1-2\mu, 0, 2\mu) = |E_{2m}|. \tag{3.44}$$

This proves the claim made earlier in (3.31).

Combining (3.37) and (3.43) gives
$$|E_{2m}| \phi^{4m+2}(q)$$
$$= |E_{2m}| - 4(-1)^m \sum_{j=1}^{\infty} \frac{(-1)^j (2j-1)^{2m} q^{2j-1}}{1 - q^{2j-1}} + 2^{2m+2} \sum_{j=1}^{\infty} \frac{j^{2m} q^j}{1 + q^{2j}}$$
$$+ \phi^{4m+2}(q) \sum_{j=1}^{\lfloor m/2 \rfloor} d_j (kk')^{2j}.$$

Substituting the expressions for k and k' given in (2.48) and (2.49), we have proved

TABLE 2. Sums of $4m+2$ squares and $4m+2$ triangular numbers

m	$4m+2$	$|E_{2m}|$	$d_1, d_2, \cdots, d_{\lfloor m/2 \rfloor}$
1	6	1	—
2	10	5	2
3	14	61	91
4	18	1385	3052, -2
5	22	50521	138677, -7381
6	26	2702765	8782962, -1907349, 2
7	30	199360981	747599583, -331261698, 597871
8	34	19391512145	82413910232, -56726638356, 1192092896, -2
9	38	2404879675441	11423178392809, -10710303333531, 795134798974, -48427561
10	42	370371188237525	1944448737984862, -2309050055554463, 371179666042442, -745058059693, 2
11	46	69348874393137901	398756027759494355, -573212064779230960, 155636292404293285, -1909093442764355, 3922632451
12	50	15514534163557086905	96965838522227598852, -163629856904292739110, 65144309725506894520, -2434405987601451345, 465661287307740, -2

THEOREM 3.3 (Sums of $4m+2$ squares).

$$|E_{2m}|\phi^{4m+2}(q)$$
$$= |E_{2m}| - 4(-1)^m \sum_{j=1}^{\infty} \frac{(-1)^j(2j-1)^{2m}q^{2j-1}}{1-q^{2j-1}} + 2^{2m+2}\sum_{j=1}^{\infty}\frac{j^{2m}q^j}{1+q^{2j}}$$
$$+ \frac{(-q;-q)_\infty^{8m+4}}{(q^2;q^2)_\infty^{4m+2}} \sum_{j=1}^{\lfloor m/2 \rfloor} d_j 16^j q^j \frac{(q^2;q^2)_\infty^{24j}}{(-q;-q)_\infty^{24j}} \quad (3.45)$$

for some integers d_j. Values of the integers d_j can be readily calculated by comparing the coefficients of the first few powers of q on both sides. The first few values are given in Table 2.

This is equivalent to Ramanujan's formulas [**16**, pp. 158-9, (133), (134), (146), (147)], which are stated without proof. Our results confirm an observation of Ramamani [**14**, pp. 6–7] that in Ramanujan's Table VI in [**16**, p. 159], the numbers 1103272 and 821888 should be 1109416 and 944768, respectively.

Taking $m=1$ in Table 2 gives

$$\left(\sum_{j=-\infty}^{\infty} q^{j^2}\right)^6 = 1 + 4\sum_{j=1}^{\infty}\frac{(-1)^j(2j-1)^2 q^{2j-1}}{1-q^{2j-1}} + 16\sum_{j=1}^{\infty}\frac{j^2 q^j}{1+q^{2j}},$$

which is due to Jacobi [**9**, p. 164]. Equating coefficients of q^n on both sides gives

$$r_6(n) = 4\sum_{\substack{de=n \\ d \equiv 1(4)}}(4e^2 - d^2) - 4\sum_{\substack{de=n \\ d \equiv 3(4)}}(4e^2 - d^2). \quad (3.46)$$

Further details may be found in [**4**].

3.5. Sums of $4m$ squares. Formulas for sums of $4m$ squares, $m \geq 1$, can be obtained by evaluating

$$2^{2m}\left[(f_1^2)^{(2m-2)}(\pi) + (-1)^m (f_1^2)^{(2m-2)}(\pi\tau)\right]$$

in two different ways. From (3.23) and (3.24) we have

$$
2^{2m}\left[(f_1^2)^{(2m-2)}(\pi) + (-1)^m (f_1^2)^{(2m-2)}(\pi\tau)\right]
$$
$$
= \frac{2^{2m}(2^{2m}-1)}{2m}|B_{2m}| + 2^{2m+1}(-1)^m \sum_{j=1}^{\infty}\frac{j^{2m-1}q^{2j}(-1)^j}{1-q^{2j}} + 2^{2m+1}\sum_{j=1}^{\infty}\frac{j^{2m-1}q^j}{1-q^{2j}}
$$
$$
= \frac{2^{2m}(2^{2m}-1)}{2m}|B_{2m}| + 2^{2m+1}\sum_{j=1}^{\infty}\frac{j^{2m-1}q^j}{1-(-1)^{m+j}q^j}.
$$
(3.47)

Next, from (3.32) we have

$$
2^{2m}\left[(f_1^2)^{(2m-2)}(\pi) + (-1)^m (f_1^2)^{(2m-2)}(\pi\tau)\right]
$$
$$
= \phi^{4m}(q)\sum_{\mu=1}^{m-1} c(2m-2;0,2m-2\mu,2\mu)(k^{2\mu}+k'^{2\mu})
$$
$$
= \phi^{4m}(q)\left[d_0 - \sum_{j=1}^{\lfloor(m-1)/2\rfloor} d_j(kk')^{2j}\right]
$$
(3.48)

for some integers $d_0, d_1, \cdots, d_{\lfloor(m-1)/2\rfloor}$, where

$$
d_0 = \sum_{\mu=1}^{m-1} c(2m-2;0,2m-2\mu,2\mu).
$$

Comparing the coefficients of q^0 in (3.47) and (3.48) gives

$$
d_0 = \sum_{\mu=1}^{m-1} c(2m-2;0,2m-2\mu,2\mu) = \frac{2^{2m}(2^{2m}-1)}{2m}|B_{2m}|.
$$
(3.49)

This proves the claim made earlier in (3.35).

Combining (3.47), (3.48) and (3.49) gives

$$
\frac{2^{2m}(2^{2m}-1)}{2m}|B_{2m}|\phi^{4m}(q) = \frac{2^{2m}(2^{2m}-1)}{2m}|B_{2m}| + 2^{2m+1}\sum_{j=1}^{\infty}\frac{j^{2m-1}q^j}{1-(-1)^{m+j}q^j}
$$
$$
+ \phi^{4m}(q)\sum_{j=1}^{\lfloor(m-1)/2\rfloor} d_j(kk')^{2j}.
$$

Thus we have proved

THEOREM 3.4 (Sums of $4m$ squares).

$$
\frac{2^{2m}(2^{2m}-1)}{2m}|B_{2m}|\phi^{4m}(q) = \frac{2^{2m}(2^{2m}-1)}{2m}|B_{2m}| + 2^{2m+1}\sum_{j=1}^{\infty}\frac{j^{2m-1}q^j}{1-(-1)^{m+j}q^j}
$$
$$
+ \frac{(-q;-q)_\infty^{8m}}{(q^2;q^2)_\infty^{4m}}\sum_{j=1}^{\lfloor(m-1)/2\rfloor} d_j 16^j q^j \frac{(q^2;q^2)_\infty^{24j}}{(-q;-q)_\infty^{24j}}.
$$
(3.50)

Values of d_j can be obtained by comparing the coefficients of the first few powers of q on both sides. The first few values are given in Table 3.

TABLE 3. Sums of $4m$ squares and $4m$ triangular numbers

m	$4m$	$\dfrac{2^{2m}(2^{2m}-1)}{2m}\|B_{2m}\|$	$d_1, d_2, \cdots, d_{\lfloor (m-1)/2 \rfloor}$
1	4	1	–
2	8	2	–
3	12	16	16
4	16	272	512
5	20	7936	19712, -256
6	24	353792	1060864, -131072
7	28	22368256	78286848, -25509888, 4096
8	32	1903757312	7615021056, -4300210176, 33554432
9	36	209865342976	944394010624, -767588499456, 33059897344, -65536
10	40	29088885112832	145444425433088, -154538996662272, 17594065092608, -8589934592
11	44	4951498053124096	27233239291658240, -35743695288401920, 7638097545134080, -42845607034880, 1048576
12	48	1015423886506852352	6092543319039016960, -9519600035503800320, 3172431722939678720, -72058212513218560, 2199023255552

This is equivalent to Ramanujan's formulas [**16**, pp. 158-9, (131), (132), (146), (147)].

The cases $m=1$, $m=2$ and $m=6$ of (3.50) give, respectively

$$\left(\sum_{j=-\infty}^{\infty} q^{j^2}\right)^4 = 1 + 8\sum_{j=1}^{\infty} \frac{jq^j}{1+(-q)^j}, \tag{3.51}$$

$$\left(\sum_{j=-\infty}^{\infty} q^{j^2}\right)^8 = 1 + 16\sum_{j=1}^{\infty} \frac{j^3 q^j}{1-(-q)^j}, \tag{3.52}$$

$$691\left(\sum_{j=-\infty}^{\infty} q^{j^2}\right)^{24} = 691 + 16\sum_{j=1}^{\infty} \frac{j^{11} q^j}{1-(-q)^j}$$
$$+ 33152 q(-q;-q)_\infty^{24} - 65536 q^2 (q^2;q^2)_\infty^{24}. \tag{3.53}$$

Equating coefficients of q^n in these equations gives

$$r_4(n) = 8 \sum_{d|n,\ 4\nmid d} d, \tag{3.54}$$

$$r_8(n) = 16(-1)^n \sum_{d|n} (-1)^d d^3, \tag{3.55}$$

$$691 r_{24}(n) = 16(-1)^n \sum_{d|n} (-1)^d d^{11} + 33152(-1)^{n-1}\tau(n) - 65536\tau(n/2). \tag{3.56}$$

The function $\tau(n)$ is defined by

$$\sum_{n=1}^{\infty} \tau(n) q^n = q(q;q)_\infty^{24} \tag{3.57}$$

and $\tau(x)$ is defined to be zero if x is not a positive integer. Equations (3.53) and (3.56) are due to Ramanujan [**16**, p. 159].

3.6. Sums of $4m+2$ triangular numbers. Ramanujan [**16**, pp. 190 – 191] obtains formulas for the number of representations of an integer as a sum of an even number of triangular numbers from the corresponding formulas for sums of squares by applying the modular transformation. In keeping with our promise to give self contained, elementary proofs, we give proofs for results on sums of triangular numbers which are similar to the ones given above for sums of squares.

For sums of $4m+2$ triangular numbers, $m \geq 1$, we evaluate

$$\frac{1}{2^{2m+1}}\left[f_2^{(2m)}(\pi+\pi\tau)+i(-1)^m f_3^{(2m)}(\pi\tau)\right] \tag{3.58}$$

in two different ways.

First, using (3.19) and (3.21) gives

$$\frac{1}{2^{2m+1}}\left[f_2^{(2m)}(\pi+\pi\tau)+i(-1)^m f_3^{(2m)}(\pi\tau)\right]$$
$$= \frac{q^{m+1/2}}{2^{4m}}\sum_{j=1}^{\infty}(2j-1)^{2m}q^{j-m-1}\left\{\frac{1}{1+q^{2j-1}}-\frac{(-1)^{m+j}}{1-q^{2j-1}}\right\}. \tag{3.59}$$

Second, using (3.27) and (3.28) gives

$$\frac{1}{2^{2m+1}}\left[f_2^{(2m)}(\pi+\pi\tau)+i(-1)^m f_3^{(2m)}(\pi\tau)\right]$$
$$= q^{m+1/2}\psi^{4m+2}(q^2)$$
$$\times \sum_{\mu=1}^{m} c(2m; 2m+1-2\mu, 2\mu, 0)\frac{\phi^{4\mu}(q)+(-1)^\mu \phi^{4\mu}(-q)}{(2q^{1/4}\psi(q^2))^{4\mu}}$$
$$= q^{m+1/2}\psi^{4m+2}(q^2)\sum_{\mu=1}^{m} c(2m; 2m+1-2\mu, 2\mu, 0)\frac{1+(-k'^2)^\mu}{k^{2\mu}}. \tag{3.60}$$

Next, observe that

$$\frac{1}{k^2}-\frac{k'^2}{k^2} = 1,$$
$$\frac{1}{k^4}+\frac{k'^4}{k^4} = \frac{(1-k'^2)^2+2k'^2}{k^4} = 1-2\left(\frac{-k'^2}{k^4}\right),$$
$$\frac{1}{k^{2m}}+(-1)^m\frac{k'^{2m}}{k^{2m}} = \frac{1}{k^{2m-2}}+(-1)^{m-1}\frac{k'^{2m-2}}{k^{2m-2}}$$
$$-\left(\frac{-k'^2}{k^4}\right)\left\{\frac{1}{k^{2m-4}}+(-1)^{m-2}\frac{k'^{2m-4}}{k^{2m-4}}\right\} \quad m=3,4,\cdots, \tag{3.61}$$

and so by induction $1/k^{2m}+(-1)^m k'^{2m}/k^{2m}$ is a polynomial in $-k'^2/k^4$ of degree $\lfloor m/2 \rfloor$, with integer coefficients and constant term 1. That is

$$k^{2m}+k'^{2m} = 1+\sum_{j=1}^{\lfloor m/2 \rfloor} a_j \left(-\frac{k'^2}{k^4}\right)^4 \tag{3.62}$$

for some integers a_j. Comparison of the recurrences and initial conditions in (3.41) and (3.61) shows that the coefficients a_j in (3.62) are identical to those in Section 3.4. Using this in (3.60) gives

$$\frac{1}{2^{2m+1}}\left[f_2^{(2m)}(\pi+\pi\tau)+i(-1)^m f_3^{(2m)}(\pi\tau)\right]$$
$$= q^{m+1/2}\psi^{4m+2}(q^2)\left[\sum_{\mu=1}^{m} c(2m;2m+1-2\mu,2\mu,0) - \sum_{j=1}^{\lfloor m/2 \rfloor} d_j\left(\frac{-k'^2}{k^4}\right)^j\right]$$
$$= q^{m+1/2}|E_{2m}|\psi^{4m+2}(q^2)$$
$$- q^{m+1/2}\frac{(q^4;q^4)_\infty^{8m+4}}{(q^2;q^2)_\infty^{4m+2}}\sum_{j=1}^{\lfloor m/2 \rfloor}\frac{(-1)^j d_j}{(256q^2)^j}\frac{(q^2;q^2)_\infty^{24j}}{(q^4;q^4)_\infty^{24j}}, \tag{3.63}$$

where $d_1, d_2, \cdots, d_{\lfloor m/2 \rfloor}$ are the same numbers as in Section 3.4.

Combining (3.59) and (3.63) gives

THEOREM 3.5 (Sums of $4m+2$ triangular numbers).

$$|E_{2m}|\psi^{4m+2}(q^2) = \frac{1}{2^{4m}}\sum_{j=1}^{\infty}(2j-1)^{2m}q^{j-m-1}\left\{\frac{1}{1+q^{2j-1}}-\frac{(-1)^{m+j}}{1-q^{2j-1}}\right\}$$
$$+\frac{(q^4;q^4)_\infty^{8m+4}}{(q^2;q^2)_\infty^{4m+2}}\sum_{j=1}^{\lfloor m/2 \rfloor}\frac{(-1)^j d_j}{(256q^2)^j}\frac{(q^2;q^2)_\infty^{24j}}{(q^4;q^4)_\infty^{24j}}. \tag{3.64}$$

The values of d_j corresponding to $m=1,2,\cdots,12$ are given in Table 2.

This is equivalent to Ramanujan's formulas [16, p. 191, (12.6), (12.63), (12.64)]. The special case $m=1$ gives

$$\left(\sum_{j=0}^{\infty}q^{j(j+1)}\right)^6 = \frac{1}{16}\sum_{j=1}^{\infty}(2j-1)^2 q^{j-2}\left\{\frac{1}{1+q^{2j-1}}+\frac{(-1)^j}{1-q^{2j-1}}\right\}. \tag{3.65}$$

This has the arithmetic interpretation

$$t_6(n) = \frac{1}{8}\sum_{\substack{d|4n+3\\d\equiv 3(4)}}d^2 - \frac{1}{8}\sum_{\substack{d|4n+3\\d\equiv 1(4)}}d^2. \tag{3.66}$$

Full details of the calculation and references may be found in [4].

3.7. Sums of $4m$ triangular numbers. Formulas for sums of $4m$ triangular numbers, $m \geq 1$, can be obtained by evaluating

$$\frac{1}{2^{2m}}\left[(-1)^m(f_1^2)^{(2m-2)}(\pi\tau)+(f_1^2)^{(2m-2)}(\pi+\pi\tau)\right]$$

in two different ways. From (3.24) and (3.25) we have

$$\frac{1}{2^{2m}}\left[(-1)^m (f_1^2)^{(2m-2)}(\pi\tau) + (f_1^2)^{(2m-2)}(\pi+\pi\tau)\right]$$
$$= \frac{1}{2^{2m-1}}\sum_{j=1}^{\infty}\frac{j^{2m-1}q^j}{1-q^{2j}} + \frac{1}{2^{2m-1}}\sum_{j=1}^{\infty}\frac{(-1)^{m+j}j^{2m-1}q^j}{1-q^{2j}}$$
$$= \frac{1}{2^{2m-2}}\sum_{j=1}^{\infty}\left(\frac{1+(-1)^{m+j}}{2}\right)\frac{j^{2m-1}q^j}{1-q^{2j}}. \tag{3.67}$$

Next, using (3.32) and simplifying gives

$$\frac{1}{2^{2m}}\left[(-1)^m (f_1^2)^{(2m-2)}(\pi\tau) + (f_1^2)^{(2m-2)}(\pi+\pi\tau)\right]$$
$$= q^m \psi^{4m}(q^2)\left[\sum_{\mu=1}^{m-1} c(2m-2;2\mu,2m-2\mu,0) - \sum_{j=1}^{\lfloor (m-1)/2 \rfloor} d_j \left(\frac{-k'^2}{k^4}\right)^j\right], \tag{3.68}$$

where $d_1, \cdots, d_{\lfloor(m-1)/2\rfloor}$ have the same values as in Section 3.5. Combining (3.67) and (3.68) and using (3.35) gives

THEOREM 3.6 (Sums of $4m$ triangular numbers).

$$\frac{2^{2m}(2^{2m}-1)|B_{2m}|}{2m}\psi^{4m}(q^2)$$
$$= \frac{1}{4^{m-1}}\sum_{j=1}^{\infty}\left(\frac{1+(-1)^{m+j}}{2}\right)\frac{j^{2m-1}q^{j-m}}{1-q^{2j}}$$
$$+ \frac{(q^4;q^4)_\infty^{8m}}{(q^2;q^2)_\infty^{4m}}\sum_{j=1}^{\lfloor(m-1)/2\rfloor}\frac{(-1)^j d_j}{(256q^2)^j}\frac{(q^2;q^2)_\infty^{24j}}{(q^4;q^4)_\infty^{24j}}. \tag{3.69}$$

The values of d_j corresponding to $m=1,2,\cdots,12$ are given in Table 3.

This is equivalent to Ramanujan's formulas [16, p. 191, (12.6), (12.61), (12.62)]. The cases $m=1$, $m=2$ and $m=6$ of (3.69) give respectively, after replacing q^2 with q,

$$\left(\sum_{j=0}^{\infty}q^{j(j+1)/2}\right)^4 = \sum_{j=1}^{\infty}\frac{(2j-1)q^{j-1}}{1-q^{2j-1}}, \tag{3.70}$$

$$\left(\sum_{j=0}^{\infty}q^{j(j+1)/2}\right)^8 = \sum_{j=1}^{\infty}\frac{j^3 q^{j-1}}{1-q^{2j}}, \tag{3.71}$$

$$176896\left(\sum_{j=0}^{\infty}q^{j(j+1)/2}\right)^{24} = \sum_{j=1}^{\infty}\frac{j^{11}q^{j-3}}{1-q^{2j}} - 2072q^{-1}(q^2;q^2)_\infty^{24} - q^{-2}(q;q)_\infty^{24}. \tag{3.72}$$

Equating coefficients of powers of q on both sides gives gives

$$t_4(n) = \sum_{d|2n+1} d, \tag{3.73}$$

$$t_8(n-1) = \sum_{\substack{d|n \\ d \text{ odd}}} \left(\frac{n}{d}\right)^3, \tag{3.74}$$

$$176896 t_{24}(n-3) = \sum_{\substack{d|n \\ d \text{ odd}}} \left(\frac{n}{d}\right)^{11} - 2072\tau(n/2) - \tau(n). \tag{3.75}$$

A proof of (3.70) using Ramanujan's $_1\psi_1$ summation formula was given by Adiga [1]. An even simpler proof, also using the $_1\psi_1$, was given by Lam [10]. Song Heng Chan [6] has pointed out that both (3.52) and (3.71) can be deduced easily from (2.40). Equations (3.72) and (3.75) were given by Ono et. al. [13].

References

[1] C. Adiga, *On the representations of an integer as a sum of two or four triangular numbers*, Nihonkai Math. J., **3** (1992), no. 2, 125–131.
[2] R. Askey, *Ramanujan's Extensions of the Gamma and Beta Functions*, American Mathematical Monthly, **87** (1980), no. 5, 346–359.
[3] B. C. Berndt, *Ramanujan's Notebooks, Part III*, Springer-Verlag, 1991.
[4] B. C. Berndt, *Fragments by Ramanujan on Lambert Series*, Number Theory and Its Applications, 35 – 49. S. Kanemitsu and K. Gÿory (eds.), Kluwer Academic Publishers, 1999.
[5] S. Bhargava, *On Ramanujan's remarkable summation formula*, Math. Student **63** (1994), 181–192.
[6] S. H. Chan, *Private communication* via H. H. Chan, 1999.
[7] E. Grosswald, *Representations of integers as sums of squares*, Springer-Verlag, New York, 1985.
[8] G. H. Hardy, *Ramanujan. Twelve lectures on subjects suggested by his life and work*, Cambridge Univ. Press, 1940.
[9] C. G. J. Jacobi, *Gesammelte Werke. Bnde I* Chelsea Publishing Co., New York 1969.
[10] H. Y. Lam, *The Development of Elliptic Functions According to Ramanujan*, MInfSc Thesis, Massey University - Albany, New Zealand, 2001.
[11] S. C. Milne, *Infinite families of exact sums of squares formulas, Jacobi elliptic functions, continued fractions, and Schur functions*, The Ramanujan Journal, **6** (2002), no. 1, 1–151, to appear.
[12] L. J. Mordell, *On the representation of numbers as the sum of $2r$ squares*, Quart. J. Pure and Appl. Math., Oxford **48** (1917), 93–104.
[13] K. Ono, S. Robins and P. T. Wahl, *On the representation of integers as sums of triangular numbers*, Aequationes Math. **50** (1995), no. 1-2, 73–94.
[14] V. Ramamani, *Some identities conjectured by Srinivasa Ramanujan found in his lithographed notes connected with partition theory and elliptic modular functions - their proofs - interconnection with various other topics in the theory of numbers and some generalisations thereon*, PhD Thesis, University of Mysore (1971).
[15] S. Ramanujan, *Notebooks (2 volumes)*, Tata Institute of Fundamental Research, Bombay, 1957.
[16] S. Ramanujan, *Collected papers*, AMS Chelsea Publishing, Providence, Rhode Island, 2000.
[17] K. Venkatachaliengar, *Development of Elliptic Functions according to Ramanujan*, Department of Mathematics, Madurai Kamaraj University, Technical Report 2, 1988.

INSTITUTE OF INFORMATION AND MATHEMATICAL SCIENCES, MASSEY UNIVERSITY – ALBANY, PRIVATE BAG 102904, NORTH SHORE MAIL CENTRE, AUCKLAND, NEW ZEALAND
E-mail address: `s.cooper@massey.ac.nz`

Some remarks on multiple Sears transformations

Yasushi Kajihara

1. Introduction

The Sears transformation for terminating balanced $_4\phi_3$ series (equation (8.3) in [**8**] and (2.10.4) in [**2**])

$$(1.1) \quad {}_4\phi_3\left[\begin{matrix} q^{-N}, a, b, c \\ d, e, f \end{matrix}; q, q\right]$$
$$= a^N \frac{(e/a)_N (f/a)_N}{(e)_N (f)_N} {}_4\phi_3\left[\begin{matrix} q^{-N}, a, d/b, d/c \\ d, aq^{1-N}/e, aq^{1-N}/f \end{matrix}; q, q\right], \quad (abc = defq^{N-1})$$

has various applications in q-analysis. On the other hand, in [**3**], the author derived some types of multiple Sears transformations by using the q-Euler transformation formula (the equation (2.3) below). The purpose of this note is to discuss some properties including some special and limiting case. We also derive a new transformation formula for multiple $_4\phi_3$ series from a multiple Sears transformation formula.

Notation. We follow the notation of Gasper and Rahman's book [**2**] for q-shifted factorials and basic hypergeometric series. In this note, we usually omit the basis q in q-shifted factorials unless stated otherwise; namely we denote $(a;q)_k$ as $(a)_k$ for $k = 0, 1, 2, \cdots$ and $(a;q)_\infty$ as $(a)_\infty$. Throughout of this note we assume that $0 < q < 1$.

2. Multiple Sears' transformations

In this section, we give some of our multiple Sears transformations (see [**3**]).

1991 *Mathematics Subject Classification.* 33D70.

The most general form of our multiple Sears transformation formula for multiple basic hypergeometric series of type A ((7.1) in [**3**]) is as follows:

$$(2.1) \quad \sum_{\gamma \in \mathbb{N}^n} q^{|\gamma|} \frac{\Delta(xq^{\gamma})}{\Delta(x)} \prod_{1 \leq i,j \leq n} \frac{(b_j x_i/x_j)_{\gamma_i}}{(qx_i/x_j)_{\gamma_i}} \prod_{1 \leq i \leq n, 1 \leq k \leq m} \frac{(c_k x_i y_k/x_n y_m)_{\gamma_i}}{(dx_i y_k/x_n y_m)_{\gamma_i}}$$

$$\times \frac{(q^{-N})_{|\gamma|}(a)_{|\gamma|}}{(e)_{|\gamma|}(f)_{|\gamma|}}$$

$$= \frac{(e/a)_N (f/a)_N}{(e)_N (f)_N} a^N$$

$$\times \sum_{\delta \in \mathbb{N}^m} q^{|\delta|} \frac{\Delta(yq^{\delta})}{\Delta(y)} \prod_{1 \leq k,l \leq m} \frac{((d/c_l)y_k/y_l)_{\delta_k}}{(qy_k/y_l)_{\delta_k}} \prod_{1 \leq i \leq n, 1 \leq k \leq m} \frac{((d/b_i)x_i y_k/x_n y_m)_{\delta_k}}{(dx_i y_k/x_n y_m)_{\delta_k}}$$

$$\times \frac{(q^{-N})_{|\gamma|}(a)_{|\gamma|}}{(q^{1-N}a/e)_{|\gamma|}(q^{1-N}a/f)_{|\gamma|}}$$

when $ab_1 \cdots b_n c_1 \cdots c_m q^{1-N} = d^m ef$. Here $\Delta(x) = \prod_{1 \leq i < j \leq n}(x_i - x_j)$.

The following formula is a special case of (2.1) ($m = n$ and $y_i = x_i^{-1}$ for $1 \leq i \leq n$)

$$(2.2) \quad \sum_{\gamma \in \mathbb{N}^n} q^{|\gamma|} \frac{\Delta(xq^{\gamma})}{\Delta(x)} \prod_{1 \leq i,j \leq n} \frac{(b_j x_i/x_j)_{\gamma_i} (c_j x_i/x_j)_{\gamma_i}}{(qx_i/x_j)_{\gamma_i} (dx_i/x_j)_{\gamma_i}} \frac{(q^{-N})_{|\gamma|}(a)_{|\gamma|}}{(e)_{|\gamma|}(f)_{|\gamma|}}$$

$$= \frac{(e/a)_N (f/a)_N}{(e)_N (f)_N} a^N \sum_{\delta \in \mathbb{N}^n} q^{|\delta|} \frac{\Delta(x^{-1}q^{\delta})}{\Delta(x^{-1})}$$

$$\times \prod_{1 \leq i,j \leq n} \frac{((d/c_i)x_j/x_i)_{\delta_i} ((d/b_i)x_j/x_i)_{\delta_i}}{(qx_j/x_i)_{\delta_i} (dx_j/x_i)_{\delta_i}} \frac{(q^{-N})_{|\delta|}(a)_{|\delta|}}{(q^{1-N}a/e)_{|\delta|}(q^{1-N}a/f)_{|\delta|}}$$

when $ab_1 \cdots b_n c_1 \cdots c_n q^{1-N} = d^n ef$.

The multiple Sears transformation (2.1) is obtained from multiple q-Euler transformation [**3**]

$$(2.3) \quad \sum_{\gamma \in \mathbb{N}^n} u^{|\gamma|} \frac{\Delta(xq^{\gamma})}{\Delta(x)} \prod_{1 \leq i,j \leq n} \frac{(A_j x_i/x_j)_{\gamma_i}}{(qx_i/x_j)_{\gamma_i}} \prod_{1 \leq i \leq n, 1 \leq k \leq m} \frac{(B_k x_i y_k/x_n y_m)_{\gamma_i}}{(Cx_i y_k/x_n y_m)_{\gamma_i}}$$

$$= \frac{(A_1 \cdots A_n B_1 \cdots B_m u/C^m)_{\infty}}{(u)_{\infty}}$$

$$\times \sum_{\delta \in \mathbb{N}^m} (A_1 \cdots A_n B_1 \cdots B_m u/C^m)^{|\delta|} \frac{\Delta(yq^{\delta})}{\Delta(y)}$$

$$\times \prod_{1 \leq k,l \leq m} \frac{((C/B_l)y_k/y_l)_{\delta_k}}{(qy_k/y_l)_{\delta_k}} \prod_{1 \leq i \leq n, 1 \leq k \leq m} \frac{((C/A_i)x_i y_k/x_n y_m)_{\delta_k}}{(Cx_i y_k/x_n y_m)_{\delta_k}}$$

for $A_1, \ldots, A_n, B_1, \ldots, B_m, C \in \mathbb{C}$, and the third Heine transformation for $_2\phi_1$ series

$$(2.4) \quad {}_2\phi_1 \begin{bmatrix} D, E \\ F \end{bmatrix}; q; u = \frac{(DEu/F)_{\infty}}{(u)_{\infty}} {}_2\phi_1 \begin{bmatrix} F/D, F/E \\ F \end{bmatrix}; q; DEu/F.$$

We consider the product of (2.3) and (2.4)

$$
(2.5) \quad \left\{ \sum_{\gamma \in \mathbb{N}^n} u^{|\gamma|} \frac{\Delta(xq^\gamma)}{\Delta(x)} \prod_{1 \le i,j \le n} \frac{(A_j x_i/x_j)_{\gamma_i}}{(qx_i/x_j)_{\gamma_i}} \prod_{1 \le i \le n, 1 \le k \le m} \frac{(B_k x_i y_k/x_n y_m)_{\gamma_i}}{(C x_i y_k/x_n y_m)_{\gamma_i}} \right\}
$$

$$
\times \frac{(DEu/F)_\infty}{(u)_\infty} {}_2\phi_1 \left[\begin{matrix} F/D, F/E \\ F \end{matrix}; q, DEu/F \right]
$$

$$
= \frac{(A_1 \cdots A_n B_1 \cdots B_m u/C^m)_\infty}{(u)_\infty}
$$

$$
\times \left\{ \sum_{\delta \in \mathbb{N}^m} (A_1 \cdots A_n B_1 \cdots B_m u/C^m)^{|\delta|} \frac{\Delta(yq^\delta)}{\Delta(y)} \right.
$$

$$
\times \prod_{1 \le k,l \le m} \frac{((C/B_l) y_k/y_l)_{\delta_k}}{(q y_k/y_l)_{\delta_k}} \prod_{1 \le i \le n, 1 \le k \le m} \left. \frac{((C/A_i) x_i y_k/x_n y_m)_{\delta_k}}{(C x_i y_k/x_n y_m)_{\delta_k}} \right\}
$$

$$
\times {}_2\phi_1 \left[\begin{matrix} D, E \\ F \end{matrix}; q, u \right].
$$

Suppose that $A_1 \cdots B_1 \cdots B_m/C = DE/F$. Next, take the coefficients of u^N in both sides of the equation above. Then by changing the parameters appropriately, we have (2.1).

Similarly, we obtain another type of multiple Sears transformation (Proposition 7.2 in [**3**]):

$$
(2.6) \quad \sum_{\gamma \in \mathbb{N}^n} q^{|\gamma|} \frac{\Delta(xq^\gamma)}{\Delta(x)} \prod_{1 \le i,j \le n} \frac{(b_j x_i/x_j)_{\gamma_i}}{(q x_i/x_j)_{\gamma_i}} \prod_{1 \le i \le n} \frac{(c x_i/x_n)_{\gamma_i}}{(d x_i/x_n)_{\gamma_i}}
$$

$$
\times \frac{(q^{-N})_{|\gamma|}}{(f)_{|\gamma|}} \prod_{1 \le k \le m} \frac{(a y_m/y_k)_{|\gamma|}}{(e_k y_m/y_k)_{|\gamma|}}
$$

$$
= \frac{(d/c)_N}{(de_1 \cdots e_m/a^m b_1 \cdots b_n c)_N} \prod_{1 \le i \le n} \frac{((d/b_i) x_i/x_n)_N}{(d x_i/x_n)_N}
$$

$$
\times \prod_{1 \le k \le m} \frac{(a y_m/y_k)_N}{(e_k y_m/y_k)_N} \left(\frac{e_1 \cdots e_m}{a^m} \right)^N
$$

$$
\times \sum_{\delta \in \mathbb{N}^m} q^{|\delta|} \frac{\Delta(yq^\delta)}{\Delta(y)} \prod_{1 \le k,l \le m} \frac{((e_l/a) y_k/y_l)_{\delta_k}}{(q y_k/y_l)_{\delta_k}} \prod_{1 \le k \le m} \frac{((f/a) y_k/y_m)_{\delta_k}}{(q^{1-N} a^{-1} y_k/y_m)_{\delta_k}}
$$

$$
\times \frac{(q^{-N})_{|\delta|}}{(q^{1-N} c/d)_{|\delta|}} \prod_{1 \le i \le n} \frac{(q^{1-N} d^{-1} x_n/x_i)_{|\delta|}}{(q^{1-N} (b_i/d) x_n/x_i)_{|\delta|}}
$$

when $a^m b_1 \cdots b_n c q^{1-N} = de_1 \cdots e_m f$. Note that the $m = n = 1$ case of (2.6) is itself an opposite version of Sears transformation obtained by reversing the order of summation. In the remaining part of this note, we deal only with transformation of the type in (2.1).

3. Some limiting cases of the multiple Sears transformation

In this section, we discuss some special and limiting cases of the multiple Sears transformation (2.1).

It is convenient to rewrite equation (2.1) as

$$
\begin{aligned}
(3.1) \quad & \sum_{\gamma \in \mathbb{N}^n} q^{|\gamma|} \frac{\Delta(xq^\gamma)}{\Delta(x)} \prod_{1 \le i,j \le n} \frac{(b_j x_i/x_j)_{\gamma_i}}{(q x_i/x_j)_{\gamma_i}} \prod_{1 \le i \le n, 1 \le k \le m} \frac{(c_k x_i y_k / x_n y_m)_{\gamma_i}}{(d x_i y_k / x_n y_m)_{\gamma_i}} \\
& \times \frac{(q^{-N})_{|\gamma|}(a)_{|\gamma|}}{(e)_{|\gamma|}(ab_1 \cdots b_n c_1 \cdots c_m q^{1-N}/d^m e)_{|\gamma|}} \\
= & \frac{(e/a)_N (d^m e / b_1 \cdots b_n c_1 \cdots c_m)_N}{(e)_N (d^m e / a b_1 \cdots b_n c_1 \cdots c_m)_N} \\
& \times \sum_{\delta \in \mathbb{N}^m} q^{|\delta|} \frac{\Delta(y q^\delta)}{\Delta(y)} \prod_{1 \le k,l \le m} \frac{((d/c_l) y_k / y_l)_{\delta_k}}{(q y_k / y_l)_{\delta_k}} \prod_{1 \le i \le n, 1 \le k \le m} \frac{((d/b_i) x_i y_k / x_n y_m)_{\delta_k}}{(d x_i y_k / x_n y_m)_{\delta_k}} \\
& \times \frac{(q^{-N})_{|\delta|}(a)_{|\delta|}}{(q^{1-N} a/e)_{|\delta|}(d^m e / b_1 \cdots b_n c_1 \cdots c_m)_{|\delta|}}.
\end{aligned}
$$

Then by taking the limit $N \to \infty$ of (3.1), we obtain the multiple non-terminating transformation formula for ${}_3\phi_2$

$$
\begin{aligned}
(3.2) \quad & \sum_{\gamma \in \mathbb{N}^n} \left(\frac{d^m e}{ab_1 \cdots b_n c_1 \cdots c_m} \right)^{|\gamma|} \frac{\Delta(xq^\gamma)}{\Delta(x)} \prod_{1 \le i,j \le n} \frac{(b_j x_i/x_j)_{\gamma_i}}{(q x_i/x_j)_{\gamma_i}} \\
& \times \prod_{1 \le i \le n, 1 \le k \le m} \frac{(c_k x_i y_k / x_n y_m)_{\gamma_i}}{(d x_i y_k / x_n y_m)_{\gamma_i}} \frac{(a)_{|\gamma|}}{(e)_{|\gamma|}} \\
= & \frac{(e/a)_\infty (d^m e / b_1 \cdots b_n c_1 \cdots c_m)_\infty}{(e)_\infty (d^m e / a b_1 \cdots b_n c_1 \cdots c_m)_\infty} \\
& \times \sum_{\delta \in \mathbb{N}^m} \left(\frac{e}{a}\right)^{|\delta|} \frac{\Delta(yq^\delta)}{\Delta(y)} \prod_{1 \le k,l \le m} \frac{((d/c_l) y_k / y_l)_{\delta_k}}{(q y_k / y_l)_{\delta_k}} \\
& \times \prod_{1 \le i \le n, 1 \le k \le m} \frac{((d/b_i) x_i y_k / x_n y_m)_{\delta_k}}{(d x_i y_k / x_n y_m)_{\delta_k}} \frac{(a)_{|\delta|}}{(d^m e / b_1 \cdots b_n c_1 \cdots c_m)_{|\delta|}},
\end{aligned}
$$

REMARK 3.1. In the case when $n = m = 1$, this formula reduces

$$
\begin{aligned}
(3.3) \quad & {}_3\phi_2 \left[\begin{matrix} a,b,c \\ d,e \end{matrix} ; q; \frac{de}{abc} \right] \\
& = \frac{(e/a)_\infty (de/bc)_\infty}{(e)_\infty (de/abc)_\infty} {}_3\phi_2 \left[\begin{matrix} a, d/b, d/c \\ d, de/bc \end{matrix} ; q; \frac{e}{a} \right],
\end{aligned}
$$

which is due to Sears [8].

Next, letting $a \to 0$ in (3.1), we get

$$
(3.4) \quad \sum_{\gamma \in \mathbb{N}^n} q^{|\gamma|} \frac{\Delta(xq^\gamma)}{\Delta(x)} \prod_{1 \le i,j \le n} \frac{(b_j x_i/x_j)_{\gamma_i}}{(q x_i/x_j)_{\gamma_i}} \prod_{1 \le i \le n, 1 \le k \le m} \frac{(c_k x_i y_k / x_n y_m)_{\gamma_i}}{(d x_i y_k / x_n y_m)_{\gamma_i}} \frac{(q^{-N})_{|\gamma|}}{(e)_{|\gamma|}}
$$

$$= \frac{(d^m e/b_1 \cdots b_n c_1 \cdots c_m)_N}{(e)_N} \left(\frac{b_1 \cdots b_n c_1 \cdots c_m}{d^m}\right)^N$$

$$\times \sum_{\delta \in \mathbb{N}^m} q^{|\delta|} \frac{\Delta(yq^\delta)}{\Delta(y)} \prod_{1 \le k,l \le m} \frac{((d/c_l)y_k/y_l)_{\delta_k}}{(qy_k/y_l)_{\delta_k}} \prod_{1 \le i \le n, 1 \le k \le m} \frac{((d/b_i)x_i y_k/x_n y_m)_{\delta_k}}{(dx_i y_k/x_n y_m)_{\delta_k}}$$

$$\times \frac{(q^{-N})_{|\delta|}}{(d^m e/b_1 \cdots b_n c_1 \cdots c_m)_{|\delta|}}.$$

Note that all the same results of the previous paragraph are restricted in the case of (2.2).

In the case when $m = 1$, the right-hand side of (2.1) reduces to a ${}_{n+3}\phi_{n+2}$ series

$$(3.5) \qquad \sum_{\gamma \in \mathbb{N}^n} q^{|\gamma|} \frac{\Delta(xq^\gamma)}{\Delta(x)} \prod_{1 \le i,j \le n} \frac{(b_j x_i/x_j)_{\gamma_i}}{(qx_i/x_j)_{\gamma_i}} \prod_{1 \le i \le n} \frac{(cx_i/x_n)_{\gamma_i}}{(dx_i/x_n)_{\gamma_i}}$$

$$\times \frac{(q^{-N})_{|\gamma|}(a)_{|\gamma|}}{(e)_{|\gamma|}(ab_1 \cdots b_n c q^{1-N}/de)_{|\gamma|}}$$

$$= \frac{(e/a)_N (de/b_1 \cdots b_n c)_N}{(e)_N (de/ab_1 \cdots b_n c)_N}$$

$$\times {}_{n+3}\phi_{n+2} \left[\begin{array}{c} q^{-N}, a, (d/b_1)x_1/x_n, \ldots (d/b_{n-1})x_{n-1}/x_n, d/b_n, d/c \\ dx_1/x_n, \ldots dx_{n-1}/x_n, d, q^{1-N}a/e, de/b_1 \cdots b_n c \end{array}; q, q\right].$$

Note also that the $c = d$ case of (3.5) essentially reduces to the classical Pfaff-Saalschutz summation formula for terminating balanced ${}_3\phi_2$ basic hypergeometric series because of the Bailey summation formula for very-well-poised ${}_6\phi_5$ basic hypergeometric series in $U(n+1)$ due to S.C. Milne [4]

$$(3.6) \qquad \frac{(a_1 \cdots a_n)_N}{(q)_N} = \sum_{\beta \in \mathbb{N}^n, |\beta|=N} \frac{\Delta(xq^\beta)}{\Delta(x)} \prod_{1 \le i,j \le n} \frac{(a_j x_i/x_j)_{\beta_i}}{(qx_i/x_j)_{\beta_i}}.$$

Letting $c \to 0$ in (3.5), we get

$$(3.7) \qquad \sum_{\gamma \in \mathbb{N}^n} q^{|\gamma|} \frac{\Delta(xq^\gamma)}{\Delta(x)} \prod_{1 \le i,j \le n} \frac{(b_j x_i/x_j)_{\gamma_i}}{(qx_i/x_j)_{\gamma_i}} \prod_{1 \le i \le n} \frac{1}{(dx_i/x_n)_{\gamma_i}} \frac{(a)_{|\gamma|}}{(e)_{|\gamma|}} (q^{-N})_{|\gamma|}$$

$$= \frac{(e/a)_N}{(e)_N} a^N$$

$$\times {}_{n+3}\phi_{n+2} \left[\begin{array}{c} q^{-N}, a, (d/b_1)x_1/x_n, \ldots (d/b_{n-1})x_{n-1}/x_n, d/b_n, d/c \\ dx_1/x_n, \ldots dx_{n-1}/x_n, d, q^{1-N}a/e, q^{1-N}a/f \end{array}; q, \frac{b_1 \cdots b_n q}{e}\right].$$

4. Transformation property of multiple Sears transformations

In this section, we discuss the transformation property of multiple Sears transformations, especially (3.5).

The classical Sears transformation (1.1)

$${}_4\phi_3 \left[\begin{array}{c} q^{-N}, a, b, c \\ d, e, abcq^{1-N}/deppp \end{array}; q, q\right]$$

$$= \frac{(e/a)_N(de/bc)_N}{(e)_N(de/abc)_N} {}_4\phi_3 \left[\begin{array}{c} q^{-N}, a, d/b, d/c \\ d, aq^{1-N}/e, de/bc \end{array} ; q, q \right]$$

is a transformation for parameters of terminating balanced ${}_4\phi_3$ basic hypergeometric series with one of the numerator parameters (a) and one of the denominator parameters (d) fixed. If we iterate Sears transformation by choosing a parameter to be fixed in various ways, we obtain other types of transformation formulas for terminating balanced ${}_4\phi_3$ series.

When $m = 1$, we can derive a transformation formula for multiple ${}_4\phi_3$ series by using the same idea.

By interchanging a and d/c in right-hand side of (3.5) and iterating the inversion of (3.5)

$${}_{n+3}\phi_{n+2} \left[\begin{array}{c} q^{-N}, a, b_1 x_1/x_n, \ldots b_{n-1}x_{n-1}/x_n, b_n, c \\ dx_1/x_n, \ldots dx_{n-1}/x_n, d, e, ab_1\cdots b_n c q^{1-N}/de \end{array} ; q, q \right]$$

$$= \frac{(e)_N(d^n e/ab_1\cdots b_n c)_N}{(e/a)_N(d^n e/b_1\cdots b_n c)_N} \sum_{\gamma \in \mathbb{N}^n} q^{|\gamma|} \frac{\Delta(xq^\gamma)}{\Delta(x)} \prod_{1\le i,j \le n} \frac{((d/b_j)x_i/x_j)_{\gamma_i}}{(qx_i/x_j)_{\gamma_i}}$$

(4.1)
$$\times \prod_{1\le i \le n} \frac{((d/c)x_i/x_n)_{\gamma_i}}{(dx_i/x_n)_{\gamma_i}} \frac{(q^{-N})_{|\gamma|}(a)_{|\gamma|}}{(q^{1-N}a/e)_{|\gamma|}(d^n e/b_1\cdots b_n c)_{|\gamma|}},$$

we have an another multiple terminating balanced ${}_4\phi_3$ transformation formula

(4.2)
$$\sum_{\gamma \in \mathbb{N}^n} q^{|\gamma|} \frac{\Delta(xq^\gamma)}{\Delta(x)} \prod_{1\le i,j \le n} \frac{(b_j x_i/x_j)_{\gamma_i}}{(qx_i/x_j)_{\gamma_i}} \prod_{1\le i \le n} \frac{(cx_i/x_n)_{\gamma_i}}{(dx_i/x_n)_{\gamma_i}}$$

$$\times \frac{(q^{-N})_{|\gamma|}(a)_{|\gamma|}}{(e)_{|\gamma|}(ab_1\cdots b_n c q^{1-N}/de)_{|\gamma|}}$$

$$= \frac{(e/b_1\cdots b_n)_N(de/ac)_N}{(e)_N(de/b_1\cdots b_n c)_N}$$

$$\times \sum_{\delta \in \mathbb{N}^n} q^{|\delta|} \frac{\Delta(xq^\delta)}{\Delta(x)} \prod_{1\le i,j \le n} \frac{(b_j x_i/x_j)_{\delta_i}}{(qx_i/x_j)_{\delta_i}} \prod_{1\le i \le n} \frac{((d/a)x_i/x_n)_{\delta_i}}{(dx_i/x_n)_{\delta_i}}$$

$$\times \frac{(q^{-N})_{|\delta|}(d/c)_{|\delta|}}{(de/ac)_{|\delta|}(q^{1-N}b_1\cdots b_n/e)_{|\delta|}}.$$

REMARK 4.1. In the case when $n = 1$, the equation above reduces to

(4.3)
$${}_4\phi_3 \left[\begin{array}{c} a, b, c, q^{-N} \\ d, e, abcq^{1-N}/de \end{array} ; q; q \right]$$

$$= \frac{(e/b)_N(de/ac)_N}{(e)_N(de/bc)_N} {}_4\phi_3 \left[\begin{array}{c} d/c, b, d/a, q^{-N} \\ d, de/ac, q^{1-N}b/e \end{array} ; q; c \right].$$

By taking the limit $N \to \infty$ in the equation above, we obtain

(4.4)
$$\sum_{\gamma \in \mathbb{N}^n} \left(\frac{de}{ab_1\cdots b_n c} \right)^{|\gamma|} \frac{\Delta(xq^\gamma)}{\Delta(x)} \prod_{1\le i,j \le n} \frac{(b_j x_i/x_j)_{\gamma_i}}{(qx_i/x_j)_{\gamma_i}}$$

$$\times \prod_{1\le i \le n} \frac{(cx_i/x_n)_{\gamma_i}}{(dx_i/x_n)_{\gamma_i}} \frac{(a)_{|\gamma|}}{(e)_{|\gamma|}}$$

$$= \frac{(e/b_1\cdots b_n)_\infty(de/ac)_\infty}{(e)_\infty(de/b_1\cdots b_n c)_\infty}$$

$$\times \sum_{\delta \in \mathbb{N}^n} \left(\frac{e}{b_1 \cdots b_n}\right)^{|\delta|} \frac{\Delta(xq^\delta)}{\Delta(x)} \prod_{1 \leq i,j \leq n} \frac{(b_j x_i/x_j)_{\delta_i}}{(qx_i/x_j)_{\delta_i}}$$

$$\times \prod_{1 \leq i \leq n} \frac{((d/a)x_i/x_n)_{\delta_i}}{(dx_i/x_n)_{\delta_i}} \frac{(d/c)_{|\delta|}}{(de/ac)_{|\delta|}}.$$

In the case when $n = 1$, this formula is an incomplete form of Hall's formula (equation (3.2.10) in [2])

(4.5) $$\quad {}_3\phi_2\left[\begin{matrix}a,b,c\\d,e\end{matrix};q;\frac{de}{abc}\right]$$
$$= \frac{(c)_\infty (de/ac)_\infty (de/bc)_\infty}{(d)_\infty (e)_\infty (de/abc)_\infty} {}_3\phi_2\left[\begin{matrix}d/b, e/b, de/abc\\de/ac, de/bc\end{matrix};q;c\right].$$

REMARK 4.2. In principle, likewise in the case of the classical Sears transformation, we can give other multiple ${}_4\phi_3$ transformations by fixing other pairs of parameters in numerators and denominators in ${}_{n+3}\phi_{n+2}$ series in (4.1). But the formula (4.2) is a unique multiple ${}_4\phi_3$ transformation formula that preserves the symmetry of parameters of interaction $x = (x_1, \ldots, x_n)$ and the parameters $b_1, \ldots, b_n$.

References

[1] G.E. Andrews, R. Askey, R. Roy, *Special functions*. Cambridge University Press, 1999.
[2] G. Gasper, M. Rahman, *Basic hypergeometric series*. Encyclopedia of Mathematics and Its Applications, (G.C.Rota, ed.), vol. 35, Cambridge Univ. Press, Cambridge, 1990.
[3] Y. Kajihara, *Euler transformation formula for multiple basic hypergeometric series of type A and some applications.* preprint.
[4] S.C. Milne. *An elementary proof of Macdonald identities for $A_l^{(1)}$.* Adv. in Math. **57**, (1985), 34–70.
[5] D.B. Sears, *On the transformation theory of hypergeometric functions*. Proc. London Math. Soc. (2) **52**, (1950), p14–35.
[6] D.B. Sears, *Transformations of basic hypergeometric functions of special type*. Proc. London Math. Soc. (2) **52**, (1951), p467–483.
[7] D.B. Sears, *On the transformation theory of hypergeometric functions and cognate trigonometric series*. Proc. London Math. Soc. (2) **53**, (1951), p138–157.
[8] D.B. Sears, *On the transformation theory of basic hypergeometric functions*. Proc. London Math. Soc. (2) **53**, (1951), p158–180.
[9] D.B. Sears, *Transformations of basic hypergeometric functions of any order*. Proc. London Math. Soc. (2) **53**, (1951), p181–191.

DEPARTMENT OF MATHEMATICS, KOBE UNIVERSITY, ROKKO, KOBE 657-8501, JAPAN
E-mail address: kaji@math.kobe-u.ac.jp

Contemporary Mathematics
Volume **291**, 2001

Another Way to Count Colored Frobenius Partitions

Louis W. Kolitsch

ABSTRACT. In this paper we look at sorting colored generalized Frobenius partitions using the color difference. Some interesting symmetries and patterns associated with sorting the partitions in this manner are developed.

1. What is a colored generalized Frobenius partition?

A generalized Frobenius partition of n using k colors (denoted by subscripts $0, 1, \ldots, k-1$) is a two-lined array of the form $\begin{pmatrix} a_1 & a_2 & \cdots & a_r \\ b_1 & b_2 & \cdots & b_r \end{pmatrix}$, where the entries in each row of the array are taken from k copies of the nonnegative integers distinguished by the k colors, the entries in each row are distinct and arranged in descending order (ordered first with respect to size and then with respect to color), and $\sum_{i=1}^{r}(a_i + b_i + 1) = n[A]$. For example, $\begin{pmatrix} 2_3 & 1_2 & 1_1 \\ 0_3 & 0_2 & 0_0 \end{pmatrix}$ is a partition of 7 using 4 colors.

The color difference for a colored generalized Frobenius partition is the sum of the color subscripts on the top row minus the sum of the color subscripts on the bottom row. For example, the color difference for $\begin{pmatrix} 2_3 & 1_2 & 1_1 \\ 0_3 & 0_2 & 0_0 \end{pmatrix}$ is 1. We let $c\phi_k(n, m)$ denote the number of generalized Frobenius partitions of n using k colors whose color difference is m. A partial table of values for $k = 3$ is given below.

$n \backslash m$	-4	-3	-2	-1	0	1	2	3	4
0					1				
1			1	2	3	2	1		
2			3	6	9	6	3		
3		2	9	18	24	18	9	2	
4	1	6	24	44	57	44	24	6	1
5	3	18	57	102	126	102	57	18	3
6	9	44	126	216	265	216	126	44	9

We immediately see that the columns for m and $-m$ are identical. This is a consequence of the fact that interchanging the entries on the top and bottom rows of our array does not change the number being partitioned but changes the sign on

---

1991 *Mathematics Subject Classification.* Primary 11P81; Secondary 05A17.

© 2001 American Mathematical Society

the color difference. We can also see that if we look from left to right starting with the column for $m = 0$, the columns begin to repeat but are shifted down. In fact, for $k = 3$ colors we see that the columns repeat in pairs.

In the table shown below for $k = 5$ colors we see that the columns repeat in blocks of ten. Because of the symmetry for positive and negative values of m, we see that we only need to know 6 columns and how the columns are shifted in order to complete the entire table for $k = 5$.

$n\backslash m$	0	1	2	3	4	5	6	7	8	9	10	11
0	1											
1	5	4	3	2	1							
2	26	22	18	12	7	2	1					
3	103	94	77	56	34	16	7	2				
4	350	324	273	202	132	70	34	12	3			
5	1062	1000	851	650	439	256	132	56	18	4	1	
6	2955	2796	2412	1874	1308	800	439	202	77	22	5	
7	7678	7314	6359	5028	3590	2290	1308	650	273	94	26	4

In general, as we will show in the next section, the columns repeat in blocks of $c = (4\ell^3 - \ell)/3$ if $k = 2\ell$ and $d = (2\ell^3 + 3\ell^2 + \ell)/6$ if $k = 2\ell + 1$.

2. Why do the colors repeat in blocks?

The generating function for $c\phi_k(n,m)$ is the coefficient of z^0 in

$$f(t,q,z) = \prod_{j=0}^{k-1}\prod_{i=0}^{\infty}(1+zq^{i+1}t^j)(1+z^{-1}q^i t^{-j})$$

[B]. In this generating function the exponent on q keeps track of the number being partitioned and the exponent on t keeps track of the color difference. We will write

$$\sum_{m=-\infty}^{\infty}\sum_{n=0}^{\infty} c\phi_k(n,m) q^n t^m$$

as coeff $(z^0, f(t,q,z))$.

For $k = 2\ell$ we can replace z by zq^{-k+1} and t by tq^2 to get

$$\sum_{m=-\infty}^{\infty}\sum_{n=0}^{\infty} c\phi_k(n,m) q^{n+2m} t^m = \text{coeff}\left(z^0, f\left(tq^2, q, zq^{-k+1}\right)\right).$$

Now

$$f(tq^2, q, zq^{-k+1}) = \prod_{j=0}^{k-1}\prod_{i=0}^{\infty}(1+zq^{i+2j-k+2}t^j)(1+z^{-1}q^{i-2j+k-1}t^{-j}).$$

We can rewrite the factors of the form $1 + zq^{i+2j-k+2}t^j$ where $i+2j-k+2 \leq 0$ (that is, when $i \leq k-2-2j$ and $j < \ell$) as

$$zq^{i+2j-k+2}t^j(1 + z^{-1}q^{-i-2j+k-2}t^{-j})$$

and the factors of the form $1 + z^{-1}q^{i-2j+k-1}t^{-j}$ where $i - 2j + k - 1 < 0$ (that is, when $i < 2j - k + 1$ and $j > \ell - 1$) as

$$z^{-1}q^{i-2j+k-1}t^{-j}(1 + zq^{-i+2j-k-1}t^j).$$

The exponent on z from these factors will be

$$\sum_{j=0}^{\ell-1}(k-2-2j+1) - \sum_{j=\ell}^{k-1}(2j-k+1) = 0.$$

The exponent on q from these factors will be

$$\sum_{j=0}^{\ell-1}\sum_{i=0}^{k-2-2j}(i+2j-k+2) + \sum_{j=\ell}^{k-1}\sum_{i=0}^{2j-k}(i-2j+k-1) = -(4\ell^3 - \ell)/3$$

and the exponent on t will be

$$\sum_{j=0}^{\ell-1}\sum_{i=0}^{k-2-2j}j + \sum_{j=\ell}^{k-1}\sum_{i=0}^{2j-k}(-j) = -(4\ell^3 - \ell)/3.$$

The binomials from these factors can be combined with the remaining factors of $f(tq^2, q, zq^{-k+1})$ to yield $f(t, q, z)$. Hence we see that

$$\sum_{m=-\infty}^{\infty}\sum_{n=0}^{\infty} c\phi_k(n,m)q^{n+2m}t^m = q^{-c}t^{-c} \cdot \text{coeff}\left(z^0, f(t,q,z)\right),$$

where $c = (4\ell^3 - \ell)/3$. This gives

$$\sum_{m=-\infty}^{\infty}\sum_{n=0}^{\infty} c\phi_k(n,m)q^{n+2m+c}t^{m+c} = \sum_{m=-\infty}^{\infty}\sum_{n=0}^{\infty} c\phi_k(n,m)q^n t^m.$$

We immediately obtain the following theorem.

THEOREM 2.1. *For k even, $c\phi_k(n + 2m + c, m + c) = c\phi_k(n,m)$ where $c = (4\ell^3 - \ell)/3$.*

In a similar manner for $k = 2\ell + 1$ we can replace z by $zq^{-\ell}$ and t by tq to get the following theorem.

THEOREM 2.2. *For $k > 1$ and odd, $c\phi_k(n + m + d, m + 2d) = c\phi_k(n,m)$, where $d = (2\ell^3 + 3\ell^2 + \ell)/6$.*

From these two theorems we see that when k is even the rth column ($r \geq c$) is the $(r-c)$th column shifted down $2r - c$ rows. Similarly when k is odd the rth column ($r \geq 2d$) is the $(r - 2d)$th column shifted down $r - d$ rows.

3. A bijection between partitions in corresponding columns

If we carefully study the transformations that we performed on the generating functions in section 2 we can develop bijections between the partitions enumerated by $c\phi_k(n,m)$ and $c\phi_k(n + 2m + c, m + c)$ for k even and between the partitions enumerated by $c\phi_k(n,m)$ and $c\phi_k(n + m + d, m + 2d)$ for k odd. The bijections can be described as follows:

Bijection for $k = 2\ell$: To transform a partition enumerated by $c\phi_k(n,m)$ into a partition enumerated by $c\phi_k(n + 2m + c, m + c)$ do the following:

(1) Replace each part a_j in the top row with $(a - k + 1 + 2j)_j$.
(2) Replace each part b_j in the bottom row with $(b + k - 1 - 2j)_j$.
(3) If none of the parts created in (1) and (2) are negative, insert parts of size $0, 1, \ldots, 2j - k$ for colors $\ell \leq j \leq k - 1$ in the top row and insert parts of size $0, 1, \ldots, k - 2j - 2$ for colors $0 \leq j \leq \ell - 1$ in the bottom row.
(4) If some of the parts created in (1) and (2) are negative, then
  (a) replace $(-a)_j$ on the top row with $(a-1)_{k-1-j}$ if $(-a)_j$ appears in the top row and $(-a)_{k-1-j}$ does not appear in the bottom row,
  (b) replace $(-b)_j$ with $(b-1)_{k-1-j}$ if $(-b)_j$ appears in the bottom row and $(-b)_{k-1-j}$ does not appear in the top row,
  (c) delete parts $(-a)_j$ and $(-a)_{k-1-j}$ if $(-a)_j$ appears in the top row and $(-a)_{k-1-j}$ appears in the bottom row,
  (d) for $0 \leq j \leq \ell - 1$ and $1 \leq a \leq k - 2j - 1$ insert $(a-1)_{k-1-j}$ in the top row and $(a-1)_j$ in the bottom row if $(-a)_j$ does not appear in the top row and $(-a)_{k-1-j}$ does not appear in the bottom row.
(5) Put the parts in the top and bottom rows in descending order.

Bijection for $k = 2\ell + 1$: To transform a partition counted by $c\phi_k(n, m)$ into a partition enumerated by $c\phi_k(n + m + d, m + 2d)$ do the following:
(1) Replace each part a_j in the top row with $(a - \ell + j)_j$.
(2) Replace each part b_j in the bottom row with $(b + \ell - j)_j$.
(3) If none of the parts created in (1) and (2) are negative, insert parts of size $0, 1, \ldots, j - \ell - 1$ for colors $\ell + 1 \leq j \leq k - 1$ in the top row and insert parts of size $0, 1, \ldots, \ell - j - 1$ for colors $0 \leq j \leq \ell - 1$.
(4) If some of the parts created in (1) and (2) are negative, then
  (a) replace $(-a)_j$ on the top row with $(a-1)_{k-1-j}$ if $(-a)_j$ appears in the top row and $(-a)_{k-1-j}$ does not appear in the bottom row,
  (b) replace $(-b)_j$ with $(b-1)_{k-1-j}$ if $(-b)_j$ appears in the bottom row and $(-b)_{k-1-j}$ does not appear in the top row,
  (c) delete parts $(-a)_j$ and $(-a)_{k-1-j}$ if $(-a)_j$ appears in the top row and $(-a)_{k-1-j}$ appears in the bottom row,
  (d) for $0 \leq j \leq \ell - 1$ and $1 \leq a \leq \ell - j$ insert $(a-1)_{k-1-j}$ in the top row and $(a-1)_j$ in the bottom row if $(-a)_j$ does not appear in the top row and $(-a)_{k-1-j}$ does not appear in the bottom row.
(5) Put the parts in the top and bottom rows in descending order.

We illustrate these bijections with the following examples.

<u>Example 1</u> Consider the partition $\begin{pmatrix} 2_3 & 1_2 & 1_1 \\ 0_3 & 0_2 & 0_0 \end{pmatrix}$ enumerated by $c\phi_4(7, 1)$. Under the bijection described above this partition becomes $\begin{pmatrix} 5_3 & 2_2 & 1_3 & 0_3 & 0_1 \\ 3_0 & 2_0 & 1_0 & 0_1 & 0_0 \end{pmatrix}$ which is a partition enumerated by $c\phi_4(19, 11)$.

<u>Example 2</u> Consider the partition $\begin{pmatrix} 2_4 & 1_2 & 1_1 & 1_0 & 0_0 \\ 1_3 & 1_2 & 0_4 & 0_1 & 0_0 \end{pmatrix}$ enumerated by $c\phi_5(12, -3)$. Under the bijection described above this partition

becomes
$\begin{pmatrix} 4_4 & 1_2 & 0_4 & 0_3 & 0_1 \\ 2_0 & 1_2 & 1_1 & 0_3 & 0_1 \end{pmatrix}$ which is a partition enumerated by $c\phi_5(14, 7)$.

4. Some other consequences

The following theorem is an immediate consequence of Theorems 2.1 and 2.2 and gives another way that $c\phi_k(n, m)$ can be calculated.

THEOREM 4.1. (1) *For k even,*

$$c\phi_k(n) = \sum_{r=-\infty}^{\infty} c\phi_k(n + 2r - c, r).$$

(2) *For k odd,*

$$c\phi_k(n) = \sum_{r=-\infty}^{\infty} c\phi_k(n + r - d, r).$$

This theorem is illustrated in the following tables.

TABLE 1. Values for $c\phi_2(n, m)$

$n \backslash m$	-3	-2	-1	0	1	2	3
0	0	0	0	1	0	0	0
1	0	0	1	2	1	0	0
2	0	0	2	5	2	0	0
3	0	0	5	10	5	0	0
4	0	1	10	20	10	1	0
5	0	2	20	36	20	2	0
6	0	5	36	65	36	5	0
7	0	10	65	110	65	10	0
8	0	20	110	185	110	20	0
9	1	36	185	300	185	36	1
10	2	65	300	481	300	65	2
11	5	110	481	752	481	110	5

TABLE 2. Values for $c\phi_3(n,m)$

$n\backslash m$	-6	-5	-4	-3	-2	-1	0	1	2	3	4	5	6
0	0	0	0	0	0	0	1	0	0	0	0	0	0
1	0	0	0	0	1	$\boxed{2}$	3	2	1	0	0	0	0
2	0	0	0	0	3	6	$\boxed{9}$	6	3	0	0	0	0
3	0	0	0	$\boxed{2}$	$\boxed{9}$	$\boxed{18}$	$\boxed{24}$	$\boxed{18}$	$\boxed{9}$	$\boxed{2}$	0	0	0
4	0	0	1	6	24	44	57	44	$\boxed{24}$	6	1	0	0
5	0	0	3	18	57	102	126	102	57	$\boxed{18}$	3	0	0
6	0	0	9	44	126	216	265	216	126	44	$\boxed{9}$	0	0
7	0	2	24	102	265	444	531	444	265	102	24	$\boxed{2}$	0
8	0	6	57	216	531	864	1026	864	531	216	57	6	0

References

[A] G. E. Andrews, *Generalized Frobenius Partitions*, Memoirs of the American Mathematical Society, Vol. 301, May 1984.

[B] L. Kolitsch, *A simple proof of some congruences for colored generalized Frobenius partitions*, Discrete Math. **81** (1990), 259–261.

Current address: Department of Mathematics and Statistics, University of Tennessee at Martin, Martin, TN 38238

E-mail address: lkolitsc@utm.edu

Proof of a summation formula for an $\tilde{A}_n$ basic hypergeometric series conjectured by Warnaar

C. Krattenthaler[†]

ABSTRACT. A proof of an unusual summation formula for a basic hypergeometric series associated to the affine root system $\tilde{A}_n$ that was conjectured by Warnaar is given. It makes use of Milne's A_n extension of Watson's transformation, Ramanujan's $_1\psi_1$-summation, and a determinant evaluation of the author. In addition, a transformation formula between basic hypergeometric series associated to the affine root systems $\tilde{A}_n$ respectively $\tilde{A}_m$, which generalizes at the same time the above summation formula and an identity due to Gessel and the author, is proposed as a conjecture.

1. Introduction, statement of the result, and of the conjecture

The purpose of this note is to prove a summation formula for a basic hypergeometric series associated to the affine root system $\tilde{A}_{n-1}$ that was conjectured by Warnaar (private communication). (Another frequently used term for such series is 'basic hypergeometric series in $SU(n)$.' We follow however the terminology for multiple basic hypergeometric series associated to root systems as laid down in [4, Sec. 7] and [1, Sec. 1]. For an overview of the state of the art of this theory and of its relevance we refer the reader to [10, 1, 2, 8] and the references cited therein.)

THEOREM. *Let n be a positive integer, let M_1 and M_2 be nonnegative integers, and let S be an integer with $-M_1 \leq S \leq M_2$. Then*

$$(1) \quad \sum_{k_1+\cdots+k_n=S} (-1)^{(n-1)S} q^{\binom{n+1}{2}\sum_{i=1}^n k_i^2 + \sum_{i=1}^n ik_i} \prod_{1 \leq i < j \leq n} (1 - q^{nk_j - nk_i + j - i})$$

$$\times \prod_{i=1}^n \frac{(q;q)_{M_1+M_2+i-1}}{(q;q)_{M_1+nk_i+i-1}(q;q)_{M_2-nk_i+n-i}} = q^{(n+1)\binom{S+1}{2}} \frac{(q;q)_{M_1+M_2}}{(q;q)_{M_1+S}(q;q)_{M_2-S}},$$

where, as usual, the shifted q-factorial $(a;q)_n$ is defined by

$$(a;q)_k := (1-a)(1-aq)\cdots(1-aq^{k-1})$$

1991 *Mathematics Subject Classification.* Primary 33D67; Secondary 05A19 05A30.

Key words and phrases. basic hypergeometric series associated to root systems, basic hypergeometric series in $SU(n)$.

[†] Research partially supported by the Austrian Science Foundation FWF, grant P12094-MAT.

if $k > 0$, $(a;q)_0 := 1$, and
$$(a;q)_k := 1/(1-a/q)(1-a/q^2)\cdots(1-aq^k)$$
if $k < 0$.

This identity is remarkable, because it essentially[1] reduces to an identity originally due to Milne [**9**, Theorem 1.9] if we let M_2 tend to infinity. The proof of Milne's identity in [**9**] uses a great deal of machinery (in fact a large part of his paper [**9**] is devoted to the proof of this identity), which, apparently, does not allow any generalization or extension. On the other hand, an elementary, combinatorial proof of Milne's identity has been given in [**4**, Theorem 22]. But, again, it seems impossible to extend this combinatorial approach to a proof of the above Theorem.

I will prove the above Theorem by an unusual combination of, on the one hand, classical and, on the other hand, more recent results in classical analysis. The proof will require Milne's A_n extension of Watson's transformation [**11**, Theorem 6.1], Ramanujan's classical $_1\psi_1$-summation (see e.g. [**3**, Eq. (5.2.1); Appendix (II.29)]), and a determinant evaluation of the author [**6**, Lemma 2.2] which is ubiquitous in classcial and combinatorial analysis (cf. [**7**, Theorem 26 and the subsequent paragraphs] for a list of occurrences).

An independent proof of the above Theorem results from an identity for supernomial coefficients due to Schilling and Shimozono [**13**, Eq. (6.6)] (cf. [**14**, remarks preceding Eq. (6.6)]). I believe that the proof of this paper is still of interest, because variations of this approach will certainly turn out to be useful in other cases as well.

A test candidate for the above judgement may be the following conjectural generalization of the Theorem. Before I state it precisely, let me recall that in [**4**, Theorem 26] it is shown that Milne's identity (i.e., the $M_2 \to \infty$ case of the above Theorem) is in fact part of an infinite *hierarchy* of *transformation* formulas between multiple basic hypergeometric *of different dimension*. (Such transformations are, up to now, very rare. Except for Section 8 of [**4**], the only occurrence of such transformations that I am aware of is [**5**].) Since Milne's identity admits the generalization stated in the above Theorem, an immediate question is whether or not it is possible to also introduce an additional parameter into this infinite hierarchy of transformation formulas. On the basis of computer experiments, there is overwhelming evidence that this is indeed the case. We state the formula in the Conjecture below.

CONJECTURE. *Let n and m be positive integers, let M_1 and M_2 be nonnegative integers, and let S_1 and S_2 be integers with $-M_1 \leq S_1 \leq M_2$ and $-M_1 \leq S_2 \leq M_2$. Then*

$$(2) \quad \sum_{k_1+\cdots+k_n=S_1} (-1)^{(n-1)S_1} q^{\frac{n(n+m)}{2}\sum_{i=1}^n k_i^2 + m\sum_{i=1}^n ik_i - m\binom{S_1+1}{2} - nS_1(S_1+m)/2}$$

$$\times \prod_{1\leq i<j\leq n}(1-q^{nk_j-nk_i+j-i})\prod_{i=1}^n \frac{(q;q)_{M_1+M_2+i-1}}{(q;q)_{M_1-S_1+nk_i+i-1}(q;q)_{M_2+S_1-nk_i+n-i}}$$

---

[1]In fact, Milne's identity is the $M_2 \to \infty$, $M_1 = 0$ case of (1). However, it is shown in [**4**, paragraph before Theorem 22], that, by what is called there the "rotation trick", Milne's identity does also imply the $M_2 \to \infty$ case of (1) (i.e., with M_1 arbitrary). The rotation trick will also be used in our proof of the Theorem.

$$= \sum_{l_1+\cdots+l_m=S_2} (-1)^{(m-1)S_2} q^{\frac{m(m+n)}{2}\sum_{i=1}^{m} l_i^2 + n\sum_{i=1}^{m} il_i - n\binom{S_2+1}{2} - mS_2(S_2+n)/2}$$

$$\times \prod_{1\le i<j\le m} (1-q^{ml_j-ml_i+j-i}) \prod_{i=1}^{m} \frac{(q;q)_{M_1+M_2+i-1}}{(q;q)_{M_1-S_2+ml_i+i-1}\,(q;q)_{M_2+S_2-ml_i+m-i}}.$$

Clearly, our Theorem is the $m=1$ case of this conjecture. Even more evidence in favour of the conjecture comes from the fact that for $M_2 \to \infty$ it reduces to Theorem 26 in [4].

By means of the "rotation trick" (see [4, paragraph before Theorem 22] and the first paragraph of the next section), it can be seen that it suffices to prove the Conjecture for $S_1 = S_2 = 0$. However, in contrast to our proof of the Theorem, for a proof of the Conjecture it will not be sufficient to apply Milne's A_{n-1} extension of Watson's transformation. Perhaps one has to start with a higher order transformation formula, for example, with one of the A_{n-1} extensions of Bailey's very-well-poised $_{10}\phi_9$-transformation formula from [12].

2. Proof of the Theorem

First of all, analogously to the remark of the last paragraph of the previous section, I claim that it is enough to prove (1) for $S=0$, i.e.,

$$(3) \quad \sum_{k_1+\cdots+k_n=0} q^{\binom{n+1}{2}\sum_{i=1}^{n} k_i^2 + \sum_{i=1}^{n} ik_i} \prod_{1\le i<j\le n} (1-q^{nk_j-nk_i+j-i})$$

$$\times \prod_{i=1}^{n} \frac{(q;q)_{M_1+M_2+i-1}}{(q;q)_{M_1+nk_i+i-1}\,(q;q)_{M_2-nk_i+n-i}} = \frac{(q;q)_{M_1+M_2}}{(q;q)_{M_1}(q;q)_{M_2}}.$$

This is seen by resorting to the "rotation trick" [4, paragraph before Theorem 22]. Let us assume that we already proved (3). Let S be some fixed integer. Division of S by n gives a unique representation $S = Qn + R$ where Q, R are integers with $0 \le R < n$. Then in (3) replace k_1 by $k_{1+R} - Q$, ..., k_{n-R} by $k_n - Q$, k_{n-R+1} by $k_1 - Q - 1$, ..., k_n by $k_R - Q - 1$. So the effect is a rotation of the summation indices, combined with a certain shift. If we rewrite (3) after these replacements and finally replace M_1 by $M_1 + S$ and M_2 by $M_2 - S$, we obtain (1) after some simplification.

Next, I claim that it is enough to prove (3) for $M_1 \equiv 0 \bmod n$. To see this, suppose that M_2 is given. Multiply both sides of (3) by $\prod_{i=1}^{n}(q^{M_1+M_2+i};q)_n$ and write the result in the form

$$(4) \quad \sum_{k_1+\cdots+k_n=0} q^{\binom{n+1}{2}\sum_{i=1}^{n} k_i^2 + \sum_{i=1}^{n} ik_i} \prod_{1\le i<j\le n} (1-q^{nk_j-nk_i+j-i})$$

$$\times \prod_{i=1}^{n} \frac{(q^{M_1+nk_i+i};q)_{M_2-nk_i+n}}{(q;q)_{M_2-nk_i+n-i}} = \frac{(q^{M_1+1};q)_{M_2}}{(q;q)_{M_2}} \prod_{i=1}^{n}(q^{M_1+M_2+i};q)_n.$$

Both sides are most obviously polynomials in q^{M_1}, of degree at most $n^2(n+M_2)$, because, in the summation, each k_i is bounded above by $1 + M_2/n$, and, hence, bounded below by $-(n-1)(1+M_2/n)$. A polynomial is uniquely determined by its evaluation at enough points, certainly at infinitely many points. Therefore, if

(4) is true for all $M_1 \equiv 0 \mod n$ then it is true for all M_1. Since (4) and (3) are equivalent, the same applies to (3).

Now, choose some $M_1 \equiv 0 \mod n$. If we want to prove (3) for this particular M_1, then an analogous argument shows that it is enough to prove it for all $M_2 \equiv 0 \mod n$.

Summarizing, it is sufficient to prove (3) for $M_1 \equiv M_2 \equiv 0 \mod n$. Therefore, for the rest of the proof, we assume that this congruence condition is satisfied.

To begin with, let us rewrite the left-hand side of (3) by replacing k_i by $k_i - M_1/n$, $i = 1, 2, \ldots, n$, and performing some rearrangement of terms,

$$(5) \quad (-1)^{nM_1} q^{M_1(M_1 n - M_1 + 2nM_2 + 2n^2 - 1)/2}$$
$$\times \sum_{k_1 + \cdots + k_n = M_1} q^{\frac{n}{2} \sum_{i=1}^{n} k_i^2 - (n-1) \sum_{i=1}^{n} i k_i}$$
$$\times \prod_{1 \leq i < j \leq n} \frac{(1 - q^{nk_j - nk_i + j - i})}{1 - q^{j-i}} \prod_{i=1}^{n} \frac{(q^{-M_1 - M_2 - n + i}; q)_{nk_i}}{(q^i; q)_{nk_i}}.$$

Next we want to apply a limiting case of Milne's A_n Watson transformation [**11**, Theorem 6.1],

$$(6) \quad \sum_{k_1, \ldots, k_l \geq 0} \left(\prod_{1 \leq r < s \leq l} \frac{1 - \frac{x_r}{x_s} q^{k_r - k_s}}{1 - \frac{x_r}{x_s}} \right) \left(\prod_{i=1}^{l} \frac{1 - \frac{x_i}{x_l} a q^{k_i + (k_1 + \cdots + k_l)}}{1 - \frac{x_i}{x_l} a} \right)$$
$$\times \left(\prod_{r=1}^{l} \prod_{s=1}^{l} \frac{(\frac{x_r}{x_s} q^{-N_s}; q)_{k_r}}{(q \frac{x_r}{x_s}; q)_{k_r}} \right) \left(\prod_{i=1}^{l} \frac{(\frac{x_i}{x_l} a; q)_{k_1 + \cdots + k_l}}{(\frac{x_i}{x_l} a q^{1+N_i}; q)_{k_1 + \cdots + k_l}} \right)$$
$$\times \left(\prod_{i=1}^{l} \frac{(\frac{x_i}{x_l} c; q)_{k_i} (\frac{x_i}{x_l} d; q)_{k_i}}{(\frac{x_i}{x_l} \frac{aq}{b}; q)_{k_i} (\frac{x_i}{x_l} \frac{aq}{e}; q)_{k_i}} \right) \frac{(b; q)_{k_1 + \cdots + k_l} (e; q)_{k_1 + \cdots + k_l}}{(\frac{aq}{c}; q)_{k_1 + \cdots + k_l} (\frac{aq}{d}; q)_{k_1 + \cdots + k_l}}$$
$$\times \left(\frac{a^2 q^{1+N_1+\cdots+N_l}}{bcde} \right)^{k_1+\cdots+k_l} q^{\sum_{i=1}^{l} i k_i}$$
$$= \frac{(aq/de; q)_{N_1+\cdots+N_l}}{(aq/d; q)_{N_1+\cdots+N_l}} \left(\prod_{i=1}^{l} \frac{(\frac{x_i}{x_l} aq; q)_{N_i}}{(\frac{x_i}{x_l} aq/e; q)_{N_i}} \right)$$
$$\times \sum_{k_1, \ldots, k_l \geq 0} q^{\sum_{i=1}^{l} i k_i} \left(\prod_{1 \leq r < s \leq l} \frac{1 - \frac{x_r}{x_s} q^{k_r - k_s}}{1 - \frac{x_r}{x_s}} \right) \left(\prod_{r=1}^{l} \prod_{s=1}^{l} \frac{(\frac{x_r}{x_s} q^{-N_s}; q)_{k_r}}{(q \frac{x_r}{x_s}; q)_{k_r}} \right)$$
$$\times \left(\prod_{i=1}^{l} \frac{(\frac{x_i}{x_l} d; q)_{k_i}}{(\frac{x_i}{x_l} \frac{aq}{b}; q)_{k_i}} \right) \frac{(\frac{aq}{bc}; q)_{k_1+\cdots+k_l} (e; q)_{k_1+\cdots+k_l}}{(\frac{aq}{c}; q)_{k_1+\cdots+k_l} (\frac{de}{a} q^{-N_1-\cdots-N_l}; q)_{k_1+\cdots+k_l}},$$

where $N_1, \ldots, N_l$ are nonnegative integers. For convenience, let us set $k_{l+1} = M - k_1 - \cdots - k_l$ and $a = x_l/q^M x_{l+1}$, so that (6) becomes

$$(7)$$
$$q^M \left(\prod_{i=1}^{l} \frac{(q \frac{x_{l+1}}{x_i}; q)_M}{(\frac{x_{l+1}}{x_i} q^{-N_i}; q)_M} \right) \frac{(q^{1-M}/b; q)_M (q^{1-M}/e; q)_M}{(\frac{x_{l+1}}{x_l} c; q)_M (\frac{x_{l+1}}{x_l} d; q)_M} \prod_{i=1}^{l} \frac{(1 - \frac{x_i}{x_{l+1}})}{(1 - \frac{x_i}{q^M x_{l+1}})}$$
$$\times \sum_{k_1 + \cdots + k_l + k_{l+1} = M} q^{-\sum_{i=1}^{l+1} i k_i} \left(\prod_{1 \leq r < s \leq l+1} \frac{1 - \frac{x_s}{x_r} q^{k_s - k_r}}{1 - \frac{x_s}{x_r}} \right) \left(\prod_{r=1}^{l+1} \prod_{s=1}^{l} \frac{(\frac{x_r}{x_s} q^{-N_s}; q)_{k_r}}{(q \frac{x_r}{x_s}; q)_{k_r}} \right)$$

$$\times \left(\prod_{i=1}^{l+1} \frac{(\frac{x_i}{x_l}c;q)_{k_i} (\frac{x_i}{x_l}d;q)_{k_i}}{(\frac{x_i}{x_{l+1}} \frac{q^{1-M}}{b};q)_{k_i} (\frac{x_i}{x_{l+1}} \frac{q^{1-M}}{e};q)_{k_i}} \right)$$

$$= \frac{(q^{1-M}x_l/x_{l+1}de;q)_{N_1+\cdots+N_l}}{(q^{1-M}x_l/x_{l+1}d;q)_{N_1+\cdots+N_l}} \left(\prod_{i=1}^{l} \frac{(\frac{x_i}{x_{l+1}} q^{1-M};q)_{N_i}}{(\frac{x_i}{x_{l+1}} q^{1-M}/e;q)_{N_i}} \right)$$

$$\times \sum_{k_1,\ldots,k_l \geq 0} q^{\sum_{i=1}^{l} i k_i} \left(\prod_{1 \leq r < s \leq l} \frac{1 - \frac{x_r}{x_s} q^{k_r - k_s}}{1 - \frac{x_r}{x_s}} \right) \left(\prod_{r=1}^{l} \prod_{s=1}^{l} \frac{(\frac{x_r}{x_s} q^{-N_s};q)_{k_r}}{(q \frac{x_r}{x_s};q)_{k_r}} \right)$$

$$\times \left(\prod_{i=1}^{l} \frac{(\frac{x_i}{x_l}d;q)_{k_i}}{(\frac{x_i}{x_{l+1}} \frac{q^{1-M}}{b};q)_{k_i}} \right) \frac{(\frac{q^{1-M}x_l}{x_{l+1}bc};q)_{k_1+\cdots+k_l} (e;q)_{k_1+\cdots+k_l}}{(\frac{q^{1-M}x_l}{x_{l+1}c};q)_{k_1+\cdots+k_l} (\frac{deq^M x_{l+1}}{x_l} q^{-N_1-\cdots-N_l};q)_{k_1+\cdots+k_l}}.$$

In this identity we replace q by q^n. Then we set $l = n-1$, $M = M_1$, $x_i = q^i$ for $i = 1, 2, \ldots, l+1$, $b = q^{-nM_1}$, $d = \delta q^{-M_1 - M_2 - 1}$, $N_i = (M_1 + M_2)/n$. Next we multiply both sides by $(1-\delta)$ (this cancels one factor in the term $(x_{l+1}d/x_l;q)_M \sim (\delta q^{-M_1-M_2};q^n)_{M_1}$ in the denominator of the left-hand side of (7) and one factor in the term $(q^{1-M}x_l/x_{l+1}d;q)_{N_1+\cdots+N_l} \sim (q^{n-nM_1+M_1+M_2}/\delta;q^n)_{(M_1+M_2)/n}$ in the denominator of the right-hand side of (7)). Finally, we let $\delta \to 1$, $c \to \infty$, and $e \to \infty$. This reduces (7) to the following transformation formula,

(8) $\quad q^{nM_1(n-M_1/2-1)} \dfrac{(q;q)_{nM_1}}{(q^{-M_1-M_2};q)_{M_1+M_2} (q;q)_{nM_1-M_1-M_2-1}}$

$$\times \sum_{k_1+\cdots+k_n=M_1} q^{\frac{n}{2}\sum_{i=1}^{n} k_i^2 - (n-1)\sum_{i=1}^{n} i k_i} \left(\prod_{1 \leq r < s \leq n} \frac{1 - q^{nk_s - nk_r + s - r}}{1 - q^{s-r}} \right)$$

$$\times \left(\prod_{r=1}^{n} \frac{(q^{-M_1-M_2-n+r};q)_{nk_r}}{(q^r;q)_{nk_r}} \right)$$

$$= -\frac{(q^{-nM_1};q)_{M_1+M_2+1}}{(q^{-nM_1};q^n)_{M_1} (q^n;q^n)_{M_2}} \sum_{k_1,\ldots,k_{n-1} \geq 0} \left(\prod_{1 \leq r < s \leq n-1} \frac{1 - q^{nk_r - nk_s + r - s}}{1 - q^{r-s}} \right)$$

$$\times \left(\prod_{r=1}^{n-1} \frac{(q^{-M_1-M_2-n+r};q)_{nk_r}}{(q^r;q)_{nk_r}} (q^{M_2})^{nk_r} \right) q^{n \sum_{i=1}^{n-1} i k_i}.$$

The series on the left-hand side of (8) is exactly the series in (5). What the transformation (8) does with this series is, in some sense which will become more transparent below, that it "entangles" the summation indices. Thus, we obtain the following expression for the left-hand side of (3),

(9) $\quad (-1)^{M_1} q^{-\binom{M_1+1}{2}} \dfrac{(q;q)_{M_1+M_2}}{(q^{-nM_1};q^n)_{M_1} (q^n;q^n)_{M_2}}$

$$\times \sum_{k_1,\ldots,k_{n-1} \geq 0} \left(\prod_{1 \leq r < s \leq n-1} \frac{1 - q^{nk_r - nk_s + r - s}}{1 - q^{r-s}} \right)$$

$$\times \left(\prod_{r=1}^{n-1} \frac{(q^{-M_1-M_2-n+r};q)_{nk_r}}{(q^r;q)_{nk_r}} (q^{M_2})^{nk_r} \right) q^{n \sum_{i=1}^{n-1} i k_i}.$$

(The sign $(-1)^{M_1}$ is no misprint since our assumption $M_1 \equiv 0 \mod n$ implies $nM_1 \equiv M_1 \mod 2$.)

The next task is to split the sum in (9) into many pieces, each of which being a product of $n-1$ one-dimensional summations. This is done by replacing the product over $1 \le s < r \le n-1$ by a Vandermonde determinant. More precisely, we have

$$\prod_{1 \le r < s \le n-1} (1 - q^{nk_r - nk_s + r - s}) = q^{-\sum_{i=1}^{n-1}(i-1)(nk_i+i)} \prod_{1 \le r < s \le n-1} (q^{nk_s+s} - q^{nk_r+r})$$

$$= q^{-\sum_{i=1}^{n-1}(i-1)(nk_i+i)} \det_{1 \le i,j \le n-1}\left(\left(q^{nk_i+i}\right)^{j-1} \right)$$

$$= q^{-n\sum_{i=1}^{n-1} ik_i + n\sum_{i=1}^{n-1} k_i - 2\binom{n}{3}} \sum_{\sigma \in S_{n-1}} \operatorname{sgn} \sigma \prod_{i=1}^{n-1} q^{(\sigma(i)-1)(nk_i+i)}.$$

Hence, the sum in (9) equals

$$(10) \quad (-1)^{\binom{n-1}{2}} \left(\prod_{i=1}^{n-1} \frac{1}{(q;q)_{i-1}} \right) \sum_{\sigma \in S_{n-1}} \operatorname{sgn} \sigma \, q^{-\binom{n}{3}} q^{\sum_{i=1}^{n-1} i(\sigma(i)-1)}$$

$$\times \prod_{i=1}^{n-1} \left(\sum_{k_i \ge 0} \frac{(q^{-M_1-M_2-n+i};q)_{nk_i}}{(q^i;q)_{nk_i}} \left(q^{M_2+\sigma(i)}\right)^{nk_i} \right).$$

The next ingredient is Ramanujan's $_1\psi_1$-summation (see [**3**, (5.2.1)]),

$$(11) \quad \sum_{k=-\infty}^{\infty} \frac{(a;q)_k}{(b;q)_k} z^k = \frac{(q;q)_\infty (b/a;q)_\infty (az;q)_\infty (q/az;q)_\infty}{(b;q)_\infty (q/a;q)_\infty (z;q)_\infty (b/az;q)_\infty}.$$

Each of the inner sums in (10) is an n-section of a special case of the left-hand side of (11). (To be precise, it is the special case $a = q^{-M_1-M_2-n+i}$, $b = q^i$, and $z = q^{M_2+\sigma(i)}$.) Thus, (10) simplifies to

$$(12) \quad (-1)^{\binom{n-1}{2}} \left(\prod_{i=1}^{n-1} \frac{1}{(q;q)_{i-1}} \right) \sum_{\sigma \in S_{n-1}} \operatorname{sgn} \sigma \, q^{-\binom{n}{3}} q^{\sum_{i=1}^{n-1} i(\sigma(i)-1)}$$

$$\times \prod_{i=1}^{n-1} \left(\frac{1}{n} \sum_{\ell_i=0}^{n-1} \frac{(q;q)_\infty (q^{M_1+M_2+n};q)_\infty}{(q^i;q)_\infty (q^{1-i+M_1+M_2+n};q)_\infty} \right.$$

$$\left. \times \frac{(q^{i+\sigma(i)-M_1-n}\omega^{\ell_i};q)_\infty (q^{1-i-\sigma(i)+M_1+n}\omega^{-\ell_i};q)_\infty}{(q^{M_2+\sigma(i)}\omega^{\ell_i};q)_\infty (q^{-\sigma(i)+M_1+n}\omega^{-\ell_i};q)_\infty} \right),$$

where ω denotes a primitive n-th root of unity. An immediate observation is that if any ℓ_i equals 0 then the corresponding summand vanishes, because of the term

$$(q^{i+\sigma(i)-M_1-n}\omega^{\ell_i};q)_\infty (q^{1-i-\sigma(i)+M_1+n}\omega^{-\ell_i};q)_\infty$$

in the numerator. Hence, we may as well sum over ℓ_i from 1 to $n-1$, $i = 1, 2, \ldots, n-1$.

Some manipulation transforms (12) into

$$(13) \quad (-1)^{\binom{n-1}{2}} \frac{1}{n^{n-1}} \prod_{i=1}^{n-1} \frac{1}{(q^{1-i+M_1+M_2+n};q)_{i-1}} \sum_{\sigma \in S_{n-1}} \operatorname{sgn} \sigma$$

$$\times \left(\sum_{\ell_1,\ldots,\ell_{n-1}=1}^{n-1} \left(\prod_{i=1}^{n-1} \frac{(q^{-M_1}\omega^{\ell_i};q)_\infty}{(q^{M_2+1}\omega^{\ell_i};q)_\infty} \omega^{\ell_i(n-i-\sigma(i))} \right) \right.$$

$$\left. \times (q^{M_2+1}\omega^{\ell_i};q)_{\sigma(i)-1} (q^{M_1+1}\omega^{-\ell_i};q)_{n-\sigma(i)-1} \right).$$

Now it is not difficult to see that if $\ell_r = \ell_s$, $r \neq s$, then the summand corresponding to the permutation σ cancels with the summand corresponding to the permutation $\sigma \circ (rs)$. (Here, (rs) denotes the transposition which interchanges r and s.) Therefore the only summands which survive this cancellation are those where the summation indices $\ell_1, \ell_2, \ldots, \ell_{n-1}$ are a permutation of $\{1, 2, \ldots, n-1\}$. Thus, (13) reduces to

(14)
$$(-1)^{\binom{n-1}{2}} \frac{1}{n^{n-1}} \frac{(q^{-nM_1};q^n)_{M_1}(q^n;q^n)_\infty (q^{M_2+1};q)_\infty}{(q^{-M_1};q)_{M_1}(q;q)_\infty (q^{nM_2+n};q^n)_\infty} \prod_{i=1}^{n-1} \frac{(1-\omega^i)}{(q^{1-i+M_1+M_2+n};q)_{i-1}}$$

$$\times \sum_{\sigma,\tau \in S_{n-1}} (\operatorname{sgn} \sigma) \omega^{\tau(i)(n-i-\sigma(i))} (q^{M_2+1}\omega^{\tau(i)};q)_{\sigma(i)-1} (q^{M_1+1}\omega^{-\tau(i)};q)_{n-\sigma(i)-1}$$

$$= (-1)^{\binom{n-1}{2}} \frac{1}{n^{n-1}} \frac{(q^{-nM_1};q^n)_{M_1}(q^n;q^n)_\infty (q^{M_2+1};q)_\infty}{(q^{-M_1};q)_{M_1}(q;q)_\infty (q^{nM_2+n};q^n)_\infty} \prod_{i=1}^{n-1} \frac{(1-\omega^i)}{(q^{1-i+M_1+M_2+n};q)_{i-1}}$$

$$\times \sum_{\tau \in S_{n-1}} \omega^{\tau(i)(n-i)} \det_{1 \leq i,j \leq n-1} \left(\omega^{-j\tau(i)} (q^{M_2+1}\omega^{\tau(i)};q)_{j-1} (q^{M_1+1}\omega^{-\tau(i)};q)_{n-j-1} \right).$$

The determinant is easily evaluated with the help of the determinant lemma [**6**, Lemma 2.2],

$$(15) \quad \det_{1 \leq i,j \leq n} \left((X_i + A_n) \cdots (X_i + A_{j+1})(X_i + B_j) \cdots (X_i + B_2) \right)$$

$$= \prod_{1 \leq i < j \leq n} (X_i - X_j) \prod_{2 \leq i \leq j \leq n} (B_i - A_j),$$

where $X_1, \ldots, X_n$, $A_2, \ldots, A_n$, and $B_2, \ldots B_n$ are arbitrary indeterminates. In order to apply (15), we rewrite the determinant in (14) as

$$\det_{1 \leq i,j \leq n-1} \left(\omega^{-j\tau(i)} (q^{M_2+1}\omega^{\tau(i)};q)_{j-1} (q^{M_1+1}\omega^{-\tau(i)};q)_{n-j-1} \right)$$

$$= (-1)^{\binom{n-1}{2}} \prod_{i=1}^{n-1} \omega^{-\tau(i)} q^{(M_1+1)+(M_1+2)+\cdots+(M_1+n-i-1)}$$

$$\times \det_{1 \leq i,j \leq n-1} \left((\omega^{-\tau(i)} - q^{M_2+1})(\omega^{-\tau(i)} - q^{M_2+2}) \cdots (\omega^{-\tau(i)} - q^{M_2+j-1}) \right.$$

$$\left. \cdot (\omega^{-\tau(i)} - q^{-M_1-n+j+1})(\omega^{-\tau(i)} - q^{-M_1-n+j+2}) \cdots (\omega^{-\tau(i)} - q^{-M_1-1}) \right).$$

Now the determinant evaluation (15) applies with $X_i = \omega^{-\tau(i)}$, $A_j = -q^{-M_1-n+j}$, and $B_j = -q^{M_2+j-1}$. If the resulting expression is substituted back into (14), we obtain

$$(16) \quad \frac{1}{n^{n-1}} \frac{(q^{-nM_1};q^n)_{M_1} (q^n;q^n)_\infty (q^{M_2+1};q)_\infty}{(q^{-M_1};q)_{M_1} (q;q)_\infty (q^{nM_2+n};q^n)_\infty} \prod_{i=1}^{n-1} (1-\omega^i)$$

$$\times \sum_{\tau \in S_{n-1}} \omega^{\tau(i)(n-i-1)} \prod_{1 \le i < j \le n-1} (\omega^{-\tau(i)} - \omega^{-\tau(j)})$$

$$= \frac{1}{n^{n-1}} \frac{(q^{-nM_1};q^n)_{M_1} (q^n;q^n)_\infty (q^{M_2+1};q)_\infty}{(q^{-M_1};q)_{M_1} (q;q)_\infty (q^{nM_2+n};q^n)_\infty} \prod_{i=1}^{n-1} (1-\omega^i) \prod_{1 \le i < j \le n-1} (\omega^{-i} - \omega^{-j})$$

$$\times \sum_{\tau \in S_{n-1}} (\mathrm{sgn}\,\tau)\, \omega^{\tau(i)(n-i-1)}.$$

The sum over permutations in the last line is just a Vandermonde determinant, and as such easily evaluated. If we substitute this in (16), the resulting expression for the sum in (9), we obtain

$$(17) \quad \frac{1}{n^{n-1}} \frac{(q;q)_{M_1+M_2}}{(q;q)_{M_1} (q;q)_{M_2}} \prod_{i=1}^{n-1} (1-\omega^i) \prod_{1 \le i < j \le n-1} (\omega^{-i} - \omega^{-j})(\omega^i - \omega^j)$$

for the left-hand side of (3). Clearly, there holds

$$\prod_{i=1}^{n-1} (1-\omega^i) = n,$$

because it is the limit $\lim_{z \to 1} (1-z^n)/(1-z)$. Moreover, we have

$$\prod_{1 \le i < j \le n-1} (\omega^i - \omega^j)(\omega^{-i} - \omega^{-j})$$

$$= \prod_{i=1}^{n-1} (1-\omega) \cdots (1-\omega^{i-1}) \prod_{i=1}^{n-1} (1-\omega^{-1}) \cdots (1-\omega^{-i+1})$$

$$= \prod_{i=1}^{n-2} (1-\omega) \cdots (1-\omega^i) \prod_{i=1}^{n-2} (1-\omega^{n-1}) \cdots (1-\omega^{i+1})$$

$$= \prod_{i=1}^{n-2} (1-\omega) \cdots (1-\omega^{n-1}) = n^{n-2},$$

in view of the previous observation. Thus, (17) does indeed reduce to the right-hand side of (3). In view of the remarks of the first paragraph of this section, the proof of the theorem is complete. □

References

[1] G. Bhatnagar and M. Schlosser, C_n and D_n very-well-poised $_{10}\phi_9$ transformations, Constr. Approx. **14** (1998), 531–567.
[2] J. F. van Diejen, On certain multiple Bailey, Rogers and Dougall type summation formulas, Publ. Research Inst. Math. Sci. Kyoto Univ. **33** (1997), 483–508.
[3] G. Gasper and M. Rahman, Basic hypergeometric series, Encyclopedia of Mathematics And Its Applications 35, Cambridge University Press, Cambridge, 1990.

[4] I. M. Gessel and C. Krattenthaler, *Cylindric partitions*, Trans. Amer. Math. Soc. **349** (1997), 429–479.
[5] Y. Kajihara, *Some multiple transformation and summation formulas related to Macdonald polynomials*, preprint.
[6] C. Krattenthaler, *Generating functions for plane partitions of a given shape*, Manuscripta Math. **69**, (1990), 173–202.
[7] C. Krattenthaler, *Advanced determinant calculus*, Séminaire Lotharingien Combin. **42** ("The Andrews Festschrift") (1999), paper B42q, 67 pp.
[8] C. Krattenthaler and M. Schlosser, *A new multidimensional matrix inverse with applications to multiple q-series*, Discrete Math. (Gould Anniversary Volume) **204** (1999), 249–279.
[9] S. C. Milne, *Classical partition functions and the $U(n+1)$ Rogers–Selberg identity*, Discrete Math. **99** (1992), 199–246.
[10] S. C. Milne, *Balanced $_3\phi_2$ summation theorems for $U(n)$ basic hypergeometric series*, Adv. in Math. **131** (1997), 93–187.
[11] S. C. Milne and G. M. Lilly, *Consequences of the A_l and C_l Bailey Transform and Bailey Lemma*, Discrete Math. **139** (1995), 319–346.
[12] S. C. Milne and J. W. Newcomb, *$U(n)$ very-well-poised $_{10}\phi_9$ transformations*, J. Comput. Appl. Math. **68** (1996), 239–285.
[13] A. Schilling and M. Shimozono, *Bosonic formula for level-restricted paths*, Advanced Studies in Pure Mathematics, vol. 28, Combinatorial Methods in Representation Theory (2000), pp. 305–325.; math/9812106.
[14] S. O. Warnaar, *Supernomial coefficients, Bailey's lemma and Rogers–Ramanujan identities. A survey of results and open problems*, Séminaire Lotharingien Combin. **42** (1999), Article B42n, 22 pp.

INSTITUT FÜR MATHEMATIK DER UNIVERSITÄT WIEN, STRUDLHOFGASSE 4, A-1090 WIEN, AUSTRIA.
E-MAIL: KRATT@AP.UNIVIE.AC.AT, WWW: http://www.mat.univie.ac.at/People/kratt

On the representation of integers as sums of squares

Zhi-Guo Liu

Dedicated to the memory of Loo-Keng Hua on the 90th anniversary of his birth.

ABSTRACT. Let k be a positive integer. Let $r_k(n)$ denote the number of representations of n as a sum of k squares. Also, let $t_k(n)$ denote the number of representations of n as a sum of k triangular numbers. In this paper we utilize a trigonometric series identity of Ramanujan to derive some striking Lambert series identities. Using these identities and some identities of Jacobi, we give a new and simple derivation of Ramanujan's formula for $r_{24}(n)$. We also derive formulas for $r_{12}(n), r_{16}(n)$, and $r_{20}(n)$ in the course of our investigations. Applying some simple modular transformations to $r_{12}(n), r_{16}(n), r_{20}(n)$, and $r_{24}(n)$, we derive formulas for $t_{12}(n), t_{16}(n), t_{20}(n)$, and $t_{24}(n)$, respectively.

1. Introduction

Suppose throughout that $q := e^{\pi i \tau}$, $\text{Im}(\tau) > 0$; this condition ensures that all the sums and products that appear here converge. We will use the familiar notation

$$(1.1) \qquad (z;q)_\infty := \prod_{n=1}^{\infty} (1 - zq^n).$$

Let $k \geq 1$ be a positive integer. Let $r_k(n)$ denote the number of representations of n as a sum of k squares. Also, let $t_k(n)$ denote the number of representations of n as a sum of k triangular numbers. To study $r_k(n)$ and $t_k(n)$, we recall Ramanujan's definitions of the theta-functions $\phi(q)$ and $\psi(q)$ defined by

$$(1.2) \qquad \phi(q) = \sum_{n=-\infty}^{\infty} q^{n^2},$$

$$(1.3) \qquad \psi(q) = \sum_{n=0}^{\infty} q^{n(n+1)/2}.$$

1991 *Mathematics Subject Classification.* Primary 11E25, 11F11, 11F27.
Key words and phrases. theta functions, Lambert series, Dedekind eta function, squares, triangular numbers, Ramanujan $\tau(n)$-function, Jacobi's identities, modular transformations.

Consequently, the generating functions for $r_k(n)$ and $t_k(n)$ are

$$(1.4) \qquad \phi^k(q) = \sum_{n=0}^{\infty} r_k(n) q^n,$$

$$(1.5) \qquad \psi^k(q) = \sum_{n=0}^{\infty} t_k(n) q^n.$$

Formulas for $r_k(n)$ have an interesting history. In his famous paper, " On certain arithmetical functions" [16], [17, pp. 136-162], S. Ramanujan established the following trigonometric series identities by elementary methods:

$$(1.6) \quad \left\{ \frac{1}{4} \cot u + \sum_{n=1}^{\infty} \frac{q^n \sin(2nu)}{1-q^n} \right\}^2 = \left\{ \frac{1}{4} \cot(u) \right\}^2 + \sum_{n=1}^{\infty} \frac{q^n \cos(2nu)}{(1-q^n)^2} + \frac{1}{2} \sum_{n=1}^{\infty} \frac{nq^n(1-\cos(2nu))}{1-q^n},$$

and

$$(1.7) \quad \left\{ \frac{1}{8} \cot^2 u + \frac{1}{12} + \sum_{n=1}^{\infty} \frac{nq^n}{1-q^n}(1-\cos(2nu)) \right\}^2 = \left\{ \frac{1}{8} \cot^2 u + \frac{1}{12} \right\}^2 + \frac{1}{12} \sum_{n=1}^{\infty} \frac{n^3 q^n}{1-q^n}(5+\cos(2nu)).$$

Identity (1.6) was used by Ramanujan to prove Jacobi's formula for $r_4(n)$ based on Jacobi's formula for $r_2(n)$ (for the details of the proof, see [16] and [9, pp. 134-135].) An interesting and more motivated proof of (1.6) has recently been given by L.-C. Shen [20]. G. E. Andrews[1] established the Jacobi's formulas for $r_2(n), r_4(n)$ and $r_8(n)$ using Bailey $_6\psi_6$ summation. In [3], R. Askey used Ramanujan's $_1\psi_1$ summation to evaluate $r_2(n)$. Recently, G. E. Andrews, R. P. Lewis and Z. -G. Liu [2] used a striking identity relating a theta product to a sum of Lambert series to derive many theorems on sums of squares.

It is surprising that Ramanujan's identity (1.7) can also be used to evaluate $r_{2k}(n)$. In this paper we will use this identity and Jacobi's identities to provide a new and simple proof of Ramanujan's formula for $r_{24}(n)$. We also derive the formulas for $r_{12}(n), r_{16}(n)$ and $r_{20}(n)$ in the course of our investigations. Applying some simple modular transformations to $r_{12}(n), r_{16}(n), r_{20}(n)$, and $r_{24}(n)$, we obtain formulas for $t_{12}(n), t_{16}(n), t_{20}(n)$, and $t_{24}(n)$ respectively. Our method is different from those of Kac and Wakimoto [10], and Milne [12, 13], and is somewhat more transparent.

2. Some basic facts about Jacobi's theta functions

The Jacobi theta functions $\theta_2(z|\tau), \theta_3(z|\tau)$, and $\theta_4(z|\tau)$ [7, p. 63], [15, p. 166] are defined as

$$\theta_2(z|\tau) = 2q^{\frac{1}{4}} \sum_{n=0}^{\infty} q^{n(n+1)} \cos(2n+1)\pi z = q^{\frac{1}{4}} \sum_{n=-\infty}^{\infty} q^{n(n+1)} e^{(2n+1)\pi i z}, \tag{2.1}$$

$$\theta_3(z|\tau) = 1 + 2\sum_{n=1}^{\infty} q^{n^2} \cos(2n\pi z) = \sum_{n=-\infty}^{\infty} q^{n^2} e^{2n\pi i z}, \tag{2.2}$$

$$\theta_4(z|\tau) = 1 + 2\sum_{n=1}^{\infty} (-1)^n q^{n^2} \cos(2n\pi z) = \sum_{n=-\infty}^{\infty} (-1)^n q^{n^2} e^{2n\pi i z}. \tag{2.3}$$

We obviously have

$$\theta_2(0|\tau) = 2q^{\frac{1}{4}} \sum_{n=0}^{\infty} q^{n(n+1)} = 2q^{\frac{1}{4}} \psi(q^2), \tag{2.4}$$

$$\theta_3(0|\tau) = \sum_{n=-\infty}^{\infty} q^{n^2} = \phi(q), \tag{2.5}$$

$$\theta_4(0|\tau) = \sum_{n=-\infty}^{\infty} (-1)^n q^{n^2} = \phi(-q). \tag{2.6}$$

We recall the Jacobi triple product identity [1, 3]

$$\sum_{n=-\infty}^{\infty} q^{n^2} x^n = (q^2; q^2)_\infty (-qx; q^2)_\infty (-qx^{-1}; q^2)_\infty. \tag{2.7}$$

Appealing to the above identity, we find that [7, p. 70]

$$\begin{aligned} \theta_2(0|\tau) &= 2q^{\frac{1}{4}} \psi(q^2) = 2q^{\frac{1}{4}} (q^2; q^2)_\infty (-q^2; q^2)_\infty^2, \\ \theta_4(0|\tau) &= \phi(-q) = (q^2; q^2)_\infty (q; q^2)_\infty^2, \\ \theta_3(0|\tau) &= \phi(q) = (q^2; q^2)_\infty (-q; q^2)_\infty^2. \end{aligned} \tag{2.8}$$

Therefore, we have

$$\begin{aligned} \theta_2(0|\tau)\theta_3(0|\tau)\theta_4(0|\tau) &= 2q^{\frac{1}{4}} (q^2; q^2)_\infty^3, \\ \theta_2(0|\tau)\theta_3(0|\tau)\theta_4^4(0|\tau) &= 2q^{\frac{1}{4}} (q; q)_\infty^6, \\ \theta_2(0|\tau)\theta_3(0|\tau)\theta_4^2(0|\tau) &= 2q^{\frac{1}{4}} (q; q)_\infty^2 (q^2; q^2)_\infty^2. \end{aligned} \tag{2.9}$$

The imaginary transformation formulas of Jacobi [7, p. 76], [15, p. 177] are

$$\begin{aligned} \theta_3(0|-1/\tau) &= \sqrt{-\tau i}\, \theta_3(0|\tau), \\ \theta_2(0|-1/\tau) &= \sqrt{-\tau i}\, \theta_4(0|\tau), \\ \theta_4(0|-1/\tau) &= \sqrt{-\tau i}\, \theta_2(0|\tau). \end{aligned} \tag{2.10}$$

The Dedekind eta-function [7, p. 122] is defined by

$$\eta(\tau) = e^{\pi i \tau / 12} \prod_{n=1}^{\infty} (1 - e^{2\pi i n \tau}) = q^{\frac{1}{12}} (q^2; q^2)_\infty. \tag{2.11}$$

It is well-known that [7, p. 126]

$$\eta(-1/\tau) = \sqrt{-\tau i}\, \eta(\tau). \tag{2.12}$$

The Bernoulli numbers B_n are defined as the coefficients in the power series

(2.13) $$\frac{x}{e^x - 1} = \sum_{k=0}^{\infty} B_k \frac{x^k}{k!}, \quad |x| < 2\pi.$$

It is to show that $B_{2k+1} = 0$ for $k \geq 1$, and the following additional values may be listed:

(2.14) $$\begin{array}{llll} B_0 = 1, & B_1 = -\frac{1}{2}, & B_2 = \frac{1}{6}, & B_4 = -\frac{1}{30}, \\ B_6 = \frac{1}{42}, & B_8 = -\frac{1}{30}, & B_{10} = \frac{5}{66}, & B_{12} = -\frac{691}{2730}. \end{array}$$

The normalized Eisenstein series $E_{2k}(\tau)$ are defined by [11, p. 111, eq. (2.11)]

$$E_{2k}(\tau) := 1 - \frac{4k}{B_{2k}} \sum_{n=1}^{\infty} \frac{n^{2k-1} e^{2\pi i n \tau}}{1 - q^{2\pi i n \tau}} = 1 - \frac{4k}{B_{2k}} \sum_{n=1}^{\infty} \frac{n^{2k-1} q^{2n}}{1 - q^{2n}}$$

(2.15) $$= 1 - \frac{4k}{B_{2k}} \sum_{n=1}^{\infty} \sigma_{2k-1}(n) q^{2n},$$

where $\sigma_k(n)$ is the standard function and is defined as

(2.16) $$\sigma_k(n) = \sum_{d|n} d^k.$$

It is also understood that $\sigma_k(x) = 0$ if x is not an integer.

The first few $E_{2k}(\tau)$ are

(2.17)
$$\begin{aligned}
E_2(\tau) &= 1 - 24 \sum_{n=1}^{\infty} \frac{n q^{2n}}{1 - q^{2n}} = 1 - 24 \sum_{n=1}^{\infty} \sigma(n) q^{2n}, \\
E_4(\tau) &= 1 + 240 \sum_{n=1}^{\infty} \frac{n^3 q^{2n}}{1 - q^{2n}} = 1 + 240 \sum_{n=1}^{\infty} \sigma_3(n) q^{2n}, \\
E_6(\tau) &= 1 - 504 \sum_{n=1}^{\infty} \frac{n^5 q^{2n}}{1 - q^{2n}} = 1 - 504 \sum_{n=1}^{\infty} \sigma_5(n) q^{2n}, \\
E_8(\tau) &= 1 + 480 \sum_{n=1}^{\infty} \frac{n^7 q^{2n}}{1 - q^{2n}} = 1 + 480 \sum_{n=1}^{\infty} \sigma_7(n) q^{2n}, \\
E_{10}(\tau) &= 1 - 264 \sum_{n=1}^{\infty} \frac{n^9 q^{2n}}{1 - q^{2n}} = 1 - 264 \sum_{n=1}^{\infty} \sigma_9(n) q^{2n}, \\
E_{12}(\tau) &= 1 + \frac{65520}{691} \sum_{n=1}^{\infty} \frac{n^{11} q^{2n}}{1 - q^{2n}} = 1 + \frac{65520}{691} \sum_{n=1}^{\infty} \sigma_{11}(n) q^{2n}.
\end{aligned}$$

We need the modular transformation formula [11, p. 110]

(2.18) $$E_{2k}(-1/\tau) = \tau^{2k} E_{2k}(\tau), \quad k \geq 2.$$

3. Some identities of Jacobi

In this section we will first prove the identity

(3.1) $$\theta_2^4(0|\tau) + \theta_3^4(0|\tau) = 1 + 24 \sum_{n=1}^{\infty} \frac{n q^n}{1 + q^n}.$$

Proof. The Lambert series expansion formula for $\theta_3^4(q)$ is

$$(3.2) \quad \theta_3^4(0|\tau) = 1 + 8\sum_{n=1}^{\infty}\frac{nq^n}{1+(-q)^n} = 1 + 8\sum_{n=1}^{\infty}\frac{nq^n}{1-q^n} - 32\sum_{n=1}^{\infty}\frac{nq^{4n}}{1-q^{4n}},$$

and the Lambert series expansion formula for $\theta_2^4(0|\tau)$ is

$$(3.3) \quad \sum_{n=0}^{\infty}\frac{(2n+1)q^{2n+1}}{1-q^{2(2n+1)}} = \frac{1}{16}\theta_2^4(0|\tau).$$

The identity (3.2) implying Jacobi's formula for $r_4(n)$, [5, Theorem 5.5], [6], and the identity (1.3) is Example (iii) in Section 17 of Chapter 17 in Ramanujan's second notebook [4, p. 123], [6]. Hence, we have

$$\begin{aligned}\theta_2^4(0|\tau) + \theta_3^4(0|\tau) &= 1 + 8\sum_{n=1}^{\infty}\frac{nq^n}{1+(-q)^n} + 16\sum_{n=0}^{\infty}\frac{(2n+1)q^{2n+1}}{1-q^{2(2n+1)}}\\ &= 1 + 8\sum_{n=1}^{\infty}\frac{2nq^{2n}}{1+q^{2n}} + 8\sum_{n=0}^{\infty}\frac{(2n+1)q^{2n+1}}{1-q^{2n+1}}\\ &\quad + 8\sum_{n=0}^{\infty}\frac{(2n+1)q^{2n+1}}{1+q^{2n+1}} + 8\sum_{n=0}^{\infty}\frac{(2n+1)q^{2n+1}}{1-q^{2n+1}}\\ &= 1 + 8\sum_{n=1}^{\infty}\frac{nq^n}{1+q^n} + 16\sum_{n=1}^{\infty}\frac{(2n+1)q^{2n+1}}{1-q^{2n+1}}\\ &= 1 + 8\sum_{n=1}^{\infty}\frac{nq^n}{1+q^n} + 16\sum_{n=0}^{\infty}\left\{\frac{nq^n}{1-q^n} - \frac{2nq^{2n}}{1-q^{2n}}\right\}\\ &= 1 + 24\sum_{n=1}^{\infty}\frac{nq^n}{1+q^n}.\end{aligned}$$

We also require the Jacobi identity for eight squares [5, Theorem 12.9], namely,

$$(3.4) \quad \theta_4^8(0|\tau) = 1 + 16\sum_{n=1}^{\infty}\frac{n^3(-q)^n}{1-q^n},$$

and the following identity due to Jacobi [18, p. 467, eq. (1.1.25)],

$$(3.5) \quad \theta_2^4(0|\tau) + \theta_4^4(0|\tau) = \theta_3^4(0|\tau).$$

An interesting proof of the above identity is given in [8]. Employing the above identity we obtain

$$(3.6) \quad \theta_2^8(0|\tau) + \theta_4^8(0|\tau) = \theta_3^8(0|\tau) - 2\theta_2^4(0|\tau)\theta_4^4(0|\tau),$$
$$(3.7) \quad \theta_2^{12}(0|\tau) + \theta_4^{12}(0|\tau) = \theta_3^{12}(0|\tau) - 3\theta_2^4(0|\tau)\theta_3^4(0|\tau)\theta_4^4(0|\tau),$$
$$\theta_2^{16}(0|\tau) + \theta_4^{16}(0|\tau) = \theta_3^{16}(0|\tau) - 4\theta_3^8(0|\tau)\theta_2^4(0|\tau)\theta_4^4(0|\tau)$$
$$(3.8) \quad\quad\quad\quad\quad\quad\quad + 2\theta_2^8(0|\tau)\theta_4^8(0|\tau).$$

4. Some Lambert series Identities

Multiplying both sides of (1.7) by 576 and then replacing u by $u+\frac{\pi}{2}$, we obtain

$$\left\{3\tan^2 u + 2 + 24\sum_{n=1}^{\infty}\frac{nq^n}{1-q^n}(1+(-1)^{n-1}\cos(2nu))\right\}^2$$

(4.1)
$$= \{3\tan^2 u + 2\}^2 + 48\sum_{n=1}^{\infty}\frac{n^3 q^n}{1-q^n}(5+(-1)^n\cos 2nu).$$

Using

$$\tan^2 u = u^2 + \frac{2}{3}u^4 + \frac{17}{45}u^6 + \frac{62}{315}u^8 + O(u^{10}),$$

$$\{3\tan^2 u + 2\}^2 = 4 + 12u^2 + 17u^4 + \frac{248}{15}u^6 + \frac{1382}{105}u^8 + O(u^{10}),$$

and

$$\cos(2nu) = 1 - 2n^2 u^2 + \frac{2}{3}n^4 u^4 - \frac{4}{45}n^6 u^6 + \frac{2}{315}n^8 u^8 + O(u^{10}),$$

we find that

$$3\tan^2 u + 2 + 24\sum_{n=1}^{\infty}\frac{nq^n}{1-q^n}(1+(-1)^{n-1}\cos(2nu))$$

$$= 2\left\{1+24\sum_{n=1}^{\infty}\frac{nq^n}{1+q^n}\right\} + 3\left\{1+16\sum_{n=1}^{\infty}\frac{n^3(-q)^n}{1-q^n}\right\}u^2$$

$$+2\left\{1-8\sum_{n=1}^{\infty}\frac{n^5(-q)^n}{1-q^n}\right\}u^4 + \frac{1}{15}\left\{17+32\sum_{n=1}^{\infty}\frac{n^7(-q)^n}{1-q^n}\right\}u^6$$

(4.2)
$$+\frac{2}{105}\left\{31-8\sum_{n=1}^{\infty}\frac{n^9(-q)^n}{1-q^n}\right\}u^8 + O(u^{10}),$$

and

$$\{3\tan^2 u + 2\}^2 + 48\sum_{n=1}^{\infty}\frac{n^3 q^n}{1-q^n}(5+(-1)^n\cos(2nu)).$$

$$= 4\left\{1+60\sum_{n=1}^{\infty}\frac{n^3 q^n}{1-q^n}+12\sum_{n=1}^{\infty}\frac{n^3(-q)^n}{1-q^n}\right\} + 12\left\{1-8\sum_{n=1}^{\infty}\frac{n^5(-q)^n}{1-q^n}\right\}u^2$$

$$+\left\{17+32\sum_{n=1}^{\infty}\frac{n^7(-q)^n}{1-q^n}\right\}u^4 + \frac{8}{15}\left\{31-8\sum_{n=1}^{\infty}\frac{n^9(-q)^n}{1-q^n}\right\}u^6$$

(4.3)
$$+\frac{2}{105}\left\{691+16\sum_{n=1}^{\infty}\frac{n^{11}(-q)^n}{1-q^n}\right\}u^8 + O(u^{10}).$$

Substituting (4.2) and (4.3) into (4.1) and then equating the coefficients of u^2, u^4, u^6 and u^8 respectively in the resulting equation we obtain the striking Lambert series identities

(4.4) $$1-8\sum_{n=1}^{\infty}\frac{n^5(-q)^n}{1-q^n} = \left\{1+24\sum_{n=1}^{\infty}\frac{nq^n}{1+q^n}\right\}\left\{1+16\sum_{n=1}^{\infty}\frac{n^3(-q)^n}{1-q^n}\right\},$$

$$17 + 32\sum_{n=1}^{\infty}\frac{n^7(-q)^n}{1-q^n} = 8\left\{1 + 24\sum_{n=1}^{\infty}\frac{nq^n}{1+q^n}\right\}\left\{1 - 8\sum_{n=1}^{\infty}\frac{n^5(-q)^n}{1-q^n}\right\}$$

(4.5)
$$+ 9\left\{1 + 16\sum_{n=1}^{\infty}\frac{n^3(-q)^n}{1-q^n}\right\}^2,$$

$$2\left\{31 - 8\sum_{n=1}^{\infty}\frac{n^9(-q)^n}{1-q^n}\right\}$$
$$= \left\{1 + 24\sum_{n=1}^{\infty}\frac{nq^n}{1+q^n}\right\}\left\{17 + 32\sum_{n=1}^{\infty}\frac{n^7(-q)^n}{1-q^n}\right\}$$

(4.6)
$$+ 45\left\{1 + 16\sum_{n=1}^{\infty}\frac{n^3(-q)^n}{1-q^n}\right\}\left\{1 - 8\sum_{n=1}^{\infty}\frac{n^5(-q)^n}{1-q^n}\right\},$$

and

$$691 + 16\sum_{n=1}^{\infty}\frac{n^{11}(-q)^n}{1-q^n}$$
$$= \left\{1 - 8\sum_{n=1}^{\infty}\frac{n^5(-q)^n}{1-q^n}\right\}^2$$
$$+ 4\left\{1 + 24\sum_{n=1}^{\infty}\frac{nq^n}{1+q^n}\right\}\left\{31 - 8\sum_{n=1}^{\infty}\frac{n^9(-q)^n}{1-q^n}\right\}$$

(4.7)
$$+ 21\left\{1 + 16\sum_{n=1}^{\infty}\frac{n^3(-q)^n}{1-q^n}\right\}\left\{17 + 32\sum_{n=1}^{\infty}\frac{n^7(-q)^n}{1-q^n}\right\}.$$

5. Formulas for $r_{12}(n)$ and $t_{12}(n)$

Substituting (3.1) and (3.4) into (4.4), we obtain the following Lambert series identity associated with $\theta_2(0|\tau), \theta_3(0|\tau)$, and $\theta_4(0|\tau)$

(5.1)
$$1 - 8\sum_{n=1}^{\infty}\frac{n^5(-q)^n}{1-q^n} = \theta_4^8(0|\tau)\left(\theta_2^4(0|\tau) + \theta_3^4(0|\tau)\right).$$

From (2.17), we find that

$$1 - 8\sum_{n=1}^{\infty}\frac{n^5(-q)^n}{1-q^n} = 1 + 8\sum_{n=1}^{\infty}\frac{n^5 q^n}{1-q^n} - 512\sum_{n=1}^{\infty}\frac{n^5 q^{2n}}{1-q^{2n}}$$

(5.2)
$$= \frac{64}{63}E_6(\tau) - \frac{1}{63}E_6(\tau/2).$$

Therefore, identity (5.1) can be written as

(5.3)
$$\frac{64}{63}E_6(\tau) - \frac{1}{63}E_6(\tau/2) = \theta_4^8(0|\tau)\left(\theta_2^4(0|\tau) + \theta_3^4(0|\tau)\right).$$

Replacing τ by $-1/\tau$ and then using modular transformation formulas, (2.10) and (2.18), we obtain

(5.4)
$$\frac{64}{63}E_6(2\tau) - \frac{64}{63}E_6(\tau) = \theta_2^8(0|\tau)\left(\theta_3^4(0|\tau) + \theta_4^4(0|\tau)\right).$$

Adding the above two equations and using (3.5) and (3.6) in the resulting equation, we have

$$\frac{64}{63}E_6(2\tau) - \frac{1}{63}E_6(\tau/2)$$
$$= \left\{\frac{64}{63}E_6(2\tau) - \frac{64}{63}E_6(\tau)\right\} + \left\{\frac{64}{63}E_6(\tau) - \frac{1}{63}E_6(\tau/2)\right\}$$
$$= \theta_4^8(0|\tau)\theta_2^4(0|\tau) + \theta_4^8(0|\tau)\theta_3^4(0|\tau) + \theta_2^8(0|\tau)\theta_3^4(0|\tau) + \theta_2^8(0|\tau)\theta_4^4(0|\tau)$$
$$= \theta_3^4(0|\tau)\left(\theta_2^8(0|\tau) + \theta_4^8(0|\tau)\right) + \theta_2^4(0|\tau)\theta_4^4(0|\tau)(\theta_2^4(0|\tau) + \theta_4^4(0|\tau))$$
$$= \theta_3^4(0|\tau)\left(\theta_3^8(0|\tau) - 2\theta_2^4(0|\tau)\theta_4^4(0|\tau)\right) + \theta_2^4(0|\tau)\theta_3^4(0|\tau)\theta_4^4(0|\tau)$$
(5.5)
$$= \theta_3^{12}(0|\tau) - \theta_2^4(0|\tau)\theta_3^4(0|\tau)\theta_4^4(0|\tau).$$

Therefore, by (2.5), the first identity in (2.9), and (5.5), we have

(5.6) $$\theta_3^{12}(0|\tau) = 16q(q^2;q^2)_\infty^{12} + 1 + 8\sum_{n=1}^\infty \frac{n^5 q^n}{1-q^n} - 512\sum_{n=1}^\infty \frac{n^5 q^{4n}}{1-q^{4n}}.$$

By using (2.17), we find that

(5.7) $$1 + 8\sum_{n=1}^\infty \frac{n^5 q^n}{1-q^n} - 512\sum_{n=1}^\infty \frac{n^5 q^{2n}}{1-q^{2n}} = 1 + 8\sum_{n=1}^\infty \left(\sigma_5(n) - 64\sigma_5(\frac{n}{4})\right)q^n.$$

We define integers $a(n)$ by

(5.8) $$q(q^2;q^2)_\infty^{12} = \sum_{n=1}^\infty a(n)q^n.$$

Substituting (5.7), (5.8), and $\theta_3^{12}(0|\tau) = \sum_{n=0}^\infty r_{12}(n)q^n$ into (5.6) and equating like powers of q to obtain Glaisher's formula for $r_{12}(n)$ [18, p. 242, eq. (7.4.30)]

(5.9) $$r_{12}(n) = 16a(n) + 8\sigma_5(n) - 512\sigma_5(\frac{n}{4}).$$

Replacing q by $-q$, we find that identity (5.6) becomes

$$\theta_4^{12}(0|\tau) = -16q(q^2;q^2)_\infty^{12} + 1 - 8\sum_{n=1}^\infty \frac{n^5 q^n}{1-q^n}$$
(5.10)
$$+ 528\sum_{n=1}^\infty \frac{n^5 q^{2n}}{1-q^{2n}} - 1024\sum_{n=1}^\infty \frac{n^5 q^{4n}}{1-q^{4n}}.$$

Using (2.11) and (2.17), we find the identity above can be rewritten as

(5.11) $$\theta_4^{12}(0|\tau) = -16\eta^{12}(\tau) + \frac{1}{63}E_6(\tau/2) - \frac{66}{63}E_6(\tau) + \frac{128}{63}E_6(2\tau).$$

Replacing τ by $-1/2\tau$ in the identity above, using (2.10), (2.12), and (2.18), we obtain

(5.12) $$2^6\theta_2^{12}(0|2\tau) = -2^{10}\eta^{12}(2\tau) - \frac{2^{12}}{63}E_6(4\tau) + \frac{33 \times 2^7}{63}E_6(2\tau) - \frac{2^7}{63}E_6(\tau).$$

Replacing q^2 as q, and performing a considerable simplification, using (2.4) and (2.11), we obtain

(5.13) $$256q^3\psi^{12}(q^2) = -q(q^2;q^2)_\infty^{12} + \sum_{n=0}^\infty \frac{(2n+1)^5 q^{2n+1}}{1-q^{2(2n+1)}}.$$

By expanding the summands into geometric series and inverting the order of summation, we find that

$$\sum_{n=0}^{\infty} \frac{(2n+1)^5 q^{2n+1}}{1-q^{2(2n+1)}} = \sum_{n=0}^{\infty} \sigma_5(2n+1) q^{2n+1}. \tag{5.14}$$

Substituting (2.4), (5.8), and $q^3 \psi^{12}(q^2) = \sum_{n=0}^{\infty} t_{12}(n) q^{2n+3}$ into the equation above and then equating the coefficients of q^n in the resulting equation we obtain the following result of Ono, Robins and Wahl [14]

$$256 t_{12}(n) = \sigma_5(2n+3) - a(2n+3). \tag{5.15}$$

6. Formulas for $r_{16}(n)$ and $t_{16}(n)$

Substituting (3.1), (3.4), and (5.1) into (4.5) and using Jacobi's identity, (3.5), in the resulting equation, we obtain the following striking Lambert series identity associated with theta functions $\theta_2(0|\tau), \theta_3(0|\tau)$, and $\theta_4(0|\tau)$:

$$17 + 32 \sum_{n=1}^{\infty} \frac{n^7 (-q)^n}{1-q^n}$$
$$= \theta_4^8(0|\tau) \left\{ 8(\theta_2^4(0|\tau) + \theta_3^4(0|\tau))^2 + 9\theta_4^8(0|\tau) \right\}$$
$$= \theta_4^8(0|\tau) \left\{ 8(\theta_3^4(0|\tau) - \theta_2^4(0|\tau))^2 + 32\theta_2^4(0|\tau)\theta_3^4(0|\tau) + 9\theta_4^8(0|\tau) \right\}$$
$$= \theta_4^8(0|\tau) \left\{ 17\theta_4^8(0|\tau) + 32\theta_2^4(0|\tau)\theta_3^4(0|\tau) \right\}$$
$$= 17\theta_4^{16}(0|\tau) + 32\theta_2^4(0|\tau)\theta_3^4(0|\tau)\theta_4^8(0|\tau). \tag{6.1}$$

Therefore, by the third identity in (2.9) and (6.1),

$$17 \theta_4^{16}(0|\tau) = 17 + 32 \sum_{n=1}^{\infty} \frac{n^7 (-q)^n}{1-q^n} - 512 q (q;q)_\infty^8 (q^2;q^2)_\infty^8. \tag{6.2}$$

We define $\Theta(n)$ as the coefficient of q^n in

$$q(q;q)_\infty^8 (q^2;q^2)_\infty^8 = \sum_{n=1}^{\infty} \Theta(n) q^n. \tag{6.3}$$

Substituting (2.6), (6.3), and

$$17 + 32 \sum_{n=1}^{\infty} \frac{n^7 (-q)^n}{1-q^n} = 17 + 32 \sum_{n=1}^{\infty} \left\{ 2^8 \sigma_7(\frac{n}{2}) - \sigma_7(n) \right\} q^n,$$

into (6.2) and equatinging the coefficients of q^n in resulting equation, we obtain the following formula of Glaisher [18, p. 242, eq. (7.4.32)]:

$$r_{16}(n) = \frac{32}{17}(-1)^n \left\{ 2^8 \sigma_7(\frac{n}{2}) - \sigma_7(n) \right\} + (-1)^{n-1} \frac{512}{17} \Theta(n). \tag{6.4}$$

By (2.17), identity (6.2) can be written as

$$\frac{256}{15} E_8(\tau) - \frac{1}{15} E_8(\tau/2) = 512 \eta^8(\tau/2) \eta^8(\tau) + 17 \theta_4^{16}(0|\tau). \tag{6.5}$$

Replacing τ by $-1/\tau$, and using (2.10), (2.12), and (2.18), we find that

$$\frac{1}{15}(E_8(\tau) - E_8(2\tau)) = 17 \times 2^8 q^4 \psi^{16}(q^2) + 32 \eta^8(2\tau) \eta^8(4\tau). \tag{6.6}$$

Replacing q^2 as q we obtain

(6.7) $$\sum_{n=1}^{\infty} \frac{n^7 q^n}{1-q^n} - \sum_{n=1}^{\infty} \frac{n^7 q^{2n}}{1-q^{2n}} = 136 q^2 \psi^{16}(q) + q(q;q)_\infty^8 (q^2;q^2)_\infty^8.$$

Equating coefficients of q^n on both sides of (6.7), and using (1.5), (6.3), and (2.17), we find that

(6.8) $$136 t_{16}(n-2) = \sigma_7(n) - \sigma_7\left(\frac{n}{2}\right) - \Theta(n).$$

7. Formulas for $r_{20}(n)$ and $t_{20}(n)$

Substituting (3.1), (3.4), (5.1), and (6.1) into (4.6), we obtain

(7.1) $$31 - 8 \sum_{n=1}^{\infty} \frac{n^9 (-q)^n}{1-q^n} = \theta_4^8(0|\tau) \left(\theta_2^4(0|\tau) + \theta_3^4(0|\tau)\right) \times \left(31 \theta_4^8(0|\tau) + 16 \theta_2^4(0|\tau) \theta_3^4(0|\tau)\right).$$

The above equation can be rewritten as

(7.2) $$\frac{2^{10}}{33} E_{10}(\tau) - \frac{1}{33} E_{10}(\tau/2) = \theta_4^8(0|\tau) \left(\theta_2^4(0|\tau) + \theta_3^4(0|\tau)\right) \times \left(31 \theta_4^8(0|\tau) + 16 \theta_2^4(0|\tau) \theta_3^4(0|\tau)\right).$$

Replacing τ by $-1/\tau$, and using (2.10) and (2.18), we obatin

(7.3) $$\frac{2^{10}}{33} \left(E_{10}(2\tau) - E_{10}(\tau)\right) = \theta_2^8(0|\tau) \left(\theta_3^4(0|\tau) + \theta_4^4(0|\tau)\right) \times \left(31 \theta_2^8(0|\tau) + 16 \theta_3^4(0|\tau) \theta_4^4(0|\tau)\right).$$

Adding the above equations, and using (3.5), (3.7), and (3.8) in the resulting equation, we find that

(7.4) $$\begin{aligned} & \frac{2^{10}}{33} E_{10}(2\tau) - \frac{1}{33} E_{10}(\tau/2) \\ &= 31 \theta_3^4(0|\tau) \left(\theta_2^{16}(0|\tau) + \theta_4^{16}(0|\tau)\right) \\ &\quad + 31 \theta_2^4(0|\tau) \theta_4^4(0|\tau) \left(\theta_2^{12}(0|\tau) + \theta_4^{12}(0|\tau)\right) \\ &\quad + 16 \theta_3^{12}(0|\tau) \theta_2^4(0|\tau) \theta_4^4(0|\tau) \\ &\quad + 32 \theta_3^4(0|\tau) \theta_2^8(0|\tau) \theta_4^8(0|\tau) \\ &= 31 \theta_3^{20}(0|\tau) - 77 \theta_3^{12}(0|\tau) \theta_2^4(0|\tau) \theta_4^4(0|\tau) \\ &\quad + \theta_3^4(0|\tau) \theta_2^8(0|\tau) \theta_4^8(0|\tau). \end{aligned}$$

Therefore, by (2.17), we have

(7.5) $$\begin{aligned} 31 \theta_3^{20}(0|\tau) &= 77 \theta_3^{12}(0|\tau) \theta_2^4(0|\tau) \theta_4^4(0|\tau) - \theta_3^4(0|\tau) \theta_2^8(0|\tau) \theta_4^8(0|\tau) \\ &\quad + 31 + 8 \sum_{n=1}^{\infty} \frac{n^9 q^n}{1-q^n} - 8192 \sum_{n=1}^{\infty} \frac{n^9 q^{4n}}{1-q^{4n}} \\ &= \frac{2480}{3} B(\tau) + \frac{1216}{3} B^*(\tau) \\ &\quad + 31 + 8 \sum_{n=1}^{\infty} \frac{n^9 q^n}{1-q^n} - 8192 \sum_{n=1}^{\infty} \frac{n^9 q^{4n}}{1-q^{4n}}, \end{aligned}$$

where

$$B(\tau) = \sum_{n=1}^{\infty} b(n) q^n$$

$$= \frac{1}{16} \theta_2^4(0|\tau) \theta_3^4(0|\tau) \theta_4^4(0|\tau) \left(\theta_3^8(0|\tau) - \theta_2^4(0|\tau) \theta_4^4(0|\tau) \right)$$

(7.6) $$= q + 228 q^3 + \cdots,$$

$$B^*(\tau) = \sum_{n=1}^{\infty} b^*(n) q^n$$

$$= \frac{1}{16} \theta_2^4(0|\tau) \theta_3^4(0|\tau) \theta_4^4(0|\tau) \left(\theta_3^8(0|\tau) + 2 \theta_2^4(0|\tau) \theta_4^4(0|\tau) \right)$$

(7.7) $$= q + 48 q^2 - 156 q^3 + 768 q^4 + \cdots.$$

Equating the coefficients of q^n in (7.5) gives the identity [19]

(7.8) $$r_{20}(n) = \frac{8}{31} \sigma_9(n) - \frac{8192}{31} \sigma_9\left(\frac{n}{4}\right) + \frac{2480}{93} b(n) + \frac{1216}{93} b^*(n).$$

Replacing q by $-q$ in (7.5), we have

$$31 \theta_4^{20}(0|\tau) = -1232 q (q;q)_\infty^{16} (q^2;q^2)_\infty^4 - 256 q^2 \frac{(q^2;q^2)_\infty^{28}}{(q;q)_\infty^8}$$

$$+ 31 - 8 \sum_{n=1}^{\infty} \frac{n^9 q^n}{1 - q^n} + 8202 \sum_{n=1}^{\infty} \frac{n^9 q^{2n}}{1 - q^{2n}}$$

(7.9) $$- 2^{14} \sum_{n=1}^{\infty} \frac{n^9 q^{4n}}{1 - q^{4n}}.$$

Using (2.11) and (2.17), we can write (7.9) in the form

$$\frac{1}{33} E_{10}(\tau/2) - \frac{1026}{33} E_{10}(\tau) + \frac{2048}{33} E_{10}(2\tau)$$

(7.10) $$= 31 \theta_4^{20}(0|\tau) + 1232 \eta^{16}(\tau/2) \eta^4(\tau) + 256 \frac{\eta^{28}(\tau)}{\eta^8(\tau/2)}.$$

Replacing τ by $-1/2\tau$, using (2.12) and (2.18), we find that

$$-\frac{2^{10}}{33} E_{10}(4\tau) + \frac{1026}{33} E_{10}(2\tau) - \frac{2}{33} E_{10}(\tau)$$

(7.11) $$= 31 \theta_2^{20}(0|2\tau) + 1232 \times 2^8 \eta^4(2\tau) \eta^{16}(4\tau) + 16 \frac{\eta^{28}(2\tau)}{\eta^8(4\tau)}.$$

Replacing q^2 as q and using (2.4) and (2.11) in the resulting equation, we obtain

$$31 \times 2^{16} q^5 \psi^{20}(q^2) = \sum_{n=0}^{\infty} \frac{(2n+1)^9 q^{2n+1}}{1 - q^{2(2n+1)}}$$

(7.12) $$- 77 \times 2^8 q^3 (q^2;q^2)_\infty^4 (q^4;q^4)_\infty^{16} - q \frac{(q^2;q^2)_\infty^{28}}{(q^4;q^4)_\infty^8}.$$

We define $c(n)$ as the coefficient of q^n in

(7.13) $$q^3 (q^2;q^2)^4 (q^4;q^4)_\infty^{16} = \sum_{n=0}^{\infty} c(n) q^{2n+3},$$

and $c^*(n)$ as the coefficient of q^n in

(7.14) $$q\frac{(q^2;q^2)_\infty^{28}}{(q^4;q^4)_\infty^8} = \sum_{n=0}^\infty c^*(n)q^{2n+1},$$

Equating coefficients of q^{2n+5} in (7.12) to conclude that

(7.15) $$31 \times 2^{16}t_{20}(n) = \sigma_9(2n+5) - 77 \times 2^8 c(n+1) - c^*(n+2).$$

8. formulas for $r_{24}(n)$ and $t_{24}(n)$

Substituting (3.1), (3.4), (5.1), (6.1), and (7.1) into (4.7), using (3.6) and (2.9), we find that

$$691 + 16\sum_{n=1}^\infty \frac{n^{11}(-q)^n}{1-q^n}$$
$$= 2\theta_4^8(0|\tau)\left\{\theta_2^4(0|\tau) + \theta_3^4(0|\tau)\right\}^2 \left\{167\theta_4^8(0|\tau) + 32\theta_2^4(0|\tau)\theta_3^4(0|\tau)\right\}$$
$$+ 21\theta_4^{16}(0|\tau)\left\{17\theta_4^8(0|\tau) + 32\theta_2^4(0|\tau)\theta_3^4(0|\tau)\right\}$$
$$= 2\theta_4^8(0|\tau)\left\{4\theta_2^4(0|\tau)\theta_3^4(0|\tau) + \theta_4^8(0|\tau)\right\} \left\{167\theta_4^8(0|\tau) + 32\theta_2^4(0|\tau)\theta_3^4(0|\tau)\right\}$$
$$+ 21\theta_4^{16}(0|\tau)\left\{17\theta_4^8(0|\tau) + 32\theta_2^4(0|\tau)\theta_3^4(0|\tau)\right\}$$
$$= 691\theta_4^{24}(0|\tau) + 2072\theta_2^4(0|\tau)\theta_3^4(0|\tau)\theta_4^{16}(0|\tau) + 256\theta_2^8(0|\tau)\theta_3^8(0|\tau)\theta_4^8(0|\tau)$$
(8.1) $$= 691\theta_4^{24}(0|\tau) + 33152q(q;q)_\infty^{24} + 65536q^2(q^2;q^2)_\infty^{24}.$$

Therefore, we obtain the identity [**9**, p. 153, eq. (9.16.1)], [**18**, p. 243, eq. (7.4.37)]

(8.2) $$\theta_4^{24}(0|\tau) = 1 + \frac{16}{691}\sum_{n=1}^\infty \frac{n^{11}(-q)^n}{1-q^n} - \frac{33152}{691}q(q;q)_\infty^{24} - \frac{65536}{691}q^2(q^2;q^2)_\infty^{24}.$$

Using (2.17), we find that

$$1 + \frac{16}{691}\sum_{n=1}^\infty \frac{n^{11}(-q)^n}{1-q^n}$$
$$= 1 + \frac{16}{691}\left\{2^{11}\sum_{n=1}^\infty \frac{n^{11}q^{2n}}{1-q^{2n}} - \sum_{n=1}^\infty \frac{n^{11}q^n}{1-q^n}\right\}$$
$$= 1 + \frac{16}{691}\left\{2^{12}\sum_{n=1}^\infty \sigma_{11}(n)q^{2n} - \sum_{n=1}^\infty \sigma_{11}(n)q^n\right\}$$
(8.3) $$= 1 + \frac{16}{691}\sum_{n=1}^\infty \left\{2^{12}\sigma_{11}\left(\frac{n}{2}\right) - \sigma_{11}(n)\right\}q^n.$$

We recall the Ramanujan $\tau(n)$-function [**18**, p. 197, eq. (6.1.13)] defined as the coefficient of q^n in

(8.4) $$q(q;q)_\infty^{24} = \sum_{n=1}^\infty \tau(n)q^n.$$

Then

$$(8.5) \qquad q^2(q^2;q^2)_\infty^{24} = \sum_{n=1}^\infty \tau(\frac{n}{2})q^n,$$

if we agree that $\tau(x)$ is 0 when x is not an integer. Substituting (8.3), (8.4), (8.5), and $\theta_4^{24}(0|\tau) = \sum_{n=0}^\infty r_{24}(n)(-q)^n$ into (8.2) and then equating coefficients of q^n we obtain Ramanujan's formula [9, p.155, eq.(9.17.1)], [18, p.241, eq.(7.4.37)]

$$(8.6) \qquad \begin{aligned} r_{24}(n) &= (-1)^n \frac{16}{691} \left\{ 2^{11} \sigma_{11}(\frac{n}{2}) - \sigma_{11}(n) \right\} \\ &+ \frac{128}{691} \left\{ (-1)^{n-1} 259\tau(n) - 512\tau(\frac{n}{2}) \right\}. \end{aligned}$$

Using (2.11) and (2.17), we can write (8.2) in the form

$$(8.7) \quad \theta_4^{24}(0|\tau) = \frac{2^{12}}{4095} E_{12}(\tau) - \frac{1}{4095} E_{12}(\tau/2) - \frac{33152}{691} \eta^{24}(\tau/2) - \frac{65536}{691} \eta^{24}(\tau).$$

Replacing τ by $-1/\tau$, and using (2.12) and (2.18), we find that

$$(8.8) \quad \theta_2^{24}(0|\tau) = \frac{2^{12}}{4095} E_{12}(\tau) - \frac{2^{12}}{4095} E_{12}(2\tau) - \frac{33152 \times 2^{12}}{691} \eta^{24}(2\tau) - \frac{65536}{691} \eta^{24}(\tau).$$

Using (2.4), (2.11), and (2.17), we obtain

$$(8.9) \qquad 176896 q^3 \psi^{24}(q) = \sum_{n=1}^\infty \frac{n^{11} q^n}{1-q^{2n}} - q(q;q)_\infty^{24} + 2072 q^2 (q^2;q^2)_\infty^{24}.$$

We note that

$$(8.10) \qquad \begin{aligned} \sum_{n=1}^\infty \frac{n^{11} q^n}{1-q^{2n}} &= \sum_{n=1}^\infty \frac{n^{11} q^n}{1-q^n} - \sum_{n=1}^\infty \frac{n^{11} q^{2n}}{1-q^{2n}} \\ &= \sum_{n=1}^\infty \left\{ \sigma_{11}(n) - \sigma_{11}(\frac{n}{2}) \right\} q^n. \end{aligned}$$

Substituting (8.3), (8.4), (8.10), and $\psi^{24}(q) = \sum_{n=0}^\infty t_{24}(n) q^n$ into (8.9), equating coefficients of q^n, we obtain the formula for $t_{24}(n)$, which is equivalent to the identity [14, Theorem 8]

$$(8.11) \qquad 176896 t_{24}(n-3) = \sigma_{11}(n) - \sigma_{11}(\frac{n}{2}) - \tau(n) - 2072\tau\left(\frac{n}{2}\right).$$

When $n = p^k$ where p is an odd prime, the identity above reduces to the identity [14]

$$(8.12) \qquad 176896 t_{24}(p^k - 3) = \sigma_{11}(p^k) - \tau(p^k).$$

Acknowledgements. I am grateful for Bruce C. Berndt for his many helpful criticisms and suggestions leading to an improvement of a earlier version of this paper. I also would like to thank Stephen C. Milne and Li-Chien Shen for their comments.

References

[1] G. E. Andrews, *Applications of basic hypergeometric function*, SIAM Review, 16 (1974), 441-484.
[2] G. E. Andrews, R. P. Lewis and Zhi-Guo Liu, *An identity relating a theta function to a sum of Lambert series*, Bull. London Math. Soc., 33 (2001), 25-31.
[3] R. Askey, *The number of representations of an integer as the sum of two squares*, Indian Journal of Mathematics, 32 (1990), 187-191.
[4] B. C. Berndt, *Ramanujan's Notebooks, Part III*, Springer-Verlag, New York, 1991.
[5] B. C. Berndt, *Ramanujan's theory of theta-functions*, Theta Functions, From the Classical to the Modern (M. Ram Murty, ed.), Centre de Recherches. Mathematiques Proceedings and Lecture Notes, Amer. Math. Soc., Providence, RI, 1993, pp 1-63.
[6] B. C. Berndt, *Fragments by Ramanujan on Lambert series*, Number Theory and Its Applications (Kyoto, 1997) (K. Györy and S Kanemitsu, eds.), vol. 2 of Dev. Math., Kluwer, Dordrecht, 1999, pp. 35-49.
[7] K. Chandrasekharan, *Elliptic Functions*, Springer-Verlag, New York, 1985.
[8] Paul Hammond, Richard Lewis and Zhi-Guo Liu, *Hirschhorn's Identities*, Bull. Austral. Math. Soc., 60 (1999), 73-80.
[9] G. H. Hardy, *Ramanujan*, Cambridge University Press, Cambridge, 1940.
[10] V. G. Kac and M. Wakimoto, *Integrable highest weight modules over affine superalgebras and number theory*, Lie Theory and Geometry (J. -L. Brylinski, R. Brylinski, V. Guillemin, and V. Kac, eds.), Birkhäuser, Boston, 1994.
[11] N. Koblitz, *Introduction to Elliptic Curves and Modular Forms*. Springer-Verlag, New York, 1984.
[12] S. C. Milne, *New infinite families of exact sums of squares formulas, Jacobi elliptic functions, and Ramanujan's tau function*, Proc. Natl. Acad. Sci. USA 93 (1996), 15004-15008.
[13] S. C. Milne, *Infinite families of exact sums of squares formulas, Jacobi elliptic functions, continued fractions, and Schur functions*, Preprint; arXiv:math.NT/0008068 (8-5-2000) (to appear in the Ramanujan Journal).
[14] K. Ono, S. Robins, and P. T. Wahl, *On the reprensentation of integers as sum of triangular numbers*, Aequations Mathematicae, 50 (1995), 73-94.
[15] H. Rademacher, *Topics in Analytic Number Theory*. vol. 169 of Grundlehren Math. Wiss., Springer-Verlag, New York, 1973.
[16] S. Ramanujan, *On certain arithmical functions*. Trans. Cambridge philos. Soc. 22(1916), 159-184.
[17] S. Ramanujan, *Collected Papers*, Chelsea, New York, 1962.
[18] R. A. Rankin, *Modular Forms and Functions*, Cambridge University Press, Cambridge, 1977.
[19] R. A. Rankin, *On the representation of a number as the sum of any number of squares, and in particular of twenty*, Acta Arith. 7 (1962), 399-407.
[20] L. -C. Shen, *On the logarithmic derivative of a theta function and a fundamental identity of Ramanujan*, J. Math. Anal. Appl. 177 (1993), 299-307.

XINXIANG EDUCATION COLLEGE, XINXIANG, HENAN 453000, PEOPLE'S REPUBLIC OF CHINA
 Current address: Nanjing Institute of Meteorology, Nanjing 210044, People's Republic of China
 E-mail address: liuzg18@hotmail.com

3-Regular Partitions and a Modular K3 Surface

Jeremy Lovejoy and David Penniston

1. Introduction

A k-regular partition of n ($k > 1$) is a non-increasing sequence of positive integers whose sum is n, with the condition that no summand is divisible by k. We denote the number of k-regular partitions of n by $b_k(n)$, and follow the convention that $b_k(0) = 1$. Elementary techniques in the theory of partitions [3] give the generating functions

$$(1.1) \qquad \sum_{n=0}^{\infty} b_k(n) q^n = \prod_{n=1}^{\infty} \left(\frac{1 - q^{kn}}{1 - q^n} \right).$$

In classical representation theory, k-regular partitions of n label irreducible k-modular representations of the symmetric group S_n when k is prime [8]. More recently, such partitions have been studied for their arithmetic properties in connection with the theory of modular forms and Galois representations [1, 6, 10, 11, 12]. Although one may presumably use the ideas from [1, 10] to study the k-regular partitions modulo any prime, more focus has been placed on the most straightforward case, the p-adic behavior of p^j-regular partitions. For example, we have

THEOREM 1 (Gordon-Ono [6]). *If $S(p, j, a)$ denotes the set of natural numbers n such that $b_{p^j}(n)$ is not divisible by p^a, then $S(p, j, a)$ has arithmetic density 0.*

In general there is no elementary characterization of the sets $S(p, j, a)$, but in the best cases we do have simple congruential formulas for $b_k(n)$. For example, the classical expansions

$$(1.2) \qquad \prod_{n=1}^{\infty} (1 - q^n) = \sum_{n=-\infty}^{\infty} (-1)^n q^{n(3n+1)/2}$$

and

$$(1.3) \qquad \prod_{n=1}^{\infty} (1 - q^n)^3 = \sum_{n=0}^{\infty} (-1)^n (2n+1) q^{n(n+1)/2}$$

reveal that $b_2(n)$ is even unless $24n + 1$ is a square and $b_4(n)$ is even unless $8n + 1$ is a square. The case of $b_2(n)$ has a famous combinatorial proof by Franklin [3],

---

1991 *Mathematics Subject Classification.* 11P83.

while K. Ono and the second author [11] have determined $b_2(n)$ modulo 8 in terms of the arithmetic of $\mathbb{Z}[\sqrt{-6}]$.

Here we undertake an investigation of the 3-adic behavior of $b_3(n)$. Let

$$\eta(z) := \prod_{n=1}^{\infty}(1-q^n)$$

denote Dedekind's eta funcion, where $q := e^{2\pi i z}$. From (1.1) we have

$$\sum_{n=0}^{\infty} b_3(n) q^{12n+1} \equiv \eta^2(12z) \pmod{3},$$

where $\eta^2(12z)$ is a weight 1 modular form which is the Mellin transform of an Artin L-function for $\mathbb{Q}(i)$. Modulo 9, it turns out that the generating function for $b_3(n)$ is related to an eigenform which is essentially the Mellin transform of the "complicated factor" in the Hasse-Weil L-function for a certain $K3$ surface.

THEOREM 2. *Let X be the $K3$ surface defined by*

(1.4) $$X : s^2 = x(x+1)y(y+1)(x+8y).$$

If p is a prime such that $p \equiv 1 \pmod{12}$, then

(1.5) $$b_3\left(\frac{p-1}{12}\right) \equiv \#X(\mathbb{F}_p) - (p+1)^2 \pmod{9}.$$

Using the fact that the relevant eigenform has complex multiplication, we can use Hecke theory and the arithmetic of the Gaussian integers to build a formula for the number of 3-regular partitions modulo 9.

THEOREM 3. *Given a positive integer n, write*

$$12n + 1 = N^2 M$$

with M squarefree. For every prime divisor p of $12n+1$, set

$$k_p := ord_p(12n+1).$$

If $p \equiv 1 \pmod{12}$, let d_p and e_p be integers such that $3 \mid d_p$ and

$$p = d_p^2 + e_p^2.$$

(1) If there is a prime p such that $p \mid M$ and $p \equiv 5, 7$ or $11 \pmod{12}$, then $b_3(n) \equiv 0 \pmod{9}$.

(2) If every prime divisor p of M satisfies $p \equiv 1 \pmod{12}$, then

(1.6) $$b_3(n) \equiv (3n+1) \cdot \prod_{\substack{p \mid (12n+1) \\ p \equiv 1 \pmod{12}}} (-1)^{k_p d_p} (k_p + 1) \cdot \prod_{\substack{p \mid (12n+1) \\ p \equiv 5 \pmod{12}}} (-1)^{\frac{k_p}{2}} \pmod{9}.$$

For comparison with (1.2) and (1.3) we cite the following, which is a direct consequence of Theorem 3.

COROLLARY 4. *$b_3(n)$ is divisible by 3 unless both of the following hold:*

(i) *All prime divisors $p \equiv 5, 7, 11 \pmod{12}$ of $12n+1$ divide $12n+1$ with even order.*

(ii) *All prime divisors $p \equiv 1 \pmod{12}$ of $12n+1$ divide $12n+1$ with order not congruent to 2 modulo 3.*

EXAMPLE. If $n = 5$, then $12n + 1 = 61 = 6^2 + 5^2$, so $b_3(5)$ is not divisible by 3. More specifically, $b_3(5) \equiv 16 \cdot (-1)^{1 \cdot 6} \cdot 2 \equiv 5 \pmod 9$. Indeed, the 3-regular partitions of 5 are $5, 4+1, 2+2+1, 2+1+1+1$, and $1+1+1+1+1$.

2. Proof of Theorem 2

Let

(2.1) $$\eta^6(4z) := \sum_{n=1}^{\infty} a(n) q^n,$$

a weight 3 cusp form for the congruence subgroup $\Gamma_0(16)$ with character $\chi_{-1}(d) := \left(\frac{-1}{d}\right)$. We denote the space of such forms by $S_3(\Gamma_0(16), \chi_{-1})$ (see [9] for definitions related to modular forms). It is well-known [5] that $\eta^6(4z)$ has complex multiplication by $K = \mathbb{Q}(i)$. Specifically, let O_K denote the ring of integers of K, and let χ be the character on $(O_K/(2))^*$ defined by $\chi(i) = -1$. Extending χ to the set of all elements of K^* prime to (2), we find that for $d + ei \in O_K$ with $d + e$ odd, $\chi(d + ei) = (-1)^e$. Denote by c the Hecke character on K with conductor (2) and exponent 2 given by

(2.2) $$c((d+ei)) = \chi(d+ei)(d+ei)^2.$$

Then

(2.3) $$\eta^6(4z) = \sum c(I) q^{N(I)},$$

where the sum is over ideals I of O_K prime to (2).

This form is the fundamental object in our work, as it relates 3-regular partitions, the $K3$ surface (1.4), and the arithmetic of the Gaussian integers.

Proof of Theorem 2. Let
$$F(z) := \frac{\eta^8(12z)}{\eta^2(36z)},$$
which is easily seen to be a modular form in $S_3(\Gamma_0(1296), \chi_{-1})$ (see [10], for example). From (1.1) and the fact that
$$\frac{\eta^9(z)}{\eta^3(3z)} \equiv 1 \pmod 9,$$
we have
$$\sum_{n=0}^{\infty} b_3(n) q^{12n+1} \equiv F(z) \pmod 9.$$

By definition, $a(n) = 0$ unless $n \equiv 1 \pmod 4$, and therefore

(2.4) $$\frac{1}{2} \sum_{n=1}^{\infty} \left(\left(\frac{n}{3}\right) a(n) + \left(\frac{n}{3}\right)\left(\frac{n}{3}\right) a(n) \right) q^n = \sum_{n \equiv 1 \pmod{12}} a(n) q^n.$$

From [9], p. 127, (2.4) is a modular form in $S_3(\Gamma_0(1296), \chi_{-1})$. By computation, the first 648 coefficients of $F(z)$ and (2.4) are equivalent modulo 9, and hence by a theorem of Sturm [13] we have for every n,

(2.5) $$b_3(n) \equiv a(12n+1) \pmod 9.$$

To complete the proof, we recall the modularity of the surface (1.4) [**2**]. For every prime $p \geq 5$, we have
$$\#X(\mathbb{F}_p) = 1 + p^2 + 20p + a(p). \tag{2.6}$$
□

REMARK. Since the L-series for X is the symmetric square of the L-series for the congruent number elliptic curve given by the equation $E : y^2 = x^3 - x$ [**2**], the congruence (2.5) is dictated by Galois actions on certain points on E. Specifically, let $g(n)$ denote the Fourier coefficients of the associated eigenform:

$$\eta^2(4z)\eta^2(8z) := \sum_{n=1}^{\infty} g(n)q^n.$$

Then for every prime $p \geq 5$, $a(p) = g(p)^2 - 2p$. Denote by $G_\mathbb{Q}$ the absolute Galois group of $\mathbb{Q}$, and by $E[n]$ the group of n-division points of E for any $n \geq 1$ (as a group, $E[n] \cong (\mathbb{Z}/n\mathbb{Z})^2$). If ℓ is prime, $G_\mathbb{Q}$ acts on the Tate module

$$T_\ell(E) = \varprojlim_m E[\ell^m] \cong \mathbb{Z}_\ell \times \mathbb{Z}_\ell,$$

and therefore we obtain a representation

$$\rho_\ell : G_\mathbb{Q} \to GL_2(\mathbb{Z}_\ell).$$

If frob_p denotes a Frobenius element for p ($p \neq \ell$), then $trace(\rho_\ell(\text{frob}_p)) = g(p)$. With (2.5), this shows that the behavior of $b_3(n)$ modulo 9 is determined by the Galois action on the 3-division points of E.

3. Proof of Theorem 3

Since $\eta^6(4z) = \sum_{n=1}^{\infty} a(n)q^n \in S_3(\Gamma_0(16), \chi_{-1})$ is a Hecke eigenform, we have that

$$a(mn) = a(m)a(n) \qquad \text{if } (m,n) = 1 \tag{3.1}$$

and

$$a(p^{k+1}) = a(p)a(p^k) - \chi_{-1}(p)a(p^{k-1})p^2 \qquad \text{if } p \geq 5 \text{ is prime and } k \geq 0. \tag{3.2}$$

In light of (2.5), (3.1), and (3.2), we begin by studying the $a(p)$ for p prime.

PROPOSITION 5. *Let p be an odd prime.*
(1) If $p \equiv 3 \pmod{4}$, then $a(p) = 0$.
(2) If $p \equiv 5 \pmod{12}$, then $3 \mid a(p)$.
(3) If $p \equiv 1 \pmod{12}$ and we write $p = d_p^2 + e_p^2$ with $3 \mid d_p$, then $a(p) \equiv (-1)^{d_p} \cdot 2p \pmod{9}$.

PROOF. For *(1)*, see (2.1), or recall (2.3) and note that since (p) is prime in O_K, there are no ideals of norm p in O_K.

Now suppose $p \equiv 1 \pmod{4}$. Then there are integers d_p and e_p with $p = d_p^2 + e_p^2$, and hence the prime ideals of O_K of norm p are $(d_p \pm e_p i)$. Since $\chi(d_p \pm e_p i) = (-1)^{e_p}$, (2.2) and (2.3) give us that

$$a(p) = (-1)^{e_p}(2d_p^2 - 2e_p^2) = (-1)^{e_p}(4d_p^2 - 2p). \tag{3.3}$$

If $p \equiv 5 \pmod{12}$, then since $p \equiv 2 \pmod 3$, it follows that $3 \nmid d_p e_p$. Hence $d_p^2 \equiv e_p^2 \equiv 1 \pmod 3$, and the proof of (2) is complete.

To finish the proof of (3), if $p \equiv 1 \pmod{12}$, then $3 \mid d_p e_p$. We assume without loss that $3 \mid d_p$. Then by (3.3),

$$a(p) \equiv (-1)^{e_p+1} \cdot 2p = (-1)^{d_p} \cdot 2p \pmod 9.$$

$\square$

Combining Proposition 5 with (3.2), it is straightforward induction to show

PROPOSITION 6. *Let p be an odd prime, k a positive integer.*
(1) If $p \equiv 3 \pmod 4$, then $a(p^{2k-1}) = 0$ and $a(p^{2k}) \equiv p^{2k} \pmod 9$.
(2) If $p \equiv 5 \pmod{12}$, then $3 \mid a(p^{2k-1})$ and $a(p^{2k}) \equiv (-p^2)^k \pmod 9$.
(3) If $p \equiv 1 \pmod{12}$ and $p = d_p^2 + e_p^2$ with $3 \mid d_p$, then $a(p^k) \equiv (-1)^{kd_p}(k+1)p^k$ (mod 9).

Theorem 3 follows now from (2.5), (3.1), and Proposition 6.

We have not observed any simple congruence condition which determines the parity of d_p as a function of p, which is tantamount to distinguishing between primes of the form $x^2 + 36y^2$ and those of the form $4x^2 + 9y^2$. In this direction it is known [4] that for all but finitely many primes $p \equiv 1 \pmod 4$, p is represented by $x^2 + 36y^2$ if and only if the minimal polynomial for $j(\sqrt{-36})$ has a root modulo p.

4. Concluding remarks

Since the generating functions for partition theoretic objects are typically products and quotients of the η function, connections to objects in arithmetic geometry such as that given by Theorem 2 are not unexpected. A striking example of this is in recent work of L. Guo and K. Ono [7], where it is shown that values of the ordinary partition function reveal structure of Tate-Shafarevich groups of motives of modular forms. In our case, an examination of, for instance, the five 3-regular partitions of 5 and the 4920 $\mathbb{F}_{61}$-points on our $K3$ surface gives one little reason to expect that there is something in the combinatorics of 3-regular partitions or irreducible 3-modular representations of S_n that is related to the structure of modular surfaces or the arithmetic of $\mathbb{Q}(i)$. We must for now be content that the theory of modular forms is a meeting place for diverse mathematical objects whose connections often cannot be otherwise explained.

References

[1] S. Ahlgren and J. Lovejoy, The arithmetic of the number of partitions into distinct parts, *Mathematika*, to appear.
[2] S. Ahlgren, K. Ono, and D. Penniston, Zeta functions of an infinite family of $K3$ surfaces, *Amer. J. Math.*, to appear.
[3] G. E. Andrews, "The Theory of Partitions," Cambridge Univ. Press, Cambridge, 1998.
[4] D. A. Cox, "Primes of the form $x^2 + ny^2$", John Wiley and Sons, Inc., 1989.
[5] D. Dummit, H. Kisilevsky, and J. McKay, Multiplicative products of η-functions, *Contemp. Math.* **45** (Finite groups - coming of age), 89-98.
[6] B. Gordon and K. Ono, The divisibility of certain partition functions, *Ramanujan J.* **1** (1997), 25-34.
[7] L. Guo and K. Ono, The partition function and the arithmetic of certain modular L-functions, *Internat. Math. Res. Notices* **21** (1999), 1179-1197.
[8] G. James and A. Kerber, "The Representation of the Symmetric Group," Addison-Wesley, Reading, 1979.

[9] N. Koblitz, "Introduction to Elliptic Curves and Modular Forms," Springer-Verlag, New York, 1984.
[10] J. Lovejoy, Divisibility and distribution of partitions into distinct parts, *Adv. Math.* **158** (2001), 253-263.
[11] K. Ono and D. Penniston, The 2-adic behavior of the number of partitions into distinct parts, *J. Combin. Theory Ser. A* **92** (2000), 138-157.
[12] D. Penniston, The p^a-regular partition function modulo p^j, preprint.
[13] J. Sturm, On the congruence of modular forms, *Lect. Notes in Math.* **1240**, Springer-Verlag, Ner York, 1984, 275-280.

DEPARTMENT OF MATHEMATICS, UNIVERSITY OF WISCONSIN, MADISON, WI 53706
E-mail address: lovejoy@math.wisc.edu

DEPARTMENT OF MATHEMATICS, FURMAN UNIVERSITY, GREENVILLE, SC 29613
E-mail address: david.penniston@furman.edu

A new look at Hecke's indefinite theta series

A. Polishchuk

ABSTRACT. We describe a family of modular q-series associated with indefinite binary quadratic forms. We prove that these series generate the same space of weight 1 modular forms as Hecke's indefinite theta series and study linear relations between them.

This note is devoted to the q-series of the form

$$\sum_{m\geq 0, n\geq 0} f(m,n) q^{Q(m,n)} - \sum_{m<0, n<0} f(m,n) q^{Q(m,n)}$$

where Q is an indefinite quadratic form on $\mathbb{Z}^2$, $f(m,n)$ is a doubly periodic function on $\mathbb{Z}^2$ such that the sums of $f(m,n)q^{Q(m,n)}$ over all vertical and all horizontal lines in $\mathbb{Z}^2$ vanish. Some of these series appeared as coefficients in univalued triple Massey products on elliptic curves computed via homological mirror symmetry in [**P**]. In particular, in this context the condition of vanishing of sums over vertical and horizontal lines appears to be related to the standard necessary condition of the existence of triple Massey products (the vanishing of two double products). In the present paper we generalize Theorem 3 of [**P**] which relates such series to the indefinite theta series considered by Hecke in [**H1**], [**H2**] (our approach is completely elementary and doesn't use the connection with triple products on elliptic curves). The main consequence of this relation is the modularity of our q-series. We also show that the problem of finding all linear relations between our series is related to the study of orbits of actions of dihedral groups on $(\mathbb{Z}/N\mathbb{Z})^2$.

1. Main result

1.1. Hecke's indefinite theta series. Let us recall the definition of these series. Let K be a totally real quadratic extension of $\mathbb{Q}$, i.e. K is either a field of the form $\mathbb{Q}(\sqrt{D})$ (where $D > 0$) or the algebra $\mathbb{Q} \oplus \mathbb{Q}$. We have the norm map $\mathrm{Nm}: K \to \mathbb{Q}$ (in case of $\mathbb{Q} \oplus \mathbb{Q}$ this is the product of components). Let us denote by $C \subset K$ the set of elements with positive norm. The cone C is a union of two components and we define the function $\mathrm{sign}: C \to \pm 1$ which assigns value 1 (resp. -1) on totally positive (resp. negative) elements (in the case of $\mathbb{Q} \oplus \mathbb{Q}$

1991 *Mathematics Subject Classification*. Primary 11E45, 11F11; Secondary 11E16, 11F27.
Key words and phrases. Indefinite theta series, modular form.
This work was partially supported by NSF Grant DMS-0070967.

© 2001 American Mathematical Society

"total positivity" means positivity of both components). Let us denote by $U_+(K)$ the subgroup of the multiplicative group K consisting of totally positive elements $k \in K^*$ with norm 1 (in the case of $\mathbb{Q} \oplus \mathbb{Q}$ this is the group of elements (r, r^{-1}) where $r > 0$). Note that the group of $\mathbb{Q}$-linear automorphisms of K preserving Nm decomposes as follows:

$$\operatorname{Aut}_{\mathbb{Q}}(K, \operatorname{Nm}) = \pm \operatorname{id} \times U_+(K) \times \operatorname{Gal}(K/\mathbb{Q})$$

where $U_+(K)$ acts on K by multiplication. Let $\Lambda \subset K$ be a lattice (i.e. a $\mathbb{Z}$-submodule of rank 2), $\Lambda + c$ be a coset for this lattice (where $c \in K$). Hecke's indefinite theta series is

$$\Theta_{\Lambda, c} = \sum_{\lambda \in (\Lambda + c) \cap C/G} \operatorname{sign}(\lambda) q^{d \cdot \operatorname{Nm}(\lambda)}$$

where G is the subgroup in $U_+(K)$ consisting of the elements preserving $\Lambda + c$, d is a positive rational number such that $d \operatorname{Nm}$ takes integer values on $\Lambda + c$. Hecke proved that this series is modular of weight 1 for the subgroup $\Gamma_0(n) \subset \operatorname{SL}_2(\mathbb{Z})$ with some explicit level n. [1] Note that the elements of $U_+(K)$ preserving Λ are totally positive units, hence, G is an infinite cyclic group. In particular, if we replace in the above definition G by any infinite subgroup in $U_+(K)$ preserving $\Lambda + c$ the resulting series will be an integral multiple of $\Theta_{\Lambda, c}$.

1.2. Formulation of the main theorem. Let $Q(m, n) = am^2 + 2bmn + cn^2$ be a $\mathbb{Q}$-valued indefinite quadratic form on $\mathbb{Z}^2$ (so $b^2 > ac$) which is positive on the cone $mn \geq 0$ (i.e. a, b and c are positive). Let $f(m, n)$ be a doubly periodic complex-valued function on $\mathbb{Z}^2$ (so $f(m + N, n) = f(m, n + N) = f(m, n)$ for some $N > 0$). Assume that for all m_0 and n_0 one has

$$\sum_{m \in \mathbb{Z}} f(m, n_0) q^{Q(m, n_0)} = \sum_{n \in \mathbb{Z}} f(m_0, n) q^{Q(m_0, n)} = 0$$

(i.e. all sums along horizontal and vertical lines are zero). Assume also that Q takes integer values on the support of f. Then the series

$$\Theta_{Q, f} = \sum_{m \geq 0, n \geq 0} f(m, n) q^{Q(m, n)} - \sum_{m < 0, n < 0} f(m, n) q^{Q(m, n)}$$

is modular of weight 1.

Moreover, the space of modular forms of weight 1 spanned by these series coincides with the space generated by Hecke's indefinite theta series.

1.3. Proof. Our first task is to unravel the condition that the sums along horizontal and vertical lines are zero. Let us extend the function $f(m, n)$ from $\mathbb{Z}^2$ to $\mathbb{Q}^2$ by zero. Then we claim that this condition is equivalent to the following two identities:

$$f(m, n) = -f(-\frac{2b}{a} n - m, n),$$

$$f(m, n) = -f(m, -\frac{2b}{c} m - n).$$

Indeed, this follows from the fact that Q restricted to a vertical or horizontal line assumes each value at exactly two points (sometimes coinciding, in which case the

---

[1] In the original definition of Hecke Λ was an ideal in the ring of integers, however, the same proof works for any lattice. Also, Hecke makes a concrete choice of d. For our purposes it is more convinient to allow any d such that $d \operatorname{Nm}$ takes integer values on $\Lambda + c$.

coefficient should be zero), so in order for the sum to be zero the corresponding coefficients should cancel out. Let us consider the following two operators preserving Q:

$$A = \begin{pmatrix} -1 & p \\ 0 & 1 \end{pmatrix},$$

$$B = \begin{pmatrix} 1 & 0 \\ r & -1 \end{pmatrix}.$$

where $p = -\frac{2b}{a}$, $r = -\frac{2b}{c}$. Then the conditions on f can be rewritten as

(1.1) $$f(Ax) = f(Bx) = -f(x)$$

for every $x \in \mathbb{Q}^2$. Let $S \subset \mathbb{Z}^2 \subset \mathbb{Q}^2$ be the support of f. We can assume that $f \neq 0$ so that S is non-empty. Let $\Lambda = \{x \in \mathbb{Q}^2 : S + x = S\}$. Since f is doubly periodic, Λ is a sublattice of $\mathbb{Z}^2$. On the other hand, both operators A and B preserve S, hence, they preserve Λ. It follows that $\text{Tr}(AB) = -2 + rp$ is an integer, i.e. $rp = \frac{4b^2}{ac}$ is an integer.

Making the change of variables of the form $m = m'/m_0$, $n = n'/n_0$, where m_0 and n_0 are positive integers such that $\frac{m_0}{n_0} = \frac{a}{2b}$, we can always assume that $a = 2b$. Then the above condition will imply that both matrices A and B have integer coefficients. In particular, we can consider them acting on $(\mathbb{Z}/N\mathbb{Z})^2$, where N is the (double) period of f. Let us denote by G_N the subgroup of $\text{GL}_2(\mathbb{Z}/N\mathbb{Z})$ generated by these two operators (by abuse of notation we will denote the corresponding elements of G_N also by A and B). Note that $A^2 = B^2 = 1$, so G_N is actually a dihedral group. Now clearly the space of functions f on $(\mathbb{Z}/N\mathbb{Z})^2$ satisfying the condition (1.1) is spanned by functions supported on orbits of G_N (and satisfying (1.1)). Let $O \subset (\mathbb{Z}/N\mathbb{Z})^2$ be an orbit of G_N, f be a function on O satisfying (1.1). In order for f to be non-zero the orbit O should satisfy the following condition: for every $x \in O$ one has $Ax \neq x$, $Bx \neq x$. Let us call such orbit *admissible*. Conversely, it is easy to see that for every admissible orbit O there is a unique (up to a constant) function f on O satisfying (1.1). Indeed, let $\chi : G_N \to \{\pm 1\}$ be the character defined by $\chi(A) = \chi(B) = -1$. Then the orbit is admissible if and only if χ is trivial on the stabilizer subgroup of a point in O (since every element $g \in G_N$ with $\chi(g) = -1$ is conjugate either to A or to B). Thus, for every admissible orbit $O = Gx$ we can define the function f_O on O by setting $f_O(gx) = \chi(g)$ (up to a sign f_O doesn't depend on x). It suffices to deal with the series associated with such functions. So, in the rest of the proof we will assume that f is a doubly periodic function on $\mathbb{Z}^2$ with values ± 1 satisfying (1.1). Let $S \subset \mathbb{Z}^2 \subset \mathbb{Q}^2$ be the support of f. Then $S = S_1 \cup S_{-1}$ where $S_1 = f^{-1}(1)$, $S_{-1} = f^{-1}(-1)$. Furthermore, we have $AS_1 = BS_1 = S_{-1}$. Let K be the quadratic extension of $\mathbb{Q}$ associated with the form Q. If $D = b^2 - ac$ is not a complete square then K is a real quadratic field $\mathbb{Q}(\sqrt{D})$, otherwise, $K = \mathbb{Q} \oplus \mathbb{Q}$. The usual notation $x + y\sqrt{D}$ for elements of a real quadratic field K can be extended to the case when D is a complete square and $K = \mathbb{Q} \oplus \mathbb{Q}$. Namely, in this case we set $x + y\sqrt{D} := (x + y\sqrt{D}, x - y\sqrt{D})$. We have

$$Q(m, n) = \frac{1}{c}[(bm + nc)^2 - Dm^2] = \frac{1}{c} \text{Nm}(bm + nc + m\sqrt{D}).$$

Thus, it makes sense to consider $\mathbb{Z}^2$ as a lattice in K via the map $(m, n) \mapsto (bm + nc + m\sqrt{D})$. For two non-zero elements $k_1, k_2 \in K$ let us denote $\langle k_1, k_2 \rangle = \mathbb{Q}_{>0} k_1 +$

$\mathbb{Q}_{>0}k_2$, $[k_1, k_2] = \mathbb{Q}_{\geq 0}k_1 + \mathbb{Q}_{\geq 0}k_2$, $\langle k_1, k_2 \rangle = \mathbb{Q}_{\geq 0}k_1 + \mathbb{Q}_{>0}k_2$. Using this notation we can write

$$\Theta_{Q,f} = \sum_{\lambda \in S_1 \cap [1, b+\sqrt{D}]} q^{\frac{\mathrm{Nm}(\lambda)}{c}} - \sum_{\lambda \in S_1 \cap \langle -1, -b-\sqrt{D} \rangle} q^{\frac{\mathrm{Nm}(\lambda)}{c}} - \sum_{\lambda \in S_{-1} \cap [1, b+\sqrt{D}]} q^{\frac{\mathrm{Nm}(\lambda)}{c}} + \sum_{\lambda \in S_{-1} \cap \langle -1, -b-\sqrt{D} \rangle} q^{\frac{\mathrm{Nm}(\lambda)}{c}}.$$

Let us extend the operators A and B from our lattice to K by $\mathbb{Q}$-linearity. We have $B(1) = -1$, $B(b+\sqrt{D}) = -b+\sqrt{D}$. Therefore, making the change of variables $\lambda \mapsto B\lambda$ in the last two sums we get

$$\sum_{\lambda \in S_{-1} \cap [1, b+\sqrt{D}]} q^{\frac{\mathrm{Nm}(\lambda)}{c}} = \sum_{\lambda \in S_1 \cap [-1, -b+\sqrt{D}]} q^{\frac{\mathrm{Nm}(\lambda)}{c}},$$

$$\sum_{\lambda \in S_{-1} \cap \langle -1, -b-\sqrt{D} \rangle} q^{\frac{\mathrm{Nm}(\lambda)}{c}} = \sum_{\lambda \in S_1 \cap \langle 1, b-\sqrt{D} \rangle} q^{\frac{\mathrm{Nm}(\lambda)}{c}}.$$

Hence, we can rewrite $\Theta_{Q,f}$ as follows:

$$\Theta_{Q,f} = \sum_{\lambda \in S_1 \cap \langle b-\sqrt{D}, b+\sqrt{D}]} q^{\frac{\mathrm{Nm}(\lambda)}{c}} - \sum_{\lambda \in S_1 \cap [-b+\sqrt{D}, -b-\sqrt{D} \rangle} q^{\frac{\mathrm{Nm}(\lambda)}{c}}.$$

Now it is easy to check that the operator $AB: K \to K$ coincides with multiplication by the element $\frac{b+\sqrt{D}}{b-\sqrt{D}}$ of norm 1. Therefore, we have

$$\Theta_{Q,f} = \sum_{\lambda \in S_1 \cap C/G} \mathrm{sign}(\lambda) q^{\frac{\mathrm{Nm}(\lambda)}{c}},$$

where G is the infinite cyclic group generated by AB. Note that the set S_1 is a union of a finite number of cosets $(\Lambda_1 + x_i, i = 1, \ldots, s)$ for the lattice $\Lambda_1 = \{x \in K : S_1 + x = S_1\}$. Furthermore, since Λ_1 is preserved by the action of G, there is a subgroup of finite index $G_0 \subset G$ preserving each of these cosets. Then we have

$$[G : G_0]\Theta_{Q,f} = \sum_{\lambda \in S_1 \cap C/G_0} \mathrm{sign}(\lambda) q^{\frac{\mathrm{Nm}(\lambda)}{c}} = \sum_{i=1}^{s} \sum_{\lambda \in (\Lambda_1 + x_i) \cap C/G_0} \mathrm{sign}(\lambda) q^{\frac{\mathrm{Nm}(\lambda)}{c}}.$$

Now each of the terms is a scalar multiple of Hecke's series.

Conversely, assume that we are given a lattice $\Lambda \subset K$ in a totally real quadratic extension of $\mathbb{Q}$ and a coset $\Lambda + c$. Let $G \subset U_+(K)$ be the subgroup preserving $\Lambda + c$. Recall that G is an infinite cyclic group. Let ϵ be a generator of G. Let us define the $\mathbb{Q}$-linear operators A and B on K as follows: $B(x) = -\overline{x}$ where $\overline{x}$ is the conjugate element to x (in the case $K = \mathbb{Q} \oplus \mathbb{Q}$ and $x = (x_1, x_2)$ one has $\overline{x} = (x_2, x_1)$), $A(x) = -\epsilon \cdot \overline{x}$. Note that $A^2 = B^2 = 1$ while $\det A = \det B = -1$. Let $k \in K$ be an eigenvector for A with eigenvalue -1, so that $\epsilon \overline{k} = k$. Changing k by $-k$ if necessary we can assume that k is totally positive. Then we have

$$\Theta_{\Lambda,c} = \sum_{\lambda \in (\Lambda+c) \cap C/G} \mathrm{sign}(\lambda) q^{d \cdot \mathrm{Nm}(\lambda)} = \sum_{\lambda \in (\Lambda+c) \cap [k, \overline{k}\rangle} q^{d \cdot \mathrm{Nm}(\lambda)} - \sum_{\lambda \in (\Lambda+c) \cap \langle -k, -\overline{k}]} q^{d \cdot \mathrm{Nm}(\lambda)}.$$

Note that we have $1 \in \langle k, \overline{k} \rangle$ since k is totally positive. Therefore, we can split each of the above sums into two according to decompositions $[k, \overline{k}] = [k, 1] \bigsqcup \langle 1, \overline{k} \rangle$,

$\langle -k, -\overline{k}] = \langle -k, -1\rangle \bigsqcup [-1, -\overline{k}]$. Making the change of variable $\lambda \mapsto B(\lambda)$ in the sums over $\langle 1, \overline{k}\rangle$ and over $[-1, -\overline{k}]$ we can rewrite the above sum as follows:

$$\Theta_{\Lambda, c} = \sum_{\lambda \in S \cap ([1,k] \cup \langle -k, -1\rangle)} f(\lambda) \operatorname{sign}(\lambda) q^{d \cdot \operatorname{Nm}(\lambda)},$$

where $S = (\Lambda + c) \cup B(\Lambda + c)$, the function f supported on S is defined by

$$f(x) = \delta_{\Lambda+c}(x) - \delta_{B(\Lambda+c)}(x)$$

where δ_I is the characteristic function of the set I. Note that since the operator AB preserves $\Lambda + c$ and $(AB)B = B(AB)^{-1}$, it also preserves $B(\Lambda + c)$, hence, $f(ABx) = f(x)$. On the other hand, by definition $f(Bx) = -f(x)$. Therefore, we also have $f(Ax) = -f(x)$. Now taking the coordinates with respect to the basis $(1, k)$ as variables of summation we see that the above series assumes the form

$$\sum_{(m,n)\in S, m\geq 0, n\geq 0} f(m,n) q^{Q(m,n)} - \sum_{(m,n)\in S, m<0, n<0} f(m,n) q^{Q(m,n)}$$

where S is a finite union of cosets with respect to some $\mathbb{Z}$-lattice in $\mathbb{Q}^2$, f is a periodic function on S with the property that sums of $f(m,n)q^{Q(m,n)}$ over all vertical and horizontal lines are zero. It remains to change variables (m, n) to (Mm, Mn) where $MS \subset \mathbb{Z}^2$ to rewrite this series in the form we require. $\square$

2. Remarks and examples

2.1. Linear relations. The series $\Theta_{Q,f}$ is often equal to zero. It is an important open problem to formulate the necessary and sufficient conditions for it to be zero. In other words, the problem is to describe all linear relations between such series for some basis in the space of functions f satisfying the assumptions of the main theorem. We restrict ourself to several observations. As above we assume that $p = -2b/a$ and $r = -2b/c$ are integers, so that we have an action of operators A and B on $\mathbb{Z}^2$ preserving the form Q. In the course of proof of the main theorem we introduced the subgroup $G_N \subset \operatorname{GL}_2(\mathbb{Z}/N\mathbb{Z})$ generated by these two operators modulo N. As we have seen above the space of functions on $(\mathbb{Z}/N\mathbb{Z})^2$ satisfying the condition (1.1) (further called *admissible* functions) has a basis (f_O) enumerated by admissible G_N-orbits. The change of variables $(m,n) \mapsto (-m,-n)$ shows that

$$\Theta_{Q,f} = -\Theta_{Q, f \circ [-1]},$$

where $f \circ [-1](m,n) = f(-m,-n)$. Let us call an admissible orbit O *symmetric* if $-O = O$, and *asymmetric* otherwise. Note that for an asymmetric orbit one has $O \cap -O = \emptyset$. For every symmetric orbit O the corresponding function f_O is either even or odd. We call a symmetric orbit O *even* (resp. *odd*) if f_O is even (resp. odd). Now the above equation shows that for an even symmetric orbit O one has $\Theta_{Q,f_O} = 0$, while for an asymmetric orbit O one has $\Theta_{Q,f_O} = \pm\Theta_{Q,f_{-O}}$ (the sign comes from the sign ambiguity in the definition of f_O).

The action of the operator $\tau : (m,n) \mapsto (n,m)$ gives some additional relations between Θ_{Q,f_O}. Indeed, for any Q we have

$$\Theta_{Q,f} = \Theta_{Q\circ\tau, f\circ\tau}.$$

If $Q \circ \tau = Q$ (i.e. $a = c$) then for every admissible orbit O we have $f_O \circ \tau = \pm f_{O'}$ for some other admissible orbit O', hence $\Theta_{Q,f_O} = \pm\Theta_{Q,f_{O'}}$. In particular, if $f_O \circ \tau = -f_O$ then $\Theta_{Q,f_O} = 0$.

Finally, we can make the changes of variables $(m,n) \mapsto (t_1 m, t_2 n)$, where t_1 and t_2 are positive rational numbers, in the case when this transformation sends the support of f into $\mathbb{Z}^2$. This transformation will always change the form Q (unless $t_1 = t_2 = 1$). However, combining it with the operator τ with respect to the new variables we can derive more linear relations for fixed Q (generalizing the above relations for the case $a = c$). Namely, assume that $c/a = t^2$ for some positive rational number t. Then the operator

$$\tau_t : (m,n) \mapsto (tn, t^{-1}m)$$

preserves Q and satisfies $\tau_t^2 = 1$, $\tau_t A = B\tau_t$. In particular if f is an admissible function such that τ_t sends the support of f into $\mathbb{Z}^2$ then $f \circ \tau_t$ is also admissible (perhaps with a different double period) and we have $\Theta_{Q, f \circ \tau_t} = \Theta_{Q,f}$.

We were not able to find any other linear relations between the series $(\Theta_{Q,f})$ for fixed Q. However, at present we are far from proving that these are all relations. Even the non-vanishing of Θ_{Q,f_O} for odd symmetric and for asymmetric admissible G_N-orbits (in the case when a/c is not a square in $\mathbb{Q}$) is still an open problem. Note that some non-vanishing results were proven in [**P**] using homological mirror symmetry.

In Hecke's paper one can find the following vanishing condition for an indefinite theta series $\Theta_{\Lambda,c}$ (see [**H2**], Satz 1): if there exists a totally negative element $\delta \in K^*$ with $\mathrm{Nm}(\delta) = 1$ such that $\delta(\Lambda + c) = \Lambda + c$ then $\Theta_{\Lambda,c} = 0$. Let us show that this vanishing is actually one of the linear relations considered above. We will use the notation introduced in the proof of the main theorem. Let (Q, f) be the data constructed in the second half of the proof so that $\Theta_{\Lambda,c} = \Theta_{Q,f}$. First of all, notice that $\delta^2 \in G$, hence $\delta^2 = \epsilon^n$ for some integer n. Changing δ by a power of ϵ we can assume that either $\delta^2 = 1$ or $\delta^2 = \epsilon$. In the former case $\delta = -1$ so one has $f \circ [-1] = f$. In the latter case we have $\epsilon \bar{\delta} = \delta$, so rescaling k we can assume that $k = -\delta$. It is easy to see that the operator $\delta B : x \mapsto -\delta \bar{x}$ preserves Nm and switches 1 and k, as well as $\Lambda + c$ and $B(\Lambda + c)$. Hence, the transposition $\tau : (m,n) \mapsto (n,m)$ preserves Q and satisfies $f \circ \tau = -f$. A different choice of k would lead to a similar relation with S replaced by τ_t.

2.2. Symmetric orbits. Henceforward, operators A and B are always considered modulo N. In the situation when the subgroup $G_N \subset \mathrm{GL}_2(\mathbb{Z}/N\mathbb{Z})$ ($N > 2$) contains the matrix $-\mathrm{id}$ every orbit is symmetric. Furthermore, since the character $\chi : G_N \to \{\pm 1\}$ defined by $\chi(A) = \chi(B) = -1$ coincides with $\det|_{G_N}$, we have $\chi(-\mathrm{id}) = 1$, hence every orbit is even. Thus, we get $\Theta_{Q,f} = 0$ for all admissible f. The following proposition gives a criterion allowing to recognize this situation in the case when N is an odd prime.

PROPOSITION 2.1. *Assume that N is an odd prime. Then $-\mathrm{id} \in G_N$ if and only if rp mod N is of the form $2 + \lambda + \lambda^{-1}$ where λ is an element of even order in $\mathbb{F}_{N^2}^*$ ($\mathbb{F}_{N^2}$ is the finite field of cardinality N^2). The number of such residues modulo N is equal to $N - \frac{n_1 + n_2}{2}$ where n_1 (resp. n_2) is the maximal odd divisor of $N - 1$ (resp. $N + 1$).*

PROOF. Since $G_N \cap \mathrm{SL}_2(\mathbb{Z}/N\mathbb{Z})$ is generated by AB the condition $-\mathrm{id} \in G_N$ is equivalent to $(AB)^n = -\mathrm{id}$ for some n. We have $\mathrm{Tr}(AB) = rp - 2$, $\det(AB) = 1$, so the eigenvalues λ_1, λ_2 of AB are roots of the equation

$$\lambda^2 - (rp - 2)\lambda + 1 = 0.$$

Assume first that $\lambda_1 = \lambda_2$. Then either $rp = 4$ or $rp = 0$. In the former case $\lambda_1 = \lambda_2 = 1$, hence, no power of AB equals $-\mathrm{id}$. In the latter case one can easily check that $(AB)^N = -\mathrm{id}$. On the other hand, 0 can be represented in the form $2 + \lambda + \lambda^{-1}$ for $\lambda = -1$.

Now assume that $\lambda_1 \neq \lambda_2$. Then the condition $(AB)^n = -\mathrm{id}$ is equivalent to $\lambda_1^n = -1$, i.e. λ_1 has even order in the multiplicative group of $\overline{\mathbb{F}}_N$. It remains to notice that $\lambda_1 \in \mathbb{F}_{N^2}^*$ and that we have $rp - 2 = \lambda_1 + \lambda_1^{-1}$.

To compute the number of such residues modulo N we note that the condition $\lambda + \lambda^{-1} \in \mathbb{F}_N$ means that either $\lambda^{N-1} = 1$ or $\lambda^{N+1} = 1$. The number of elements λ of even order such that $\lambda^m = 1$ (where m is either $N-1$ or $N+1$) is equal to $m - n$ where n is the maximal odd divisor of m. Therefore, the number of elements in $\mathbb{F}_N$ of the form $\lambda + \lambda^{-1}$ is equal to

$$1 + \frac{(N-1) - n_1 - 1}{2} + \frac{(N+1) - n_2 - 1}{2} = N - \frac{n_1 + n_2}{2}.$$

$\square$

Taking in the above proposition λ to be -1, ζ_4 and ζ_6 (where ζ_l is a primitive root of unity of order l) we get $rp \equiv 0 \mod (N)$, $rp \equiv 2 \mod (N)$ and $rp \equiv 3 \mod (N)$ respectively. On the other hand, we claim that if $rp \equiv 1 \mod (N)$ or $rp \equiv 4 \mod (N)$ then $-\mathrm{id} \notin G_N$. Indeed, the equation $4 = 2 + \lambda + \lambda^{-1}$ has the only solution $\lambda = 1$ while the solutions of the equation $1 = 2 + \lambda + \lambda^{-1}$ are roots of unity of order 3. These are the only cases of the above criterion which are independent of N. Here are the lists of values of $rp \mod (N)$ such that $-\mathrm{id} \notin G_N$ for small odd primes N:

$N = 3$: $rp \equiv 1 \mod (3)$;
$N = 5$: $rp \equiv 1, 4 \mod (5)$;
$N = 7$: $rp \equiv 1, 4 \mod (7)$;
$N = 11$: $rp \equiv 1, 4, 5, 9 \mod (11)$.
$N = 13$: $rp \equiv 1, 4, 9, 10, 12 \mod (13)$.

Our last general observation is that in the case when N is an odd prime, all symmetric G_N-orbits have the same parity, i.e. they are either all odd or all even.

PROPOSITION 2.2. *Assume that N is an odd prime. Then either $-\mathrm{id} \in G_N$ or every symmetric G_N-orbit is odd.*

PROOF. Assume that there exists a non-zero vector $v \in (\mathbb{Z}/N\mathbb{Z})^2$ and an element $g \in G_N$ such that $gv = -v$ and $\det(g) = 1$. Then both eigenvalues of g are -1, hence, $g^N = -\mathrm{id}$. $\square$

2.3. Examples. In all examples below we assume that a, c, $p = -\frac{2b}{a}$ and $r = -\frac{2b}{c}$ are integers (b is a half-integer). Note that we are interested only in the cases when G_N doesn't contain $-\mathrm{id}$. In particular, if N is an odd prime we can assume that $rp \not\equiv 0 \mod (N)$. In this case the conjugacy class of the subgroup $G_N \subset \mathrm{GL}_2(\mathbb{Z}/N)^2$ depends only on $rp \mod (N)$. For instance, if $rp \equiv 1 \mod (N)$ then G_N is isomorphic to the permutation group S_3. In examples 1 and 2 below we consider in details cases $N = 3$ and $N = 5$. It turns out that in these cases all admissible orbits are symmetric (they are automatically odd by proposition 2.2). The simplest example of an asymmetric admissible orbit (for prime N) occurs for $N = 7$ (see example 3 below).

1. $N = 3$, $rp \equiv 1 \mod (3)$. Then there is a unique admissible orbit: the orbit of $(1,0)$. For $r \equiv p \equiv 1 \mod (3)$ (resp. $r \equiv p \equiv -1 \mod (3)$) the corresponding admissible function is $f(m,n) = \chi_3(m+n)$ (resp. $f(m,n) = \chi_3(m-n)$) where χ_3 is the non-trivial Dirichlet character modulo 3 such that $\chi_3(\pm 1) = \pm 1$. Let us assume that $a \leq c$ (we can always achieve this using the transformation $(m,n) \mapsto (n,m)$ if necessary). Then we have

$$\Theta_{Q,f} \equiv q^a + \chi_3(r) q^c \mod (q^{a+1}).$$

It follows that this theta series doesn't vanish unless $r \equiv -1 \mod (3)$ and $a = c$. In the latter case we have $Q(n,m) = Q(m,n)$ while $f(n,m) = -f(m,n)$ so that $\Theta_{Q,f} = 0$.

2. $N = 5$.

(a) $rp \equiv 1 \mod (5)$. In this case there are two distinct admissible orbits: the orbit of $(1,0)$ and the orbit of $(2,0)$. It is easy to see that unless $a = c$ the corresponding two theta functions Θ_{Q,f_1} and Θ_{Q,f_2} are linearly independent. More precisely, the initial terms of these series look as follows (in (i) and (ii) we assume that $a \leq c$):

(i) $p \equiv r \equiv 1(5)$:

$$\Theta_{Q,f_1} \equiv q^a + q^c \mod (q^{a+1}), \quad \Theta_{Q,f_2} \equiv q^{4a} + q^{4c} \mod (q^{4a+1}).$$

(ii) $p \equiv r \equiv -1(5)$:

$$\Theta_{Q,f_1} \equiv q^a - q^c \mod (q^{a+1}), \quad \Theta_{Q,f_2} \equiv q^{4a} - q^{4c} \mod (q^{4a+1}).$$

(iii) $p \equiv 2(5)$, $r \equiv -2(5)$:

$$\Theta_{Q,f_1} \equiv q^a - q^{4c} \mod (q^{\min(a,4c)+1}), \quad \Theta_{Q,f_2} \equiv q^c - q^{4a} \mod (q^{\min(4a,c)+1}).$$

Furthermore, in the case $a = 4c$ we have

$$\Theta_{Q,f_1} \equiv q^{9c} \mod (q^{9c+1})$$

while in the case $c = 4a$ we have

$$\Theta_{Q,f_2} \equiv q^{9a} \mod (q^{9a+1}).$$

If $a = c$ then in the case (ii) we have $\Theta_{Q,f_1} = \Theta_{Q,f_2} = 0$ while in the case (iii) we have $\Theta_{Q,f_2} = \Theta_{Q,f_1}$.

(b) $rp \equiv -1 \mod (5)$. In this case AB has order 5 but there are still two admissible orbits: the orbit of $(1,0)$ and the orbit of $(2,0)$.[2] The analysis of the initial terms of these series (very similar to the case (a)) implies that the corresponding two theta series are linearly independent unless $a = c$.

3. $N = 7$, $r \equiv p \equiv 1 \mod (7)$. There are 5 admissible orbits: 3 symmetric orbits and 2 asymmetric orbits. The symmetric orbits are $O_1 = G_N \cdot (1,0)$, $2 \cdot O_1$, and $3 \cdot O_1$. The asymmetric orbits are $O_2 = G_N \cdot (1,3)$ and $-O_2$. Using the relation $\Theta_{Q,f_{-O_2}} = -\Theta_{Q,f_{O_2}}$ we can exclude the orbit $-O_2$ from our consideration. The initial terms of the remaining 4 theta series look as follows (assuming that $a \leq c$)

$$\Theta_{Q,f_{O_1}} \equiv q^a + q^c \mod (q^{a+1}),$$

$$\Theta_{Q,f_{2O_1}} \equiv q^{4a} + q^{4c} \mod (q^{4a+1}),$$

$$\Theta_{Q,f_{3O_1}} \equiv q^{9a} + q^{9c} \mod (q^{9a+1}),$$

$$\Theta_{Q,f_{O_2}} \equiv q^{9a+c+6b} + q^{a+9c+6b} \mod (q^{9a+c+6b+1}).$$

---

[2] In this case A and B have a common invariant vector which allows to have bigger admissible orbits than in case (a).

This immediately implies that they are linearly independent.

4. The indefinite theta series considered in Theorem 2 of [**P**] correspond to the following situation. Let us assume that $\frac{b}{a}$ and $\frac{b}{c}$ are integers (not just half-integers as before). In this case the discriminant $D = b^2 - ac$ is divisible by ac. We are going to take $N = \frac{4D}{ac}$. Let s_1 and s_2 be arbitrary odd numbers. It is easy to check that the G_N-orbit of the element

$$v_{s_1,s_2} = (\frac{b}{a}s_2 - s_1, \frac{b}{c}s_1 - s_2) \in (\mathbb{Z}/N\mathbb{Z})^2$$

is admissible and consists of four elements which are congruent to v_{s_1,s_2} modulo $N/2$. On the other hand, if l divides $\frac{b}{a}+1$ and $\frac{b}{c}+1$ then $\frac{D}{ac}$ is divisible by l and the $2l$-torsion element $v_l = \frac{2D}{lac}(1,1) \in (\mathbb{Z}/N\mathbb{Z})^2$ is G_N-invariant. The series considered in [**P**] correspond to the orbits of the elements $v_{s_1,s_2} + t \cdot v_l$ where $t \in \mathbb{Z}$ (these orbits depend only on $t \mod (l)$).

References

[H1] E. Hecke, *Über einen neuen Zusammenhang zwischen elliptischen Modulfunktionen und indefiniten quadratischen Formen*, no. 22 in *Mathematische Werke*, p. 418–427, Göttingen, 1983.

[H2] E. Hecke, *Zur Theorie der elliptischen Modulfunktionen*, no. 23 in *Mathematische Werke*, p. 428–460, Göttingen, 1983.

[P] A. Polishchuk, *Indefinite theta series of signature $(1,1)$ from the point of view of homological mirror symmetry*, preprint math.AG/0003076.

DEPARTMENT OF MATHEMATICS AND STATISTICS, BOSTON UNIVERSITY, BOSTON, MA 02215
E-mail address: `apolish@math.bu.edu`

A proof of a multivariable elliptic summation formula conjectured by Warnaar

Hjalmar Rosengren

ABSTRACT. We prove a multivariable elliptic analogue of Jackson's $_8W_7$ summation formula, which was recently conjectured by S. O. Warnaar.

1. Introduction

Elliptic hypergeometric series form a natural generalization of hypergeometric and basic hypergeometric (or q-) series. It is surprising that they were introduced only very recently, by Frenkel and Turaev [**FT**], who expressed the $6j$-symbols corresponding to certain elliptic solutions of the Yang–Baxter equation, cf. [**DJ**], in terms of the $_{10}\omega_9$-sums defined below. It is expected that elliptic hypergeometric series play a fundamental role in the representation theory of elliptic quantum groups, though so far there has been little work in this direction.

Recall that a series $\sum_n a_n$ is called hypergeometric if $f(n) = a_{n+1}/a_n$ is a rational function of n and basic hypergeometric if f is a rational function of q^n for some q. This can be compared with Weierstrass' theorem, stating that a meromorphic function of z which satisfies an algebraic addition theorem is either a rational function, a rational function of q^z, or, in the most general case, an elliptic function. This suggests that an elliptic hypergeometric series should be a series $\sum_n a_n$ with a_{n+1}/a_n an elliptic function of n. Actually, the series introduced by Frenkel and Turaev only fit this description if one interprets the term "elliptic" somewhat loosely. Nevertheless, their properties stem from addition theorems for elliptic functions (it is worth noting that the Yang–Baxter equation is an algebraic addition theorem for matrix-valued functions).

Let us write $[x]$ for the "elliptic number" (a Jacobi theta function, normalized so that $[1] = 1$)

$$[x] = \frac{q^{-\frac{x}{2}} \prod_{j=0}^{\infty}(1-q^x p^j)(1-q^{-x}p^{j+1})}{q^{-\frac{1}{2}} \prod_{j=0}^{\infty}(1-qp^j)(1-q^{-1}p^{j+1})},$$

1991 *Mathematics Subject Classification.* 33D67, 33E05.

© 2001 American Mathematical Society

where p and q are fixed parameters with $|p|<1$. When $p=0$, $q=e^{2ih}$, we have the trigonometric number

$$[x]=\frac{q^{\frac{x}{2}}-q^{-\frac{x}{2}}}{q^{\frac{1}{2}}-q^{-\frac{1}{2}}}=\frac{\sin(hx)}{\sin(h)},$$

which tends to the rational number $[x]=x$ as q tends to 1. Returning to the general case, we write

$$[x]_n=[x][x+1]\cdots[x+n-1]$$

for the elliptic Pochhammer symbols. The elliptic, or modular, hypergeometric series occurring in [**FT**] are finite sums of the form

$$_{r+1}\omega_r(a;-n,b_1,\ldots,b_{r-3})$$
$$=\sum_{k=0}^n\frac{[a+2k]}{[a]}\frac{[a]_k[-n]_k[b_1]_k\cdots[b_{r-3}]_k}{[1]_k[1+a+n]_k[1+a-b_1]_k\cdots[1+a-b_{r-3}]_k},$$

where

(1.1) $$(r-3)(a+1)=2\left(1-n+\sum_i b_i\right).$$

When $p=0$, this is a terminating very-well-poised balanced basic hypergeometric series [**GR**], which tends to the corresponding hypergeometric series as q tends to 1. As was pointed out in [**FT**], the series $_{r+1}\omega_r$ has remarkable invariance properties under the standard action of $\mathrm{SL}(2,\mathbb{Z})$ on (p,q)-space.

Most (or possibly all) known identities involving terminating q-series may be proved by induction, using the trigonometric addition formula

(1.2) $$[x+z]_{p=0}[x-z]_{p=0}=[x+y]_{p=0}[x-y]_{p=0}+[y+z]_{p=0}[y-z]_{p=0}.$$

However, only a tiny subset of these identities may be obtained from the elliptic addition formula

$$[x+z][x-z][y+w][y-w]=[x+y][x-y][z+w][z-w]+[x+w][x-w][y+z][y-z]$$

satisfied by the elliptic numbers. At least as a rule of thumb, these are the identities involving series which are both well-poised and balanced, and thus only these admit elliptic analogues. In particular, Frenkel and Turaev obtained the elliptic Jackson–Dougall summation formula
(1.3)
$$_8\omega_7(a;-n,b,c,d,e)=\frac{[a+1]_n[a+1-b-c]_n[a+1-b-d]_n[a+1-c-d]_n}{[a+1-b]_n[a+1-c]_n[a+1-d]_n[a+1-b-c-d]_n}$$

and (more generally) the elliptic Bailey transformation formula

$$_{10}\omega_9(a;-n,b,c,d,e,f,g)=\frac{[a+1]_n[a+1-e-f]_n[\lambda+1-e]_n[\lambda+1-f]_n}{[a+1-e]_n[a+1-f]_n[\lambda+1-e-f]_n[\lambda+1]_n}$$
$$\times\,_{10}\omega_9(\lambda;-n,\lambda+b-a,\lambda+c-a,\lambda+d-a,e,f,g),$$

where $\lambda=2a+1-b-c-d$; note that the balanced condition (1.1) is assumed.

If one wants to further develop the theory of elliptic hypergeometric series, there are two natural directions: quadratic (or higher) transformation formulas and multivariable series. In [**W**], Warnaar initiated the investigation of both topics. We will be concerned with the multivariable theory. As Warnaar pointed out, progress in this direction requires essentially new ideas, since the known proofs in the trigonometric and rational case usually depend on "lower level" identities, corresponding to the degenerate addition theorem (1.2).

The purpose of this paper is to prove an identity conjectured by Warnaar in [**W**], cf. Theorem 2.1, which is a generalization of (1.3) connected with the root system C_n. Our main tool will be a different generalization of (1.3), obtained by Warnaar [**W**] from a determinant evaluation.

We mention that one degenerate case of Theorem 2.1 is the terminating case of a multivariable $_6\psi_6$ sum due to van Diejen [**D**]. It generalizes various Macdonald–Morris-type identities for root systems, cf. [**D**] for a detailed discussion. Moreover, van Diejen's sum gives the norm evaluation for the multivariable q-Racah polynomials studied by van Diejen and Stokman [**DSt**].

When [**W**] was published, Theorem 2.1 was new even in the trigonometric case ($p = 0$). This case of the conjecture was settled by van Diejen and Spiridonov [**DS**], who deduced it from a certain multiple integral due to Gustafson [**G**], which reduces to the Nassrallah–Rahman integral [**NR**] in the one-variable case. The multiple q-series in question appears as a sum of residues of the integrand. Moreover, it was demonstrated that both sides of the equality in Theorem 2.1 are invariant under the action of $SL(2, \mathbb{Z})$. Using the theory of modular forms, it was then proved that for $q = e^{2ih}$, the two sides are equal at least up to order h^{10} around $h = 0$; a strong indication that Warnaar's conjecture is true. Finally, van Diejen and Spiridonov conjectured an elliptic generalization of Gustafson's integral, involving the elliptic gamma function introduced by Ruijsenaars [**R**]. A proof of this identity would yield another proof of Theorem 2.1, completely different from the one given here. The one-variable case of the integral is treated in [**Sp**].

Acknowledgement: I would like to thank Jan Felipe van Diejen and Vyacheslav Spiridonov for illuminating correspondence and the referee for some useful comments.

2. Notation and statement of results

In the rest of the paper we will use the "multiplicative" notation of [**W**] rather than the "additive" notation of [**FT**] used in the introduction. Since the elliptic modulus p is fixed we suppress it from the notation. Thus we write

$$E(x) = \prod_{j=0}^{\infty} \left(1 - xp^j\right)\left(1 - p^{j+1}/x\right),$$

$$E(x_1, \ldots, x_m) = E(x_1) \cdots E(x_m),$$

$$(a; q)_k = \begin{cases} \prod_{j=0}^{k-1} E(aq^j), & k \in \mathbb{Z}_{\geq 0}, \\ \prod_{j=0}^{-k-1} \dfrac{1}{E(aq^{k+j})}, & k \in \mathbb{Z}_{<0}, \end{cases}$$

$$(a_1, \ldots, a_m; q)_k = (a_1; q)_k \cdots (a_m; q)_k,$$

$$(a; q, x)_\lambda = \prod_{j=1}^{n} (ax^{1-j}; q)_{\lambda_j}, \quad \lambda \in \mathbb{Z}^n,$$

$$(a_1, \ldots, a_m; q, x)_\lambda = (a_1; q, x)_\lambda \cdots (a_m; q, x)_\lambda.$$

We will use freely standard identities such as $(a;q)_n(aq^n;q)_k = (a;q)_{n+k}$. We also mention the easily verified identity

$$(2.1) \qquad (aq;q)_n \prod_{1\leq i<j\leq n} E(aq^{i+j}) = (aq;q^2)_n \prod_{1\leq i<j\leq n} E(aq^{i+j-1}).$$

We can now state the main result of the paper, conjectured by Warnaar [**W**].

THEOREM 2.1. *In the notation above,*

$$(2.2) \quad \sum_\lambda \prod_{i=1}^n \left(\frac{E(ax^{2(1-i)}q^{2\lambda_i})}{E(ax^{2(1-i)})} q^{\lambda_i} x^{2(i-1)\lambda_i} \right)$$

$$\times \prod_{1\leq i<j\leq n} \left(\frac{E(x^{j-i}q^{\lambda_i-\lambda_j})}{E(x^{j-i})} \frac{E(ax^{2-i-j}q^{\lambda_i+\lambda_j})}{E(ax^{2-i-j})} \frac{(ax^{3-i-j};q)_{\lambda_i+\lambda_j}(x^{j-i+1};q)_{\lambda_i-\lambda_j}}{(aqx^{1-i-j};q)_{\lambda_i+\lambda_j}(qx^{j-i-1};q)_{\lambda_i-\lambda_j}} \right)$$

$$\times \frac{(ax^{1-n},b,c,d,e,q^{-N};q,x)_\lambda}{(qx^{n-1},aq/b,aq/c,aq/d,aq/e,aq^{N+1};q,x)_\lambda}$$

$$= \frac{(aq,aq/bc,aq/bd,aq/cd;q,x)_{N^n}}{(aq/b,aq/c,aq/d,aq/bcd;q,x)_{N^n}},$$

where the sum is over the partitions

$$\lambda \in \Lambda_{nN} = \{\lambda \in \mathbb{Z}^n; N \geq \lambda_1 \geq \lambda_2 \geq \cdots \geq \lambda_n \geq 0\},$$

where

$$bcdex^{n-1} = a^2 q^{N+1}$$

and where N^n denotes the partition with $\lambda_i = N$, $i = 1, \ldots, n$.

Note that the factor

$$\prod_{i=1}^{n-1} \left(\frac{1}{(q;q)_{\lambda_i-\lambda_{i+1}}} \right) \frac{(q^{-N};q,x)_\lambda}{(qx^{n-1};q,x)_\lambda}$$

vanishes for $\lambda \in \mathbb{Z}^n \setminus \Lambda_{nN}$. Therefore, we may equivalently sum over all $\lambda \in \mathbb{Z}^n$.

Our main tool for proving Theorem 2.1 will be the following identity, again due to Warnaar.

LEMMA 2.2. *In the notation above,*

$$(2.3) \quad \sum_{k_1,\ldots,k_n=0}^1 \prod_{i=1}^n \frac{(bx_i,cx_i,dx_i,ex_i;q)_{k_i}}{(aqx_i/b,aqx_i/c,aqx_i/d,aqx_i/e;q)_{k_i}} (-1)^{k_i} q^{(i-1)k_i}$$

$$\times \prod_{1\leq i<j\leq n} \frac{E(q^{k_i-k_j}x_i/x_j) E(ax_ix_jq^{k_i+k_j})}{E(x_i/x_j) E(ax_ix_jq)}$$

$$= (aq/bc,aq/bd,aq/cd;q^{-1})_n \prod_{i=1}^n \frac{E(aqx_i^2)}{E(aq^{2-n}/bcdx_i,aqx_i/b,aqx_i/c,aqx_i/d)},$$

where $a^2 q^{3-n} = bcde$.

In fact, Warnaar proved the more general identity [**W**, Theorem 5.1]

$$\sum_{k_1,\ldots,k_n=0}^{N} \prod_{i=1}^{n} \frac{E(ax_i^2 q^{2k_i})}{E(ax_i^2)} \frac{(ax_i^2, bx_i, cx_i, dx_i, ex_i, q^{-N}; q)_{k_i}}{(q, aqx_i/b, aqx_i/c, aqx_i/d, aqx_i/e, aq^{N+1}x_i^2; q)_{k_i}} q^{ik_i}$$

$$\times \prod_{1 \le i < j \le n} \frac{E(q^{k_i - k_j} x_i/x_j)}{E(x_i/x_j)} \frac{E(ax_i x_j q^{k_i + k_j})}{E(ax_i x_j q^N)}$$

$$= \prod_{i=1}^{n} \frac{(aqx_i^2, aq^{2-i}/bc, aq^{2-i}/bd, aq^{2-i}/cd; q)_N}{(aq^{2-n}/bcdx_i, aqx_i/b, aqx_i/c, aqx_i/d; q)_N},$$

where $a^2 q^{N+2-n} = bcde$. For $n = 1$, this is equivalent to (1.3) and for $N = 1$ it reduces to (2.3). The case $p = 0$ is due to Schlosser [**S**].

We will prove Theorem 2.1 by induction on the "terminator" N. However, because of a duality property for the sums in question, cf. Proposition 4.1, we can alternatively formulate the proof as an induction on the number n of variables. In this context we remark that, for $p = 0$, (2.3) is a special case not only of Schlosser's identity but also of yet another multivariable Jackson–Dougall formula due to Denis and Gustafson [**DG**] and Milne and Lilly [**ML**]. The degeneration of the latter to the $_6\psi_6$-level, together with induction on the number of variables, was used by van Diejen [**D**] to prove the trigonometric $_6\psi_6$-version of Theorem 2.1. Nevertheless, our proof is essentially different from the one in [**D**], since we only need a very special case of the (as yet unproved) elliptic Denis–Gustafson–Milne–Lilly identity.

3. Proof of Theorem 2.1

We will prove Theorem 2.1 by induction on N, starting from the trivial case $N = 0$. Assume that Theorem 2.1 holds for a fixed value of N. Let us fix parameters with

(3.1) $$bcdex^{n-1} = a^2 q^{N+2}.$$

We write the right-hand side of (2.2) with N replaced by $N + 1$ as

$$R = \frac{(aq, aq/bc, aq/bd, aq/cd; q, x)_{(N+1)^n}}{(aq/b, aq/c, aq/d, aq/bcd; q, x)_{(N+1)^n}}$$

$$= \frac{(aq, aq/bc, aq/bd, aq/cd; x^{-1})_n}{(aq/b, aq/c, aq/d, aq/bcd; x^{-1})_n} \frac{(aq^2, aq^2/bc, aq^2/bd, aq^2/cd; q, x)_{N^n}}{(aq^2/b, aq^2/c, aq^2/d, aq^2/bcd; q, x)_{N^n}},$$

where the second factor is the right-hand side of (2.2) with a replaced by aq and e by eq. Using our induction hypothesis, we have

$$R = \frac{(aq, aq/bc, aq/bd, aq/cd; x^{-1})_n}{(aq/b, aq/c, aq/d, aq/bcd; x^{-1})_n} \sum_{\lambda} \prod_{i=1}^{n} \left(\frac{E(ax^{2(1-i)} q^{2\lambda_i + 1})}{E(ax^{2(1-i)} q)} q^{\lambda_i} x^{2(i-1)\lambda_i} \right)$$

$$\times \prod_{1 \le i < j \le n} \left(\frac{E(x^{j-i} q^{\lambda_i - \lambda_j})}{E(x^{j-i})} \frac{E(ax^{2-i-j} q^{\lambda_i + \lambda_j + 1})}{E(ax^{2-i-j} q)} \frac{(aqx^{3-i-j}; q)_{\lambda_i + \lambda_j}}{(aq^2 x^{1-i-j}; q)_{\lambda_i + \lambda_j}} \right.$$

$$\left. \times \frac{(x^{j-i+1}; q)_{\lambda_i - \lambda_j}}{(qx^{j-i-1}; q)_{\lambda_i - \lambda_j}} \right) \frac{(aqx^{1-n}, b, c, d, eq, q^{-N}; q, x)_{\lambda}}{(qx^{n-1}, aq^2/b, aq^2/c, aq^2/d, aq/e, aq^{N+2}; q, x)_{\lambda}}.$$

We now apply (2.3) with

$$(a, b, c, d, e, x_i, q) \mapsto (aq/x, b, c, d, eq^{-N}, x^{1-i} q^{\lambda_i}, x),$$

which allows us to write

$$(aq/bc, aq/bd, aq/cd; x^{-1})_n \prod_{i=1}^n E(ax^{2(1-i)}q^{2\lambda_i+1})$$

$$= \prod_{i=1}^n E(ax^{i-n}q^{1-\lambda_i}/bcd, ax^{1-i}q^{\lambda_i+1}/b, ax^{1-i}q^{\lambda_i+1}/c, ax^{1-i}q^{\lambda_i+1}/d)$$

$$\times \sum_{k_1,\ldots,k_n=0}^1 \prod_{i=1}^n \frac{(bx^{1-i}q^{\lambda_i}, cx^{1-i}q^{\lambda_i}, dx^{1-i}q^{\lambda_i}, ex^{1-i}q^{\lambda_i-N}; x)_{k_i}(-1)^{k_i}x^{(i-1)k_i}}{(ax^{1-i}q^{\lambda_i+1}/b, ax^{1-i}q^{\lambda_i+1}/c, ax^{1-i}q^{\lambda_i+1}/d, ax^{1-i}q^{\lambda_i+N+1}/e; x)_{k_i}}$$

$$\times \prod_{1\leq i<j\leq n} \frac{E(x^{j-i+k_i-k_j}q^{\lambda_i-\lambda_j})}{E(x^{j-i}q^{\lambda_i-\lambda_j})} \frac{E(ax^{1-i-j+k_i+k_j}q^{\lambda_i+\lambda_j+1})}{E(ax^{2-i-j}q^{\lambda_i+\lambda_j+1})}.$$

Plugging this into the previous identity and then replacing λ by $\lambda - k$ in the summation (here we use the observation succeeding the statement of Theorem 2.1) yields

$$R = \frac{(aq; x^{-1})_n}{(aq/b, aq/c, aq/d, aq/bcd; x^{-1})_n} \sum_{\lambda, k} \prod_{i=1}^n \left((-1)^{k_i} q^{\lambda_i - k_i} x^{(i-1)(2\lambda_i - k_i)} \right.$$

$$\times \frac{E(ax^{i-n}q^{k_i-\lambda_i+1}/bcd, ax^{1-i}q^{\lambda_i-k_i+1}/b, ax^{1-i}q^{\lambda_i-k_i+1}/c, ax^{1-i}q^{\lambda_i-k_i+1}/d)}{E(aqx^{2(1-i)})}$$

$$\left. \times \frac{(bx^{1-i}q^{\lambda_i-k_i}, cx^{1-i}q^{\lambda_i-k_i}, dx^{1-i}q^{\lambda_i-k_i}, ex^{1-i}q^{\lambda_i-k_i-N}; x)_{k_i}}{(ax^{1-i}q^{\lambda_i-k_i+1}/b, ax^{1-i}q^{\lambda_i-k_i+1}/c, ax^{1-i}q^{\lambda_i-k_i+1}/d, ax^{1-i}q^{\lambda_i-k_i+N+1}/e; x)_{k_i}} \right)$$

$$\times \prod_{1\leq i<j\leq n} \left\{ \frac{E(x^{j-i+k_i-k_j}q^{\lambda_i-\lambda_j-k_i+k_j})}{E(x^{j-i})} \frac{E(ax^{1-i-j+k_i+k_j}q^{\lambda_i+\lambda_j-k_i-k_j+1})}{E(ax^{2-i-j}q)} \right.$$

$$\left. \times \frac{(aqx^{3-i-j}; q)_{\lambda_i+\lambda_j-k_i-k_j}(x^{j-i+1}; q)_{\lambda_i-\lambda_j-k_i+k_j}}{(aq^2x^{1-i-j}; q)_{\lambda_i+\lambda_j-k_i-k_j}(qx^{j-i-1}; q)_{\lambda_i-\lambda_j-k_i+k_j}} \right\}$$

$$\times \frac{(aqx^{1-n}, b, c, d, eq, q^{-N}; q, x)_{\lambda-k}}{(qx^{n-1}, aq^2/b, aq^2/c, aq^2/d, aq/e, aq^{N+2}; q, x)_{\lambda-k}}.$$

We will identify the sum with respect to k as a case of (2.3) with q replaced by x^{-1}. Since $k_i \in \{0, 1\}$, we can write

$$(ex^{1-i}q^{\lambda_i-k_i-N}; x)_{k_i} = (ex^{1-i}q^{\lambda_i-1-N}; x^{-1})_{k_i},$$

$$\frac{(aqx^{1-n}, eq, q^{-N}; q, x)_{\lambda-k}}{(qx^{n-1}, aq/e, aq^{N+2}; q, x)_{\lambda-k}} = \frac{(aqx^{1-n}, eq, q^{-N}; q, x)_\lambda}{(qx^{n-1}, aq/e, aq^{N+2}; q, x)_\lambda}$$

$$\times \prod_{i=1}^n \frac{(x^{n-i}q^{\lambda_i}, ax^{1-i}q^{\lambda_i}/e, ax^{1-i}q^{\lambda_i+N+1}; x^{-1})_{k_i}}{(ax^{2-n-i}q^{\lambda_i}, ex^{1-i}q^{\lambda_i}, x^{1-i}q^{\lambda_i-N-1}; x^{-1})_{k_i}},$$

$$(b; q, x)_{\lambda-k} \prod_{i=1}^n (bx^{1-i}q^{\lambda_i-k_i}; x)_{k_i} = (b; q, x)_\lambda,$$

$$\frac{1}{(aq/b; x^{-1})_n (aq^2/b; q, x)_{\lambda-k}} \prod_{i=1}^n \frac{E(ax^{1-i}q^{\lambda_i-k_i+1}/b)}{(ax^{1-i}q^{\lambda_i-k_i+1}/b; x)_{k_i}} = \frac{1}{(aq/b; q, x)_\lambda}$$

and similarly with b replaced by c and d. Using the reflection formula $E(x) = -xE(1/x)$ and recalling (3.1), we have

$$\frac{E(ax^{i-n}q^{k_i-\lambda_i+1}/bcd)}{(ax^{1-i}q^{\lambda_i-k_i+N+1}/e;x)_{k_i}} = q^{k_i}\frac{E(ex^{i-1}q^{-\lambda_i-N-1}/a)}{(ax^{1-i}q^{\lambda_i+N+1}/e;x^{-1})_{k_i}}.$$

Considering the four cases k_i, $k_j = 0, 1$ separately, we find that the factor in curly brackets may be written as

$$\frac{E(x^{j-i+k_j-k_i}q^{\lambda_i-\lambda_j}, ax^{3-i-j-k_i-k_j}q^{\lambda_i+\lambda_j}, aqx^{1-i-j})}{E(x^{j-i}, ax^{3-i-j}, aqx^{2-i-j})}$$

$$\times \frac{(ax^{3-i-j};q)_{\lambda_i+\lambda_j}(x^{j-i+1};q)_{\lambda_i-\lambda_j}}{(aqx^{1-i-j};q)_{\lambda_i+\lambda_j}(qx^{j-i-1};q)_{\lambda_i-\lambda_j}}.$$

Finally, by (2.1), we have

$$(aq;x^{-1})_n \prod_{i=1}^{n}\frac{1}{E(aqx^{2(1-i)})} \prod_{1\le i<j\le n}\frac{E(aqx^{1-i-j})}{E(aqx^{2-i-j})} = 1.$$

These simplifications lead to

$$R = \frac{1}{(aq/bcd;x^{-1})_n}\sum_{\lambda}\prod_{i=1}^{n}\left(E(ex^{i-1}q^{-\lambda_i-N-1}/a)q^{\lambda_i}x^{2(i-1)\lambda_i}\right)$$

$$\times \prod_{1\le i<j\le n}\left(\frac{E(x^{j-i}q^{\lambda_i-\lambda_j}, ax^{2-i-j}q^{\lambda_i+\lambda_j})}{E(x^{j-i}, ax^{3-i-j})}\frac{(ax^{3-i-j};q)_{\lambda_i+\lambda_j}}{(aqx^{1-i-j};q)_{\lambda_i+\lambda_j}}\right.$$

$$\left.\times \frac{(x^{j-i+1};q)_{\lambda_i-\lambda_j}}{(qx^{j-i-1};q)_{\lambda_i-\lambda_j}}\right)\frac{(b,c,d,aqx^{1-n},eq,q^{-N};q,x)_{\lambda}}{(qx^{n-1},aq/b,aq/c,aq/d,aq/e,aq^{N+2};q,x)_{\lambda}}$$

$$\times \sum_{k}\prod_{1\le i<j\le n}\frac{E(x^{j-i+k_j-k_i}q^{\lambda_i-\lambda_j})}{E(x^{j-i}q^{\lambda_i-\lambda_j})}\frac{E(ax^{3-i-j-k_i-k_j}q^{\lambda_i+\lambda_j})}{E(ax^{2-i-j}q^{\lambda_i+\lambda_j})}$$

$$\times \prod_{i=1}^{n}\frac{(x^{n-i}q^{\lambda_i},ax^{1-i}q^{\lambda_i}/e,ax^{1-i}q^{\lambda_i+N+1},ex^{1-i}q^{\lambda_i-N-1};x^{-1})_{k_i}(-1)^{k_i}}{(ax^{2-n-i}q^{\lambda_i},ex^{1-i}q^{\lambda_i},x^{1-i}q^{\lambda_i-N-1},ax^{1-i}q^{\lambda_i+N+1}/e;x^{-1})_{k_i}x^{(i-1)k_i}}.$$

The sum in k is the left-hand side of (2.3) with

$$(a,b,c,d,e,x_i,q) \mapsto (ax, x^{n-1}, a/e, aq^{N+1}, eq^{-N-1}, x^{1-i}q^{\lambda_i}, x^{-1}),$$

and thus equals

$$\prod_{i=1}^{n}\frac{E(ax^{2(1-i)}q^{2\lambda_i},ex^{i-n},x^{i-n}q^{-N-1},ex^{i-1}q^{-N-1}/a)}{E(ax^{2-i-n}q^{\lambda_i},ex^{1-i}q^{\lambda_i},x^{1-i}q^{\lambda_i-N-1},ex^{i-1}q^{-\lambda_i-N-1}/a)}$$

$$= \frac{(aq/bcd;x^{-1})_n(e,q^{-N-1};q,x)_{\lambda}}{(eq,q^{-N};q,x)_{\lambda}}\prod_{i=1}^{n}\frac{E(ax^{2(1-i)}q^{2\lambda_i})}{E(ax^{2-i-n}q^{\lambda_i},ex^{i-1}q^{-\lambda_i-N-1}/a)}.$$

Finally, we use (2.1) to write

$$(aqx^{1-n};q,x)_{\lambda}\prod_{i=1}^{n}\frac{1}{E(ax^{2-i-n}q^{\lambda_i})}\prod_{1\le i<j\le n}\frac{1}{E(ax^{3-i-j})}$$

$$= (ax^{1-n};q,x)_{\lambda}\prod_{i=1}^{n}\frac{1}{E(ax^{2(1-i)})}\prod_{1\le i<j\le n}\frac{1}{E(ax^{2-i-j})}.$$

Putting all this together we find that R equals

$$\sum_\lambda \prod_{i=1}^n \left(\frac{E(ax^{2(1-i)}q^{2\lambda_i})}{E(ax^{2(1-i)})} q^{\lambda_i} x^{2(i-1)\lambda_i} \right) \prod_{1\leq i<j\leq n} \left(\frac{E(x^{j-i}q^{\lambda_i-\lambda_j})}{E(x^{j-i})} \right.$$

$$\left. \times \frac{E(ax^{2-i-j}q^{\lambda_i+\lambda_j})}{E(ax^{2-i-j})} \frac{(ax^{3-i-j};q)_{\lambda_i+\lambda_j}(x^{j-i+1};q)_{\lambda_i-\lambda_j}}{(aqx^{1-i-j};q)_{\lambda_i+\lambda_j}(qx^{j-i-1};q)_{\lambda_i-\lambda_j}} \right)$$

$$\times \frac{(b,c,d,ax^{1-n},e,q^{-N-1};q,x)_\lambda}{(qx^{n-1},aq/b,aq/c,aq/d,aq/e,aq^{N+2};q,x)_\lambda},$$

which is indeed the left-hand side of (2.2) with N replaced by $N+1$. This completes the proof of Theorem 2.1.

4. Duality

In this section we prove a duality property for sums of the type occurring in Theorem 2.1. To state the result, we use the notation

$$_{r+1}\Omega_r^{(n)}(a;b_1,\ldots,b_{r-3},q^{-N};q,x) = \sum_{\lambda \in \Lambda_{nN}} \prod_{i=1}^n \left(\frac{E(ax^{2(1-i)}q^{2\lambda_i})}{E(ax^{2(1-i)})} q^{\lambda_i} x^{2(i-1)\lambda_i} \right)$$

$$\times \prod_{1\leq i<j\leq n} \left(\frac{E(x^{j-i}q^{\lambda_i-\lambda_j})}{E(x^{j-i})} \frac{E(ax^{2-i-j}q^{\lambda_i+\lambda_j})}{E(ax^{2-i-j})} \frac{(ax^{3-i-j};q)_{\lambda_i+\lambda_j}(x^{j-i+1};q)_{\lambda_i-\lambda_j}}{(aqx^{1-i-j};q)_{\lambda_i+\lambda_j}(qx^{j-i-1};q)_{\lambda_i-\lambda_j}} \right)$$

$$\times \frac{(ax^{1-n},b_1,\ldots,b_{r-3},q^{-N};q,x)_\lambda}{(qx^{n-1},aq/b_1,\ldots,aq/b_{r-3},aq^{N+1};q,x)_\lambda}.$$

It is natural to assume the balanced condition $(aq)^{r-3} = (x^{n-1}q^{1-N}\prod_i b_i)^2$, though we do not need it to prove the following proposition.

PROPOSITION 4.1. *One has*

$$_{r+1}\Omega_r^{(n)}(a;b_1,\ldots,b_{r-3},q^{-N};q,x) = {_{r+1}\Omega_r^{(N)}}(aqx;b_1,\ldots,b_{r-3},x^n;x^{-1},q^{-1}).$$

In fact, this holds as a termwise symmetry between the two sums, the change of summation variable $\Lambda_{nN} \to \Lambda_{Nn}$ being conjugation of partitions. Let us write λ' for the conjugate of a partition λ. Note that, since we consider partitions into non-negative parts, λ' depends not only on the Young diagram of λ but also on the choice of n and N. For instance, $(3,2,0) \in \Lambda_{33}$ and $(3,2,0) \in \Lambda_{34}$ has conjugate $(2,2,1)$ and $(2,2,1,0)$, respectively.

To prove Proposition 4.1, we observe that, since $(b;q,x)_\lambda = (b;x^{-1},q^{-1})_{\lambda'}$, it is enough to show that, for $\lambda \in \Lambda_{nN}$, the two quantities

$$A_\lambda = \prod_{1\leq i\leq j\leq n} \frac{E(ax^{2-i-j}q^{\lambda_i+\lambda_j})}{E(ax^{2-i-j})} \prod_{1\leq i<j\leq n} \frac{(ax^{3-i-j};q)_{\lambda_i+\lambda_j}}{(aqx^{1-i-j};q)_{\lambda_i+\lambda_j}} \frac{(ax^{1-n};q,x)_\lambda}{(aq^{N+1};q,x)_\lambda}$$

and

$$B_\lambda = \prod_{i=1}^n q^{\lambda_i} x^{2(i-1)\lambda_i} \prod_{1\leq i<j\leq n} \frac{E(x^{j-i}q^{\lambda_i-\lambda_j})}{E(x^{j-i})} \frac{(x^{j-i+1};q)_{\lambda_i-\lambda_j}}{(qx^{j-i-1};q)_{\lambda_i-\lambda_j}} \frac{(q^{-N};q,x)_\lambda}{(qx^{n-1};q,x)_\lambda}$$

are invariant under the transformation $(a,q,x,n,N,\lambda) \mapsto (aqx,x^{-1},q^{-1},N,n,\lambda')$.

We prove the invariance of A_λ, the case of B_λ being similar. We fix n and N and proceed by induction on the number of boxes in the Young diagram of λ, starting from the trivial case of zero boxes. Suppose that the invariance holds for a fixed partition λ. We will show that it also holds for any partition λ^+ obtained by adding a box to the Young diagram of λ. There exist k and l with $1 \leq k \leq n$, $1 \leq l \leq N$ such that $\lambda_j^+ = \lambda_j$ for $j \neq k$, $\lambda_k = l - 1$ and $\lambda_k^+ = l$. After straight-forward simplifications, we may write

$$\frac{A_{\lambda^+}}{A_\lambda} = \frac{E(ax^{2-2k}q^{2l}, ax^{1-2k}q^{2l-1}, ax^{2-n-k}q^{l-1})}{E(ax^{2-2k}q^{2l-1}, ax^{3-2k}q^{2l-2}, ax^{1-k}q^{l+N})}$$
$$\times \prod_{i=1}^{n} \frac{E(ax^{2-i-k}q^{\lambda_i+l}, ax^{3-i-k}q^{\lambda_i+l-1})}{E(ax^{2-i-k}q^{\lambda_i+l-1}, ax^{1-i-k}q^{\lambda_i+l})}.$$

Next we observe that

$$\prod_{i=1}^{n} \frac{E(bx^{1-i}q^{\lambda_i})}{E(bx^{-i}q^{\lambda_i})} = \prod_{i=1}^{\lambda'_N} \frac{E(bx^{1-i}q^N)}{E(bx^{-i}q^N)} \prod_{i=\lambda'_N+1}^{\lambda'_{N-1}} \frac{E(bx^{1-i}q^{N-1})}{E(bx^{-i}q^{N-1})} \times \cdots \times \prod_{i=\lambda'_1+1}^{n} \frac{E(bx^{1-i})}{E(bx^{-i})}$$
$$= \frac{E(bq^N)}{E(bx^{-\lambda'_N}q^N)} \frac{E(bx^{-\lambda'_N}q^{N-1})}{E(bx^{-\lambda'_{N-1}}q^{N-1})} \cdots \frac{E(bx^{-\lambda'_1})}{E(bx^{-n})} = \frac{E(bq^N)}{E(bx^{-n})} \prod_{i=1}^{N} \frac{E(bx^{-\lambda'_i}q^{i-1})}{E(bx^{-\lambda'_i}q^i)},$$

which gives

$$\frac{A_{\lambda^+}}{A_\lambda} = \frac{E(ax^{2-2k}q^{2l}, ax^{1-2k}q^{2l-1}, ax^{2-k}q^{l+N-1})}{E(ax^{2-2k}q^{2l-1}, ax^{3-2k}q^{2l-2}, ax^{1-k-n}q^l)}$$
$$\times \prod_{i=1}^{N} \frac{E(ax^{1-k-\lambda'_i}q^{l+i-1}, ax^{2-k-\lambda'_i}q^{l+i-2})}{E(ax^{1-k-\lambda'_i}q^{l+i}, ax^{2-k-\lambda'_i}q^{l+i-1})}.$$

This agrees with the expression obtained from the previous one by substituting $(a, q, x, n, N, \lambda, k, l) \mapsto (aqx, x^{-1}, q^{-1}, N, n, \lambda', l, k)$. Thus the invariance of A_λ implies that of A_{λ^+}.

References

[DJ] E. Date, M. Jimbo, A. Kuniba, T. Miwa and M. Okado, *Exactly solvable SOS models: local height probabilities and theta function identities*, Nuclear Phys. B 290 (1987), 231–273.

[DG] R. Y. Denis and R. A. Gustafson, *An SU(n) q-beta integral transformation and multiple hypergeometric series identities*, SIAM J. Math. Anal. 23 (1992), 552–561.

[D] J. F. van Diejen, *On certain multiple Bailey, Rogers and Dougall type summation formulas*, Publ. Res. Inst. Math. Sci. 33 (1997), 483–508.

[DS] J. F. van Diejen and V. P. Spiridonov, *An elliptic Macdonald-Morris conjecture and modular hypergeometric sums*, Math. Res. Lett. 7 (2000), 729–746.

[DSt] J. F. van Diejen and J. V. Stokman, *Multivariable q-Racah polynomials*, Duke Math. J. 91 (1998), 89–136.

[FT] I. B. Frenkel and V. G. Turaev, *Elliptic solutions of the Yang-Baxter equation and modular hypergeometric functions*, The Arnold–Gelfand mathematical seminars, 171–204, Birkhäuser, Boston, MA, 1997.

[GR] G. Gasper and M. Rahman, *Basic Hypergeometric Series*, Cambridge University Press, Cambridge, 1990.

[G] R. A. Gustafson, *Some q-beta integrals on* SU(n) *and* Sp(n) *that generalize the Askey-Wilson and Nasrallah-Rahman integrals*, SIAM J. Math. Anal. 25 (1994), 441–449.

[ML] S. C. Milne and G. M. Lilly, *Consequences of the A_l and C_l Bailey transform and Bailey lemma*, Discrete Math. 139 (1995), 319–346.

[NR] B. Nassrallah and M. Rahman, *Projection formulas, a reproducing kernel and a generating function for q-Wilson polynomials*, SIAM J. Math. Anal. 16 (1985), 186–197.

[R] S. M. Ruijsenaars, *First order analytic difference equations and integrable quantum systems*, J. Math. Phys. 38 (1997), 1069–1146.

[S] M. Schlosser, *Summation theorems for multidimensional basic hypergeometric series by determinant evaluations*, Discrete Math. 210 (2000), 151–169.
[Sp] V. P. Spiridonov, *On an elliptic beta function*, Russian Math. Surveys, to appear.
[W] S. O. Warnaar, *Summation and transformation formulas for elliptic hypergeometric series*, math.QA/0001006.

DEPARTMENT OF MATHEMATICS, CHALMERS UNIVERSITY OF TECHNOLOGY AND GÖTEBORG UNIVERSITY, SE-412 96 GÖTEBORG, SWEDEN
E-mail address: hjalmar@math.chalmers.se

Multilateral transformations of q-series with quotients of parameters that are nonnegative integral powers of q

Michael Schlosser

ABSTRACT. We give multidimensional generalizations of several transformation formulae for basic hypergeometric series of a specific type. Most of the upper parameters of the series differ multiplicatively from corresponding lower parameters by a nonnegative integral power of the base q. In one dimension, formulae for such series have been found, in the $q \to 1$ case, by B. M. Minton and P. W. Karlsson, and in the basic case by G. Gasper, by W. C. Chu, and more recently by the author. Our identities involve multilateral basic hypergeometric series associated to the root system A_r (or equivalently, the unitary group $U(r+1)$).

1. Introduction

The theory of hypergeometric and basic hypergeometric (or q-hypergeometric) series (cf. L. J. Slater [33], and G. Gasper and M. Rahman [13]) contains numerous summation and transformation formulae. Many of these appear in applications including number theory, combinatorics, physics, representation theory, and computer algebra (see e.g. G. E. Andrews [1]).

One particular example is B. M. Minton's [26] summation formula, found in 1970, which is useful for simplifying sums that arise in certain problems in theoretical physics (such as Racah coefficients). B. M. Minton's formula is of special interest since it sums a specific hypergeometric series with an arbitrary number of parameters. B. M. Minton derived his formula by expanding a hypergeometric series in terms of other hypergeometric series, exploiting an identity already obtained by C. Fox [9] in 1925 (but published in 1927). B. M. Minton iterated this expansion and suitably specialized the parameters to successively evaluate the (inner) sums. A condition on the parameters of the specific hypergeometric series considered by B. M. Minton is that most of the upper parameters differ from corresponding lower ones by a nonnegative integer. B. M. Minton's result was slightly extended by P. W. Karlsson [19] who was using the same method.

In the early 1980's, G. Gasper [10] found q-analogues of Karlsson and Minton's results. In the basic case, the condition on the parameters is that most of the upper

---

2000 *Mathematics Subject Classification.* Primary 33D15; Secondary 33D67.

Key words and phrases. bilateral basic hypergeometric series, A_r series, $U(r+1)$ series, Karlsson–Minton type identities.

parameters differ *multiplicatively* from corresponding lower ones by a nonnegative integral power of q. G. Gasper even extended his results to a transformation formula [**10**, Eq. (19)]. For the above material, see the exposition in G. Gasper and M. Rahman [**13**], in particular Section 1.9, and Exercises 1.30 and 1.34.

Note that G. Gasper and M. Rahman [**13**] use the terminology "Karlsson–Minton" and "q-Karlsson–Minton", respectively, to denote the type of the series in question. We are dropping this terminology in the present paper, since the work is based on expanding a hypergeometric function in terms of another, which has a longer history. Instead, we introduce the acronyms IPD and q-IPD, respectively, where IPD stands for "**I**ntegral **P**arameter **D**ifferences" (motivated by the title of P. W. Karlsson's [**19**] article), see Section 2. It should be mentioned that expansions of hypergeometric series in terms of other hypergeometric series have also been obtained by J. L. Fields and J. Wimp [**8**], by A. Verma [**35**], and in more generality (concerning identities between general sequences), by J. L. Fields and M. E. H. Ismail [**7**]. Thus, as pointed out to us by Mourad Ismail [**18**], one may easily write generalizations of the Karlsson–Minton formulae to series involving partly hypergeometric coefficients and partly general sequences.

By using an essentially different method, namely by partial fraction expansions, W. C. Chu [**5**] generalized G. Gasper's q-IPD type identities further to a bilateral series transformation. In another article, G. Gasper [**11**, Eq. (5.13)] found a new summation for a very-well-poised basic hypergeometric series of q-IPD type. Again, W. C. Chu [**6**] extended G. Gasper's result to a summation for a very-well-poised *bilateral* basic hypergeometric series.

Very recently, the author [**31**, Sec. 8] found even more general identities of q-IPD type, by elementary manipulations of series, using L. J. Slater's [**32**] general transformations for bilateral basic hypergeometric series. Already earlier J. Haglund [**17**, pp. 415–416] had discovered that W. C. Chu's [**5**] bilateral transformation formula can be obtained by specializing L. J. Slater's [**32**] general transformation for ${}_t\psi_t$ series.

In this article, we provide *multidimensional* extensions of several specific transformation formulae of q-IPD type, in particular, multivariate extensions of the identities in Propositions 2.1, 2.2, 2.3 and 2.4. These multivariate extensions involve multiple basic hypergeometric series associated to the root system A_{r-1} (or equivalently, the unitary group $U(r)$). Such type of series are considered in the work of R. A. Gustafson, S. C. Milne, and several other authors, see e.g. [**4**], [**14**] [**15**], [**16**], [**20**], [**21**], [**22**], [**23**], [**24**], [**25**], [**27**], [**28**], and [**29**].

As a matter of fact, there are unfortunately no suitable *multidimensional* extensions of L. J. Slater's [**32**] general transformation formulae known (yet). Thus, in higher dimensions we cannot specialize down from such higher level identities. Instead we proceed from lower level identities to systematically derive the upper level ones. In this fashion, using certain A_{r-1} summation theorems (from R. A. Gustafson [**15**] and S. C. Milne [**22**]), elementary manipulation of series, and induction, we prove two multilateral transformations of q-IPD type, namely Theorems 4.2 and 4.6. The first one of these, Theorem 4.2, involves *very-well-poised* multilateral series (over A_{r-1}), and contains r-dimensional generalizations of W. C. Chu's [**6**, Theorem 2] and G. Gasper's [**11**, Eq. (5.13)] summations as special cases, see Corollaries 4.3 and 4.4, respectively. The other transformation formula in Theorem 4.6, involves multilateral series with an arbitrary argument

z. Four other multilateral transformations of q-IPD type are derived by simpler means, using tools developed in [**25**], see Theorems 4.7, 4.8, 4.9, and 4.10.

In [**29**, Theorem 6.4], we already gave some multiple series generalizations (associated to the root systems of classical type) of W. C. Chu's [**5**] bilateral transformation. The multiple series identities in [**29**] were derived by using one-dimensional identities, combined with certain determinant evaluations. Using the same determinantal method, one could also deduce multilateral generalizations of L. J. Slater's [**32**] general transformation formulae, and in particular of the q-IPD type transformations which were found in [**31**, Sec. 8]. However, most of the results of this article, in particular Theorems 4.2 and 4.6, are deeper since they are derived by using genuine multidimensional summation theorems.

Our article is organized as follows. In Section 2, we introduce some standard notation for q-series and basic hypergeometric series, and state several important one-dimensional results. In Section 3, we consider multiple series and recollect some specific ingredients which we need in Section 4 to state and prove our multilateral identities of q-IPD type.

2. Notation and one-dimensional results

In order to state and prove our theorems, we employ some standard q-series notation (cf. G. Gasper and M. Rahman [**13**]). For a complex number q with $0 < |q| < 1$, define the *q-shifted factorial* by

$$(a;q)_\infty := \prod_{j=0}^{\infty}(1-aq^j),$$

and

(2.1) $$(a;q)_k := \frac{(a;q)_\infty}{(aq^k;q)_\infty}, \quad \text{where } k \text{ is an integer.}$$

Further, for brevity, we also employ the notation

$$(a_1,\ldots,a_m;q)_k \equiv (a_1;q)_k \ldots (a_m;q)_k,$$

where k is an integer or infinity. Further, we utilize the notations

(2.2) $${}_t\phi_{t-1}\begin{bmatrix} a_1, a_2, \ldots, a_t \\ b_1, b_2, \ldots, b_{t-1} \end{bmatrix}; q, z \end{bmatrix} := \sum_{k=0}^{\infty} \frac{(a_1,a_2,\ldots,a_t;q)_k}{(q,b_1,\ldots,b_{t-1};q)_k} z^k,$$

and

(2.3) $${}_t\psi_t\begin{bmatrix} a_1, a_2, \ldots, a_t \\ b_1, b_2, \ldots, b_t \end{bmatrix}; q, z \end{bmatrix} := \sum_{k=-\infty}^{\infty} \frac{(a_1,a_2,\ldots,a_t;q)_k}{(b_1,b_2,\ldots,b_t;q)_k} z^k,$$

for *basic hypergeometric ${}_t\phi_{t-1}$ series*, and *bilateral basic hypergeometric ${}_t\psi_t$ series*, respectively. Note that G. Gasper and M. Rahman [**13**] have more general definitions for ${}_r\phi_s$ series and for ${}_r\psi_s$ series, but in this article we are only really concerned with the case where $s = r - 1$ for the ${}_r\phi_s$ series, and where $r = s$ for the ${}_r\psi_s$ series.

Clearly, a bilateral ${}_t\psi_t$ series becomes a unilateral ${}_t\phi_{t-1}$ series if one of the lower parameters, say b_t, is q (or more generally, q^j where j is a positive integer). This is because $(q;q)_k^{-1} = 0$, for $k = -1, -2, \ldots$, by definition (2.1). In this case, the ${}_t\psi_t$ series terminates naturally from below. On the other hand, if in a ${}_t\phi_{t-1}$ series one of the upper parameters, say a_t, equals q^{-n}, where n is a nonnegative integer, then the ${}_t\phi_{t-1}$ series terminates naturally from above. This is because $(q^{-n};q)_k = 0$, for

$k = n+1, n+2, \ldots$, by definition (2.1). Such a ${}_t\phi_{t-1}$ series terminates after $n+1$ terms.

The ratio test gives simple criteria of when the above series converge, if they do not terminate. Remember that we assume $0 < |q| < 1$. The ${}_t\phi_{t-1}$ series in (2.2) converges absolutely in the radius $|z| < 1$, while the ${}_t\psi_t$ series in (2.3) converges absolutely in the annulus $|b_1 \ldots b_t / a_1 \ldots a_t| < |z| < 1$.

The classical theory of basic hypergeometric series consists of several summation and transformation formulae involving ${}_t\phi_{t-1}$ series. The classical summation theorems for terminating ${}_3\phi_2$, ${}_6\phi_5$, and ${}_8\phi_7$ series require that the parameters satisfy the additional condition of being either balanced and/or very-well-poised. A ${}_t\phi_{t-1}$ basic hypergeometric series is called *balanced* if $b_1 \cdots b_{t-1} = a_1 \cdots a_t q$ and $z = q$. An ${}_t\phi_{t-1}$ series is *well-poised* if $a_1 q = a_2 b_1 = \cdots = a_t b_{t-1}$. It is called *very-well-poised* if it is well-poised and if $a_2 = q\sqrt{a_1}$ and $a_3 = -q\sqrt{a_1}$. Note that the factor

$$(2.4) \qquad \frac{(q\sqrt{a_1}, -q\sqrt{a_1}; q)_k}{(\sqrt{a_1}, -\sqrt{a_1}; q)_k} = \frac{1 - a_1 q^{2k}}{1 - a_1}$$

appears in a very-well-poised series. The parameter a_1 is usually referred to as the *special parameter* of such a series, and we call (2.4) the *very-well-poised term* of the series. Similarly, a bilateral ${}_t\psi_t$ basic hypergeometric series is well-poised if $a_1 b_1 = a_2 b_2 \cdots = a_t b_t$ and very-well-poised if, in addition, $a_1 = -a_2 = qb_1 = -qb_2$.

In our proofs in Section 4, we often make use of some elementary identities involving q-shifted factorials, listed in G. Gasper and M. Rahman [13, Appendix I].

With the above notations for basic hypergeometric and bilateral basic hypergeometric series, we are ready to state some important (one-dimensional) summation formulae.

One of the most fundamental summation theorems in the theory of (bilateral) basic hypergeometric series is W. N. Bailey's [2] very-well-poised ${}_6\psi_6$ summation,

$$(2.5) \quad {}_6\psi_6 \left[\begin{matrix} q\sqrt{a}, -q\sqrt{a}, b, c, d, e \\ \sqrt{a}, -\sqrt{a}, aq/b, aq/c, aq/d, aq/e \end{matrix} ; q, \frac{a^2 q}{bcde} \right]$$
$$= \frac{(aq, aq/bc, aq/bd, aq/be, aq/cd, aq/ce, aq/de, q, q/a; q)_\infty}{(aq/b, aq/c, aq/d, aq/e, q/b, q/c, q/d, q/e, a^2q/bcde; q)_\infty},$$

provided the series either terminates, or $|q| < 1$ and $|a^2 q/bcde| < 1$, for convergence. For a simple proof of (2.5) using elementary manipulations of series, see [30].

Another important summation is the terminating balanced q-Pfaff–Saalschütz summation (cf. [13, Eq. (II.12)]),

$$(2.6) \qquad {}_3\phi_2 \left[\begin{matrix} a, b, q^{-n} \\ c, abq^{1-n}/c \end{matrix} ; q, q \right] = \frac{(c/a, c/b; q)_n}{(c, c/ab; q)_n}.$$

S. Ramanujan's ${}_1\psi_1$ summation (cf. [13, Eq. (5.2.1)]) reads as follows,

$$(2.7) \qquad {}_1\psi_1 \left[\begin{matrix} a \\ b \end{matrix} ; q, z \right] = \frac{(q, b/a, az, q/az; q)_\infty}{(b, q/a, z, b/az; q)_\infty},$$

provided the series either terminates, or $|q| < 1$ and $|b/a| < |z| < 1$, for convergence.

Finally, the terminating q-binomial theorem is (cf. [13, Eq. (II.4)])

$$(2.8) \qquad {}_1\phi_0 \left[\begin{matrix} q^{-n} \\ - \end{matrix} ; q, z \right] = (zq^{-n}; q)_n.$$

Note that (2.8) is just the special case $a \to q^{-n}$, $b \to q$ of (2.7).

In this article, we prove multidimensional extensions (associated to the root system A_{r-1}) of four transformations of q-IPD type, namely Propositions 2.1, 2.2, 2.3, and 2.4. We need to explain our terminology first.

We say that a basic hypergeometric series is of *q-IPD type* if there are s upper parameters $a_1, \ldots, a_s$ and s lower parameters $b_1, \ldots, b_s$ such that each a_i differs from b_i multiplicatively by a nonnegative integral power of q, i.e. $a_i = b_i q^{m_i}$, $m_i \geq 0$. G. Gasper [10] found some summation formulae for particular basic hypergeometric series of such type. These were q-analogues of formulae originally discovered by B. M. Minton [26] and P. W. Karlsson [19], using C. Fox' [9] expansion of a hypergeometric function in terms of other hypergeometric functions. We call the series considered by B. M. Minton and P. W. Karlsson to be of *IPD type*, where IPD stands for "**I**ntegral **P**arameter **D**ifferences", motivated by the title of P. W. Karlsson's [19] article. G. Gasper [10, Eq. (19)] also extended his summations to a transformation formula. Later, W. C. Chu [5] found *bilateral* summations and transformations of q-IPD type, generalizing G. Gasper's identities of [10]. In an expository paper, G. Gasper [11, Eq. (5.13)] derived a summation formula for a specific *very-well-poised* basic hypergeometric series of q-IPD type. His result was then generalized to a summation for bilateral series, again by W. C. Chu [6, Theorem 2]. (It is maybe interesting that as application W. C. Chu [6, Eq. (5.25)] applied an inverse relation to his bilateral summation and (re-)derived an important bibasic identity, actually due to G. Gasper and M. Rahman [12, Eq. (2.8)]. This shows how strongly seemingly different aspects in q-series are interconnected.)

In a recent article [31, Sec. 8], the author found formulae of q-IPD type covering all of the above q-IPD type identities as special cases. In the following, we list the four transformation formulae from [31, Sec. 8] which we extend to higher dimensions. The first one of these involves *very-well-poised* bilateral basic hypergeometric series.

PROPOSITION 2.1 (A bilateral very-well-poised q-IPD type transformation). *Let a, b, c, d, e, f, and $h_1, \ldots, h_s$ be indeterminate, let $m_1, \ldots, m_s$ be nonnegative integers, let $|m| = \sum_{i=1}^{s} m_i$, and suppose that the series in (2.9) are well-defined. Then*

$$(2.9) \quad {}_{6+2s}\psi_{6+2s}\left[\begin{array}{c} q\sqrt{a}, -q\sqrt{a}, b, c, d, e, \\ \sqrt{a}, -\sqrt{a}, \frac{aq}{b}, \frac{aq}{c}, \frac{aq}{d}, \frac{aq}{e}, \end{array}\right.$$
$$\left.\begin{array}{c} h_1, \ldots, h_s, \frac{aq^{1+m_1}}{h_1}, \ldots, \frac{aq^{1+m_s}}{h_s} \\ \frac{aq}{h_1}, \ldots, \frac{aq}{h_s}, h_1 q^{-m_1}, \ldots, h_s q^{-m_s} \end{array}; q, \frac{a^2 q^{1-|m|}}{bcde}\right]$$
$$= \frac{(a, \frac{q}{a}, \frac{fq}{b}, \frac{fq}{c}, \frac{fq}{d}, \frac{fq}{e}, \frac{aq}{bf}, \frac{aq}{cf}, \frac{aq}{df}, \frac{aq}{ef}; q)_\infty}{(\frac{q}{b}, \frac{q}{c}, \frac{q}{d}, \frac{q}{e}, \frac{aq}{b}, \frac{aq}{c}, \frac{aq}{d}, \frac{aq}{e}, \frac{f^2 q}{a}, \frac{aq}{f^2}; q)_\infty} \prod_{i=1}^{s} \frac{(\frac{fq}{h_i}, \frac{aq}{fh_i}; q)_{m_i}}{(\frac{aq}{h_i}, \frac{q}{h_i}; q)_{m_i}}$$
$$\times {}_{6+2s}\psi_{6+2s}\left[\begin{array}{c} \frac{qf}{\sqrt{a}}, -\frac{qf}{\sqrt{a}}, \frac{bf}{a}, \frac{cf}{a}, \frac{df}{a}, \frac{ef}{a}, \\ \frac{f}{\sqrt{a}}, -\frac{f}{\sqrt{a}}, \frac{fq}{b}, \frac{fq}{c}, \frac{fq}{d}, \frac{fq}{e}, \end{array}\right.$$
$$\left.\begin{array}{c} \frac{fh_1}{a}, \ldots, \frac{fh_s}{a}, \frac{fq^{1+m_1}}{h_1}, \ldots, \frac{fq^{1+m_s}}{h_s} \\ \frac{fq}{h_1}, \ldots, \frac{fq}{h_s}, \frac{fh_1 q^{-m_1}}{a}, \ldots, \frac{fh_s q^{-m_s}}{a} \end{array}; q, \frac{a^2 q^{1-|m|}}{bcde}\right],$$

where the series either terminate, or $|a^2 q^{1-|m|}/bcde| < 1$, for convergence.

Note that f does not appear on the left side of (2.9).

The special case $f \mapsto b$, $c \mapsto a/b$ of Proposition 2.1 is exactly W. C. Chu's summation in [6, Theorem 2]. If we specialize this summation then further by setting $e \mapsto a$ we arrive at G. Gasper's [11, Eq. (5.13)] summation.

The following transformation formula involves bilateral basic hypergeometric series with an independent argument z.

PROPOSITION 2.2 (A bilateral q-IPD type transformation). *Let a, b, c, z, and $h_1, \ldots, h_s$ be indeterminate, let $m_1, \ldots, m_s$ be nonnegative integers, let $|m| = \sum_{i=1}^{s} m_i$, and suppose that the series in (2.10) are well-defined. Then*

$$(2.10) \quad {}_{1+s}\psi_{1+s}\left[\begin{matrix} a, h_1 q^{m_1}, \ldots, h_s q^{m_s} \\ b, h_1, \ldots, h_s \end{matrix} ; q, z\right]$$

$$= \frac{(c/a, bq/c, az, q/az; q)_\infty}{(q/a, b, azq/c, c/az; q)_\infty} \prod_{i=1}^{s} \frac{(h_i q/c; q)_{m_i}}{(h_i; q)_{m_i}}$$

$$\times {}_{1+s}\psi_{1+s}\left[\begin{matrix} aq/c, h_1 q^{1+m_1}/c, \ldots, h_s q^{1+m_s}/c \\ bq/c, h_1 q/c, \ldots, h_s q/c \end{matrix} ; q, z\right],$$

where the series either terminate, or $|bq^{-|m|}/a| < |z| < 1$, for convergence.

Note that c does not appear on the left side of (2.10).

The next two transformations involve series whose argument depends on the parameters.

PROPOSITION 2.3 (A bilateral q-IPD type transformation). *Let a, b, c, d, e, and $h_1, \ldots, h_s$ be indeterminate, let N be an arbitrary integer, $m_1, \ldots, m_s$ be nonnegative integers, let $|m| = \sum_{i=1}^{s} m_i$, and suppose that the series in (2.11) are well-defined. Then*

$$(2.11) \quad {}_{2+s}\psi_{2+s}\left[\begin{matrix} a, b, h_1 q^{m_1}, \ldots, h_s q^{m_s} \\ c, d, h_1, \ldots, h_s \end{matrix} ; q, \frac{eq^{-N}}{ab}\right]$$

$$= \left(\frac{e}{q}\right)^N \frac{(e/a, e/b, cq/e, dq/e; q)_\infty}{(q/a, q/b, c, d; q)_\infty} \prod_{i=1}^{s} \frac{(h_i q/e; q)_{m_i}}{(h_i; q)_{m_i}}$$

$$\times {}_{2+s}\psi_{2+s}\left[\begin{matrix} aq/e, bq/e, h_1 q^{1+m_1}/e, \ldots, h_s q^{1+m_s}/e \\ cq/e, dq/e, h_1 q/e, \ldots, h_s q/e \end{matrix} ; q, \frac{eq^{-N}}{ab}\right],$$

where the series either terminate, or $|e/ab| < |q^N| < |eq^{|m|}/cd|$, for convergence.

If we reverse the ${}_{2+s}\psi_{2+s}$ series on the right side of (2.11), we obtain

PROPOSITION 2.4 (A bilateral q-IPD type transformation). *Let a, b, c, d, e, and $h_1, \ldots, h_s$ be indeterminate, let N be an arbitrary integer, $m_1, \ldots, m_s$ be nonnegative integers, let $|m| = \sum_{i=1}^{s} m_i$, and suppose that the series in (2.12) are well-defined. Then*

$$(2.12) \quad {}_{2+s}\psi_{2+s}\left[\begin{matrix} a, b, h_1 q^{m_1}, \ldots, h_s q^{m_s} \\ c, d, h_1, \ldots, h_s \end{matrix} ; q, \frac{eq^{-N}}{ab}\right]$$

$$= \left(\frac{e}{q}\right)^N \frac{(e/a, e/b, cq/e, dq/e; q)_\infty}{(q/a, q/b, c, d; q)_\infty} \prod_{i=1}^{s} \frac{(h_i q/e; q)_{m_i}}{(h_i; q)_{m_i}}$$

$$\times {}_{2+s}\psi_{2+s}\left[\begin{matrix} e/c, e/d, e/h_1, \ldots, e/h_s \\ e/a, e/b, eq^{-m_1}/h_1, \ldots, eq^{-m_s}/h_s \end{matrix} ; q, \frac{cdq^{N-|m|}}{e}\right],$$

where the series either terminate, or $|e/ab| < |q^N| < |eq^{|m|}/cd|$, for convergence.

The $e = aq$ case of Proposition 2.4 reduces to W. C. Chu's [5, Eq. (15)] transformation. If we specialize the resulting transformation further by setting $c = q$ we obtain G. Gasper's [10, Eq. (19)] transformation.

Propositions 2.1, 2.2, 2.3, and 2.4 appeared as Corollaries 8.6, 8.3, 8.2 and Equation (8.8) in [31]. They were originally derived as special cases from even more general transformations for bilateral basic hypergeometric series of q-IPD type.

3. Preliminaries on multiple series

In general, we consider multiple series of the form

$$(3.1) \qquad \sum_{k_1,\ldots,k_r=-\infty}^{\infty} S(\mathbf{k}),$$

where $\mathbf{k} = (k_1, \ldots, k_r)$, which reduce to classical (bilateral) basic hypergeometric series when $r = 1$. We call such a multiple basic hypergeometric series *balanced* if it reduces to a balanced series when $r = 1$. We define *well-poised* and *very-well-poised* series analogously. In case these series do not terminate from below, we also call such series *multilateral* basic hypergeometric series.

In our particular cases, we also have

$$(3.2) \qquad \prod_{1 \leq i < j < r} \left(\frac{z_i q^{k_i} - z_j q^{k_j}}{z_i - z_j} \right)$$

(or something similar), as a factor of $S(\mathbf{k})$. A typical example is the right side of (3.5). Since we may associate (3.2) with the product side of the Weyl denominator formula for the root system A_{r-1} (see e.g. D. Stanton [34]), we call our series A_{r-1} basic hypergeometric series, in accordance with I. M. Gessel and C. Krattenthaler [14, Eq. (7.1)]. Note that often in the literature (e.g. [3], [23], [25], [27], [28]) these r-dimensional series are called A_r series instead of A_{r-1} series.

For convenience, we frequently use the notation $|\mathbf{k}| := k_1 + \cdots + k_r$. Note that on the right side of (3.5) we have (in addition to (3.2))

$$(3.3) \qquad \prod_{i=1}^{r} \left(\frac{1 - az_i q^{k_i + |\mathbf{k}|}}{1 - az_i} \right)$$

appearing as a factor in the summand of the series. It is easy to see that the $r = 1$ case of (3.3) essentially reduces to (2.4). To clarify the special appearance of the very-well-poised term in the multidimensional case (and even in the one-dimensional) case, it is useful to view the series in one higher dimension. In particular, we can write

$$(3.4) \qquad \prod_{1 \leq i < j < r} \left(\frac{z_i q^{k_i} - z_j q^{k_j}}{z_i - z_j} \right) \prod_{i=1}^{r} \left(\frac{1 - az_i q^{k_i + (k_1 + \cdots + k_r)}}{1 - az_i} \right)$$

$$= q^{k_1 + \cdots + k_r} \prod_{1 \leq i < j \leq r+1} \left(\frac{z_i q^{k_i} - z_j q^{k_j}}{z_i - z_j} \right),$$

where $z_{r+1} = 1/a$ and $k_{r+1} = -(k_1 + \cdots + k_r)$. Thus, some A_{r-1} basic hypergeometric series identities are sometimes better viewed as identites associated to the affine root system $\tilde{A}_r$ (or, equivalently, the special unitary group $SU(r+1)$). For such an example, see Remark 3.2.

Let a, $b_1, \ldots, b_r$, c, d, $e_1, \ldots, e_r$, $z_1, \ldots, z_r$, and w be indeterminate. For purpose of compact notation, we define for $r \geq 1$

$$(3.5) \quad {}_6\Psi_6^{(r)}\left[a; b_1, \ldots, b_r; c, d; e_1, \ldots, e_r; z_1, \ldots, z_r \,\middle|\, q, w\right]$$

$$:= \sum_{k_1,\ldots,k_r=-\infty}^{\infty} \left(\prod_{1 \leq i < j \leq r} \left(\frac{z_i q^{k_i} - z_j q^{k_j}}{z_i - z_j} \right) \prod_{i=1}^{r} \left(\frac{1 - a z_i q^{k_i + |\mathbf{k}|}}{1 - a z_i} \right) \right.$$

$$\times \prod_{i,j=1}^{r} \frac{(b_j z_i / z_j; q)_{k_i}}{(a z_i q / e_j z_j; q)_{k_i}} \prod_{i=1}^{r} \frac{(e_i z_i; q)_{|\mathbf{k}|}}{(a z_i q / b_i; q)_{|\mathbf{k}|}}$$

$$\left. \times \prod_{i=1}^{r} \frac{(c z_i; q)_{k_i}}{(a z_i q / d; q)_{k_i}} \cdot \frac{(d; q)_{|\mathbf{k}|}}{(a q / c; q)_{|\mathbf{k}|}} w^{|\mathbf{k}|} \right).$$

The above ${}_6\Psi_6^{(r)}$ series is an r-dimensional ${}_6\psi_6$ series (which reduces to a classical very-well-poised ${}_6\psi_6$ when $r=1$).

For convenience, we sometimes use capital letters to abbreviate the (r-fold) products of certain variables. Specifically, in this article we use $A \equiv a_1 \cdots a_r$, $B \equiv b_1 \cdots b_r$, $C \equiv c_1 \cdots c_r$, $E \equiv e_1 \cdots e_r$, and $F \equiv f_1 \cdots f_r$, respectively.

In our derivation of the multilateral q-IPD type transformation in Theorem 4.2 we utilize the following r-dimensional generalization of W. N. Bailey's summation formula in (2.5).

THEOREM 3.1 ((Gustafson) An A_{r-1} ${}_6\psi_6$ summation). *Let a, $b_1, \ldots, b_r$, c, d, $e_1, \ldots, e_r$, and $z_1, \ldots, z_r$ be indeterminate, let $r \geq 1$, and suppose that none of the denominators in (3.6) vanishes. Then*

$$(3.6) \quad {}_6\Psi_6^{(r)}\left[a; b_1, \ldots, b_r; c, d; e_1, \ldots, e_r; z_1, \ldots, z_r \,\middle|\, q, \frac{a^{r+1} q}{BcdE}\right]$$

$$= \frac{(aq/Bc, a^r q/dE, aq/cd; q)_\infty}{(a^{r+1} q/BcdE, aq/c, q/d; q)_\infty} \prod_{i,j=1}^{r} \frac{(a z_i q / b_i e_j z_j, q z_i / z_j; q)_\infty}{(q z_i / b_i z_j, a z_i q / e_j z_j; q)_\infty}$$

$$\times \prod_{i=1}^{r} \frac{(aq/ce_i z_i, a z_i q / b_i d, a z_i q, q / a z_i; q)_\infty}{(a z_i q / b_i, q / e_i z_i, q / c z_i, a z_i q / d; q)_\infty},$$

provided $|a^{r+1} q / BcdE| < 1$.

REMARK 3.2. Using (3.4), the multilateral identity in (3.6) can also be written in a more compact form. We then have R. A. Gustafson's [**15**, Theorem 1.15] $\tilde{A}_r$ ${}_6\psi_6$ summation: Let $a_1, \ldots, a_{r+1}$, $b_1, \ldots, b_{r+1}$, and $z_1, \ldots, z_{r+1}$ be indeterminate, let $r \geq 1$, and suppose that none of the denominators in (3.7) vanishes. Then

$$(3.7) \quad \sum_{\substack{-\infty \leq k_1,\ldots,k_{r+1} \leq \infty \\ k_1 + \cdots + k_{r+1} = 0}} \prod_{1 \leq i < j \leq r+1} \left(\frac{z_i q^{k_i} - z_j q^{k_j}}{z_i - z_j} \right) \prod_{i,j=1}^{r+1} \frac{(a_j z_i / z_j; q)_{k_i}}{(b_j z_i / z_j; q)_{k_i}}$$

$$= \frac{(b_1 \ldots b_{r+1} q^{-r}, q/a_1 \ldots a_{r+1}; q)_\infty}{(q, b_1 \ldots b_{r+1} q^{-r} / a_1 \ldots a_{r+1}; q)_\infty} \prod_{i,j=1}^{r+1} \frac{(q z_i / z_j, b_j z_i / a_i z_j; q)_\infty}{(b_j z_i / z_j, z_i q / a_i z_j; q)_\infty},$$

provided $|b_1 \ldots b_{r+1} q^{-r}/a_1 \ldots a_{r+1}| < 1$. It is not difficult to see that (3.7) and (3.6) are equivalent.

We also need the following r-dimensional generalization of the terminating q-Pfaff–Saalschütz summation from S. C. Milne [**22**, Theorem 4.15].

THEOREM 3.3 ((Milne) An A_{r-1} terminating $_3\phi_2$ summation). *Let $a_1, \ldots, a_r$, b, c, and $x_1, \ldots, x_r$, be indeterminate, let N be a nonnegative integer, let $r \geq 1$, and suppose that none of the denominators in (3.8) vanishes. Then*

$$(3.8) \quad \sum_{\substack{k_1,\ldots,k_r \geq 0 \\ 0 \leq |\mathbf{k}| \leq N}} \left(\prod_{1 \leq i < j \leq r} \left(\frac{x_i q^{k_i} - x_j q^{k_j}}{x_i - x_j} \right) \prod_{i,j=1}^{r} \frac{(a_j x_i/x_j; q)_{k_i}}{(q x_i/x_j; q)_{k_i}} \right.$$

$$\left. \times \prod_{i=1}^{r} \frac{(bx_i; q)_{k_i}}{(cx_i; q)_{k_i}} \cdot \frac{(q^{-N}; q)_{|\mathbf{k}|}}{(a_1 \ldots a_r b q^{1-N}/c; q)_{|\mathbf{k}|}} q^{|\mathbf{k}|} \right)$$

$$= \frac{(c/b; q)_N}{(c/a_1 \ldots a_r b; q)_N} \prod_{i=1}^{r} \frac{(cx_i/a_i; q)_N}{(cx_i; q)_N}.$$

The $r = 1$ case of (3.8) clearly reduces to (2.6).

In our derivation of the multilateral q-IPD type transformation in Theorem 4.6 we utilize R. A. Gustafson's [**15**, Theorem 1.17] A_{r-1} extension of S. Ramanujan's $_1\psi_1$ summation (2.7).

THEOREM 3.4 ((Gustafson) An A_{r-1} $_1\psi_1$ summation). *Let $a_1, \ldots, a_r$, $b_1, \ldots, b_r$, $x_1, \ldots, x_r$, and z be indeterminate, let $r \geq 1$, and suppose that none of the denominators in (3.9) vanishes. Then*

$$(3.9) \quad \sum_{k_1,\ldots,k_r=-\infty}^{\infty} \prod_{1 \leq i < j \leq r} \left(\frac{x_i q^{k_i} - x_j q^{k_j}}{x_i - x_j} \right) \prod_{i,j=1}^{r} \frac{(a_j x_i/x_j; q)_{k_i}}{(b_j x_i/x_j; q)_{k_i}} z^{|\mathbf{k}|}$$

$$= \frac{(Az, q/Az; q)_\infty}{(z, Bq^{1-r}/Az; q)_\infty} \prod_{i,j=1}^{r} \frac{(b_j x_i/a_i x_j, q x_i/x_j; q)_\infty}{(q x_i/a_i x_j, b_j x_i/x_j; q)_\infty},$$

where $|Bq^{1-r}/A| < |z| < 1$.

Further, we make use of the following terminating q-binomial theorem from S. C. Milne [**22**, Theorem 5.46], which is a multiple extension of (2.8).

THEOREM 3.5 ((Milne) An A_{r-1} terminating q-binomial theorem). *Let $x_1, \ldots, x_r$, and z be indeterminate, let $n_1, \ldots, n_r$ be nonnegative integers, let $r \geq 1$, and suppose that none of the denominators in (3.10) vanishes. Then*

$$(3.10) \quad \sum_{\substack{0 \leq k_i \leq n_i \\ i=1,\ldots,r}} \left(\prod_{1 \leq i < j \leq r} \left(\frac{x_i q^{k_i} - x_j q^{k_j}}{x_i - x_j} \right) \prod_{i,j=1}^{r} \frac{(q^{-n_j} x_i/x_j; q)_{k_i}}{(q x_i/x_j; q)_{k_i}} \prod_{i=1}^{r} x_i^{k_i} \right.$$

$$\left. \times q^{-\binom{|\mathbf{k}|}{2}+\sum_{i=1}^{r}\binom{k_i}{2}} z^{|\mathbf{k}|} \right) = \prod_{i=1}^{r} (zx_i q^{-|\mathbf{n}|}; q)_{n_i}.$$

In Section 4, we also give two multiple series extensions each of Propositions 2.3 and 2.4, see Theorems 4.7, 4.8, 4.9, and 4.10. These A_{r-1} extensions are not as deep as those in Theorems 4.2 or 4.6. In our derivations, we make use of Lemmas 4.3 and 4.9 from [**25**], displayed as follows:

LEMMA 3.6. *Let* $b_1,\ldots,b_r$ *and* $x_1,\ldots,x_r$ *be indeterminate, let* $r \geq 1$, *and suppose that none of the denominators in* (3.11) *vanishes. Then, if* $f(n)$ *is an arbitrary function of integers* n, *we have*

$$(3.11) \quad \sum_{n=-\infty}^{\infty} \frac{f(n)}{(Bq^{1-r};q)_n} = \frac{(q;q)_\infty}{(Bq^{1-r};q)_\infty} \prod_{i,j=1}^{r} \frac{(b_j x_i/x_j;q)_\infty}{(qx_i/x_j;q)_\infty}$$

$$\times \sum_{k_1,\ldots,k_r=-\infty}^{\infty} \Bigg(\prod_{1\leq i<j\leq r} \left(\frac{x_i q^{k_i} - x_j q^{k_j}}{x_i - x_j} \right) \prod_{i,j=1}^{r} (b_j x_i/x_j;q)_{k_i}^{-1} \prod_{i=1}^{r} x_i^{rk_i-|\mathbf{k}|}$$

$$\times (-1)^{(r-1)|\mathbf{k}|} q^{-\binom{|\mathbf{k}|}{2}+r\sum_{i=1}^{r}\binom{k_i}{2}} \cdot f(|\mathbf{k}|) \Bigg),$$

provided the series converge.

LEMMA 3.7. *Let* $a_1,\ldots,a_r$, $b_1,\ldots,b_r$, *and* $x_1,\ldots,x_r$ *be indeterminate, let* $r \geq 1$, *and suppose that none of the denominators in* (3.12) *vanishes. Then, if* $g(n)$ *is an arbitrary function of integers* n, *we have*

$$(3.12) \quad \sum_{n=-\infty}^{\infty} \frac{(A;q)_n}{(Bq^{1-r};q)_n} g(n)$$

$$= \frac{(q, Bq^{1-r}/A;q)_\infty}{(Bq^{1-r}, q/A;q)_\infty} \prod_{i,j=1}^{r} \frac{(b_j x_i/x_j, x_i q/a_i x_j;q)_\infty}{(qx_i/x_j, b_j x_i/a_i x_j;q)_\infty}$$

$$\times \sum_{k_1,\ldots,k_r=-\infty}^{\infty} \prod_{1\leq i<j\leq r} \left(\frac{x_i q^{k_i} - x_j q^{k_j}}{x_i - x_j} \right) \prod_{i,j=1}^{r} \frac{(a_j x_i/x_j;q)_{k_i}}{(b_j x_i/x_j;q)_{k_i}} \cdot g(|\mathbf{k}|),$$

provided the series converge.

4. Multilateral identities of q-IPD type

Here we give six new multilateral transformations of q-IPD type, extending the q-IPD type transformations of Propositions 2.1, 2.2, 2.3, and 2.4 to higher dimensions. The transformation formula in Theorem 4.2, which generalizes Proposition 2.1, involves multiple series very-well-poised over the root system A_{r-1}. A special case of that theorem is given as Corollary 4.3, which is a multilateral summation formula extending W. C. Chu's [**6**, Theorem 2] bilateral summation to r-dimensions. A further specialization gives a multiple extension of G. Gasper's [**11**, Eq. (5.13)] very-well-poised summation, see Corollary 4.4. In Theorem 4.6 we provide an A_{r-1} extension of Proposition 2.2. The interesting feature about that transformation is that it involves multilateral series with an *independent* argument z (subject to convergence), similar to the case of S. Ramanujan's $_1\psi_1$ summation (2.7) and its extension in Theorem 3.4. We were, unfortunately, not able to give multidimensional extensions of Propositions 2.3 or 2.4 which are as deep as Theorems 4.2 and 4.6. Instead, we derive multiple extensions of a *simpler* type, using Lemmas 3.6 and 3.7. Theorems 4.7 and 4.8 are simple A_{r-1} extensions of Proposition 2.3, while Theorems 4.9 and 4.10 are simple A_{r-1} extensions of Proposition 2.4. Of course, by the same method one could also derive simple multilateral generalizations of the q-IPD type transformations in Propositions 2.1 and 2.2. However, we decided to derive the identities of simpler type only in the cases were we were unable to find corresponding deeper ones.

For our derivation of Theorem 4.2, we need the following lemma, which is easily established by applying Theorem 3.1 twice.

LEMMA 4.1. *Let $a, b_1, \ldots, b_r, c, d, e_1, \ldots, e_r, f_1, \ldots, f_r,$ and $z_1, \ldots, z_r$ be indeterminate, let $r \geq 1$, and suppose that none of the denominators in (4.1) vanishes. Then*

$$(4.1) \quad {}_6\Psi_6^{(r)}\left[a; b_1, \ldots, b_r; c, d; e_1, \ldots, e_r; z_1, \ldots, z_r \,\Big|\, q, \frac{a^{r+1}q}{BcdE}\right]$$

$$= \frac{(Fq/d, aq/cF; q)_\infty}{(aq/c, q/d; q)_\infty} \prod_{i,j=1}^{r} \frac{(qz_i/z_j, az_iq/e_jf_iz_j, f_jz_iq/b_iz_j; q)_\infty}{(qf_jz_i/f_iz_j, qz_i/b_jz_j, az_iq/e_jz_j; q)_\infty}$$

$$\times \prod_{i=1}^{r} \frac{(az_iq, q/az_i, Fq/e_iz_i, az_iq/b_iF, az_iq/df_i, f_iq/cz_i; q)_\infty}{(Ff_iq/az_i, az_iq/Ff_i, az_iq/b_i, q/e_iz_i, q/cz_i, az_iq/d; q)_\infty}$$

$$\times {}_6\Psi_6^{(r)}\left[\frac{F}{a}; \frac{e_1f_1}{a}, \ldots, \frac{e_rf_r}{a}; \frac{d}{a}, \frac{cF}{a}; \frac{b_1F}{af_1}, \ldots, \frac{b_rF}{af_r}; \frac{f_1}{z_1}, \ldots, \frac{f_r}{z_r} \,\Big|\, q, \frac{a^{r+1}q}{BcdE}\right],$$

provided $|a^{r+1}q/BcdE| < 1$.

Now, for compact notation, let us extend definition (3.5) by introducing additional indeterminates $g_1, \ldots, g_s$ and $h_1, \ldots, h_s$:

$$(4.2) \quad {}_{6+2s}\Psi_{6+2s}^{(r)}\big[a; b_1, \ldots, b_r; c, d; e_1, \ldots, e_r; z_1, \ldots, z_r;$$

$$g_1, \ldots, g_s; h_1, \ldots, h_s \,\big|\, q, w\big]$$

$$:= \sum_{k_1,\ldots,k_r=-\infty}^{\infty} \left(\prod_{1 \leq i < j \leq r}\left(\frac{z_iq^{k_i} - z_jq^{k_j}}{z_i - z_j}\right) \prod_{i=1}^{r}\left(\frac{1 - az_iq^{k_i+|\mathbf{k}|}}{1 - az_i}\right)\right.$$

$$\times \prod_{i,j=1}^{r} \frac{(b_jz_i/z_j; q)_{k_i}}{(az_iq/e_jz_j; q)_{k_i}} \prod_{i=1}^{r} \frac{(e_iz_i; q)_{|\mathbf{k}|}}{(az_iq/b_i; q)_{|\mathbf{k}|}}$$

$$\times \prod_{i=1}^{r} \frac{(cz_i, g_1z_i, \ldots, g_sz_i; q)_{k_i}}{(az_iq/d, az_iq/h_1, \ldots, az_iq/h_s; q)_{k_i}}$$

$$\left.\times \frac{(d, h_1, \ldots, h_s; q)_{|\mathbf{k}|}}{(aq/c, aq/g_1, \ldots, aq/g_s; q)_{|\mathbf{k}|}} w^{|\mathbf{k}|}\right).$$

We have

THEOREM 4.2 (A multilateral very-well-poised A_{r-1} q-IPD type transformation). *Let $a, b_1, \ldots, b_r, c, d, e_1, \ldots, e_r, f_1, \ldots, f_r, z_1, \ldots, z_r,$ and $h_1, \ldots, h_s$ be indeterminate, let $N_1, \ldots, N_s$ be nonnegative integers, let $|N| = \sum_{i=1}^{s} N_i$, $r \geq 1$, and suppose that none of the denominators in (4.3) vanishes. Then*

$$(4.3) \quad {}_{6+2s}\Psi_{6+2s}^{(r)}\bigg[a; b_1, \ldots, b_r; c, d; e_1, \ldots, e_r; z_1, \ldots, z_r;$$

$$\frac{aq^{1+N_1}}{h_1}, \ldots, \frac{aq^{1+N_s}}{h_s}; h_1, \ldots, h_s \,\Big|\, q, \frac{a^{r+1}q^{1-|N|}}{BcdE}\bigg]$$

$$= \prod_{j=1}^{s} \left[\frac{(Fq/h_j; q)_{N_j}}{(q/h_j; q)_{N_j}} \prod_{i=1}^{r} \frac{(az_iq/f_ih_j; q)_{N_j}}{(az_iq/h_j; q)_{N_j}}\right]$$

$$\times \frac{(Fq/d, aq/cF; q)_\infty}{(aq/c, q/d; q)_\infty} \prod_{i,j=1}^{r} \frac{(qz_i/z_j, az_iq/e_jf_iz_j, f_jz_iq/b_iz_j; q)_\infty}{(qf_jz_i/f_iz_j, qz_i/b_iz_j, az_iq/e_jz_j; q)_\infty}$$

$$\times \prod_{i=1}^{r} \frac{(az_iq, q/az_i, Fq/e_iz_i, az_iq/b_iF, az_iq/df_i, f_iq/cz_i; q)_\infty}{(Ff_iq/az_i, az_iq/Ff_i, az_iq/b_i, q/e_iz_i, q/cz_i, az_iq/d; q)_\infty}$$

$$\times \;_{6+2s}\Psi_{6+2s}^{(r)}\!\left[\begin{array}{c} \frac{F}{a}; \frac{e_1f_1}{a}, \ldots, \frac{e_rf_r}{a}; \frac{d}{a}, \frac{cF}{a}; \frac{b_1F}{af_1}, \ldots, \frac{b_rF}{af_r}; \frac{f_1}{z_1}, \ldots, \frac{f_r}{z_r}; \\ \frac{h_1}{a}, \ldots, \frac{h_s}{a}; \frac{Fq^{1+N_1}}{h_1}, \ldots, \frac{Fq^{1+N_s}}{h_s} \end{array} \right| \left. q, \frac{a^{r+1}q^{1-|N|}}{BcdE} \right],$$

provided $|a^{r+1}q^{1-|N|}/BcdE| < 1$.

PROOF. We proceed by induction on s. For $s = 0$ (4.3) is true by Lemma 4.1. So, suppose that the transformation is already shown for $s \mapsto s - 1$. Then, by using some elementary identities from [**13**, Appendix I],

$$\;_{6+2s}\Psi_{6+2s}^{(r)}\!\left[a; b_1, \ldots, b_r; c, d; e_1, \ldots, e_r; z_1, \ldots, z_r;\right.$$
$$\left. \frac{aq^{1+N_1}}{h_1}, \ldots, \frac{aq^{1+N_s}}{h_s}; h_1, \ldots, h_s \right| \left. q, \frac{a^{r+1}q^{1-|N|}}{BcdE}\right]$$

$$= \sum_{k_1, \ldots, k_r = -\infty}^{\infty} \left(\prod_{1 \le i < j \le r} \left(\frac{z_iq^{k_i} - z_jq^{k_j}}{z_i - z_j} \right) \prod_{i=1}^{r} \left(\frac{1 - az_iq^{k_i+|\mathbf{k}|}}{1 - az_i} \right) \right.$$

$$\times \prod_{i,j=1}^{r} \frac{(b_jz_i/z_j; q)_{k_i}}{(az_iq/e_jz_j; q)_{k_i}} \prod_{i=1}^{r} \frac{(e_iz_i; q)_{|\mathbf{k}|}}{(az_iq/b_i; q)_{|\mathbf{k}|}}$$

$$\times \prod_{i=1}^{r} \frac{(cz_i, az_iq^{1+N_1}/h_1, \ldots, az_iq^{1+N_{s-1}}/h_{s-1}; q)_{k_i}}{(az_iq/d, az_iq/h_1, \ldots, az_iq/h_{s-1}; q)_{k_i}}$$

$$\times \frac{(d, h_1, \ldots, h_{s-1}; q)_{|\mathbf{k}|}}{(aq/c, h_1q^{-N_1}, \ldots, h_{s-1}q^{-N_{s-1}}; q)_{|\mathbf{k}|}} \left(\frac{a^{r+1}q^{1-|N|}}{BcdE} \right)^{|\mathbf{k}|}$$

$$\left. \times \frac{(h_s; q)_{|\mathbf{k}|}}{(h_sq^{-N_s}; q)_{|\mathbf{k}|}} \prod_{i=1}^{r} \frac{(az_iq^{1+N_s}/h_s; q)_{k_i}}{(az_iq/h_s; q)_{k_i}} \right)$$

$$= \frac{(a^rq/Eh_s; q)_{N_s}}{(q/h_s; q)_{N_s}} \prod_{i=1}^{r} \frac{(e_iz_iq/h_s; q)_{N_s}}{(az_iq/h_s; q)_{N_s}}$$

$$\times \sum_{k_1, \ldots, k_r = -\infty}^{\infty} \left(\prod_{1 \le i < j \le r} \left(\frac{z_iq^{k_i} - z_jq^{k_j}}{z_i - z_j} \right) \prod_{i=1}^{r} \left(\frac{1 - az_iq^{k_i+|\mathbf{k}|}}{1 - az_i} \right) \right.$$

$$\times \prod_{i,j=1}^{r} \frac{(b_jz_i/z_j; q)_{k_i}}{(az_iq/e_jz_j; q)_{k_i}} \prod_{i=1}^{r} \frac{(e_iz_i; q)_{|\mathbf{k}|}}{(az_iq/b_i; q)_{|\mathbf{k}|}}$$

$$\times \prod_{i=1}^{r} \frac{(cz_i, az_iq^{1+N_1}/h_1, \ldots, az_iq^{1+N_{s-1}}/h_{s-1}; q)_{k_i}}{(az_iq/d, az_iq/h_1, \ldots, az_iq/h_{s-1}; q)_{k_i}}$$

$$\times \frac{(d, h_1, \ldots, h_{s-1}; q)_{|\mathbf{k}|}}{(aq/c, h_1q^{-N_1}, \ldots, h_{s-1}q^{-N_{s-1}}; q)_{|\mathbf{k}|}} \left(\frac{a^{r+1}q^{1-(N_1+\cdots+N_{s-1})}}{BcdE} \right)^{|\mathbf{k}|}$$

$$\times \frac{(q^{1-|\mathbf{k}|}/h_s;q)_{N_s}}{(a^r q/Eh_s;q)_{N_s}} \prod_{i=1}^{r} \frac{(az_i q^{1+k_i}/h_s;q)_{N_s}}{(e_i z_i q/h_s;q)_{N_s}}\Bigg).$$

Now we expand the last factors (those involving $(\cdot;q)_{N_s}$) by applying the $a_i \mapsto e_i q^{-k_i}/a$, $b \mapsto q^{|\mathbf{k}|}$, $c \mapsto q/h_s$, $x_i \mapsto e_i z_i$, $i = 1, \ldots, r$, and $N \mapsto N_s$ case of the A_{r-1} q-Pfaff–Saalschütz summation in Theorem 3.3. We obtain

$$\frac{(a^r q/Eh_s;q)_{N_s}}{(q/h_s;q)_{N_s}} \prod_{i=1}^{r} \frac{(e_i z_i q/h_s;q)_{N_s}}{(az_i q/h_s;q)_{N_s}}$$

$$\times \sum_{k_1,\ldots,k_r=-\infty}^{\infty} \left(\prod_{1\le i<j\le r} \left(\frac{z_i q^{k_i} - z_j q^{k_j}}{z_i - z_j} \right) \prod_{i=1}^{r} \left(\frac{1 - az_i q^{k_i+|\mathbf{k}|}}{1 - az_i} \right) \right.$$

$$\times \prod_{i,j=1}^{r} \frac{(b_j z_i/z_j;q)_{k_i}}{(az_i q/e_j z_j;q)_{k_i}} \prod_{i=1}^{r} \frac{(e_i z_i;q)_{|\mathbf{k}|}}{(az_i q/b_i;q)_{|\mathbf{k}|}}$$

$$\times \prod_{i=1}^{r} \frac{(cz_i, az_i q^{1+N_1}/h_1, \ldots, az_i q^{1+N_{s-1}}/h_{s-1};q)_{k_i}}{(az_i q/d, az_i q/h_1, \ldots, az_i q/h_{s-1};q)_{k_i}}$$

$$\times \frac{(d, h_1, \ldots, h_{s-1};q)_{|\mathbf{k}|}}{(aq/c, h_1 q^{-N_1}, \ldots, h_{s-1} q^{-N_{s-1}};q)_{|\mathbf{k}|}} \left(\frac{a^{r+1} q^{1-(N_1+\cdots+N_{s-1})}}{BcdE} \right)^{|\mathbf{k}|}$$

$$\times \sum_{\substack{l_1,\ldots,l_r\ge 0 \\ 0\le|\mathbf{l}|\le N_s}} \prod_{1\le i<j\le r} \left(\frac{e_i z_i q^{l_i} - e_j z_j q^{l_j}}{e_i z_i - e_j z_j} \right) \prod_{i,j=1}^{r} \frac{(e_i z_i q^{-k_j}/az_j;q)_{l_i}}{(qe_i z_i/e_j z_j;q)_{l_i}}$$

$$\times \prod_{i=1}^{r} \frac{(ez_i q^{|\mathbf{k}|};q)_{l_i}}{(e_i z_i q/h_s;q)_{l_i}} \cdot \frac{(q^{-N_s};q)_{|\mathbf{l}|}}{(Eh_s q^{-N_s}/a^r;q)_{|\mathbf{l}|}} q^{|\mathbf{l}|} \Bigg)$$

$$= \frac{(a^r q/Eh_s;q)_{N_s}}{(q/h_s;q)_{N_s}} \prod_{i=1}^{r} \frac{(e_i z_i q/h_s;q)_{N_s}}{(az_i q/h_s;q)_{N_s}}$$

$$\times \sum_{\substack{l_1,\ldots,l_r\ge 0 \\ 0\le|\mathbf{l}|\le N_s}} \left(\prod_{1\le i<j\le r} \left(\frac{e_i z_i q^{l_i} - e_j z_j q^{l_j}}{e_i z_i - e_j z_j} \right) \prod_{i,j=1}^{r} \frac{(e_i z_i/az_j;q)_{l_i}}{(qe_i z_i/e_j z_j;q)_{l_i}}\right.$$

$$\times \prod_{i=1}^{r} \frac{(ez_i;q)_{l_i}}{(e_i z_i q/h_s;q)_{l_i}} \cdot \frac{(q^{-N_s};q)_{|\mathbf{l}|}}{(Eh_s q^{-N_s}/a^r;q)_{|\mathbf{l}|}} q^{|\mathbf{l}|}$$

$$\times {}_{6+2(s-1)}\Psi^{(r)}_{6+2(s-1)}\left[a; b_1, \ldots, b_r; c, d; e_1 q^{l_1}, \ldots, e_r q^{l_r}; z_1, \ldots, z_r; \right.$$

$$\left. \frac{aq^{1+N_1}}{h_1}, \ldots, \frac{aq^{1+N_{s-1}}}{h_{s-1}}; h_1, \ldots, h_{s-1} \Big| q, \frac{a^{r+1} q^{1-(N_1+\cdots+N_{s-1})-|\mathbf{l}|}}{BcdE} \right]\Bigg).$$

By the $e_i \mapsto e_i q^{l_i}$, $i = 1, \ldots, r$, case of the inductive hypothesis we obtain

$$\frac{(a^r q/Eh_s;q)_{N_s}}{(q/h_s;q)_{N_s}} \prod_{i=1}^{r} \frac{(e_i z_i q/h_s;q)_{N_s}}{(az_i q/h_s;q)_{N_s}}$$

$$\times \sum_{\substack{l_1,\ldots,l_r \geq 0 \\ 0 \leq |\mathbf{l}| \leq N_s}} \Bigg(\prod_{1 \leq i < j \leq r} \left(\frac{e_i z_i q^{l_i} - e_j z_j q^{l_j}}{e_i z_i - e_j z_j} \right) \prod_{i,j=1}^{r} \frac{(e_i z_i/a z_j; q)_{l_i}}{(q e_i z_i/e_j z_j; q)_{l_i}}$$

$$\times \prod_{i=1}^{r} \frac{(e z_i; q)_{l_i}}{(e_i z_i q/h_s; q)_{l_i}} \cdot \frac{(q^{-N_s}; q)_{|\mathbf{l}|}}{(Eh_s q^{-N_s}/a^r; q)_{|\mathbf{l}|}} q^{|\mathbf{l}|}$$

$$\times \prod_{j=1}^{s-1} \left[\frac{(Fq/h_j; q)_{N_j}}{(q/h_j; q)_{N_j}} \prod_{i=1}^{r} \frac{(a z_i q/f_i h_j; q)_{N_j}}{(a z_i q/h_j; q)_{N_j}} \right]$$

$$\times \frac{(Fq/d, aq/cF; q)_\infty}{(aq/c, q/d; q)_\infty} \prod_{i,j=1}^{r} \frac{(qz_i/z_j, a z_i q^{1-l_j}/e_j f_i z_j, f_j z_i q/b_i z_j; q)_\infty}{(q f_j z_i/f_i z_j, q z_i/b_i z_j, a z_i q^{1-l_j}/e_j z_j; q)_\infty}$$

$$\times \prod_{i=1}^{r} \frac{(a z_i q, q/a z_i, Fq^{1-l_i}/e_i z_i, a z_i q/b_i F, a z_i q/d f_i, f_i q/c z_i; q)_\infty}{(F f_i q/a z_i, a z_i q/F f_i, a z_i q/b_i, q^{1-l_i}/e_i z_i, q/c z_i, a z_i q/d; q)_\infty}$$

$$\times \sum_{k_1,\ldots,k_r = -\infty}^{\infty} \prod_{1 \leq i < j \leq r} \left(\frac{f_i q^{k_i}/z_i - f_j q^{k_j}/z_j}{f_i/z_i - f_j/z_j} \right) \prod_{i=1}^{r} \left(\frac{1 - F f_i q^{k_i + |\mathbf{k}|}/a z_i}{1 - F f_i/a z_i} \right)$$

$$\times \prod_{i,j=1}^{r} \frac{(e_j f_i z_j q^{l_j}/a z_i; q)_{k_i}}{(f_i z_j q/b_j z_i; q)_{k_i}} \prod_{i=1}^{r} \frac{(b_i F/a z_i; q)_{|\mathbf{k}|}}{(Fq^{1-l_i}/e_i z_i; q)_{|\mathbf{k}|}}$$

$$\times \prod_{i=1}^{r} \frac{(d f_i/a z_i, f_i h_1/a z_i, \ldots, f_i h_{s-1}/a z_i; q)_{k_i}}{(f_i q/c z_i, f_i h_1 q^{-N_1}/a z_i, \ldots, f_i h_{s-1} q^{-N_{s-1}}/a z_i; q)_{k_i}}$$

$$\times \frac{(cF/a, Fq^{1+N_1}/h_1, \ldots, Fq^{1+N_{s-1}}/h_{s-1}; q)_{|\mathbf{k}|}}{(Fq/d, Fq/h_1, \ldots, Fq/h_{s-1}; q)_{|\mathbf{k}|}} \left(\frac{a^{r+1} q^{1-(N_1+\cdots+N_{s-1})-|\mathbf{l}|}}{BcdE} \right)^{|\mathbf{k}|} \Bigg)$$

$$= \frac{(a^r q/Eh_s; q)_{N_s}}{(q/h_s; q)_{N_s}} \prod_{i=1}^{r} \frac{(e_i z_i q/h_s; q)_{N_s}}{(a z_i q/h_s; q)_{N_s}} \prod_{j=1}^{s-1} \left[\frac{(Fq/h_j; q)_{N_j}}{(q/h_j; q)_{N_j}} \prod_{i=1}^{r} \frac{(a z_i q/f_i h_j; q)_{N_j}}{(a z_i q/h_j; q)_{N_j}} \right]$$

$$\times \frac{(Fq/d, aq/cF; q)_\infty}{(aq/c, q/d; q)_\infty} \prod_{i,j=1}^{r} \frac{(qz_i/z_j, a z_i q/e_j f_i z_j, f_j z_i q/b_i z_j; q)_\infty}{(q f_j z_i/f_i z_j, q z_i/b_i z_j, a z_i q/e_j z_j; q)_\infty}$$

$$\times \prod_{i=1}^{r} \frac{(a z_i q, q/a z_i, Fq/e_i z_i, a z_i q/b_i F, a z_i q/d f_i, f_i q/c z_i; q)_\infty}{(F f_i q/a z_i, a z_i q/F f_i, a z_i q/b_i, q/e_i z_i, q/c z_i, a z_i q/d; q)_\infty}$$

$$\times \sum_{k_1,\ldots,k_r=-\infty}^{\infty} \Bigg(\prod_{1 \leq i < j \leq r} \left(\frac{f_i q^{k_i}/z_i - f_j q^{k_j}/z_j}{f_i/z_i - f_j/z_j} \right) \prod_{i=1}^{r} \left(\frac{1 - F f_i q^{k_i + |\mathbf{k}|}/a z_i}{1 - F f_i/a z_i} \right)$$

$$\times \prod_{i,j=1}^{r} \frac{(e_j f_i z_j/a z_i; q)_{k_i}}{(f_i z_j q/b_j z_i; q)_{k_i}} \prod_{i=1}^{r} \frac{(b_i F/a z_i; q)_{|\mathbf{k}|}}{(Fq/e_i z_i; q)_{|\mathbf{k}|}}$$

$$\times \prod_{i=1}^{r} \frac{(d f_i/a z_i, f_i h_1/a z_i, \ldots, f_i h_{s-1}/a z_i; q)_{k_i}}{(f_i q/c z_i, f_i h_1 q^{-N_1}/a z_i, \ldots, f_i h_{s-1} q^{-N_{s-1}}/a z_i; q)_{k_i}}$$

$$\times \frac{(cF/a, Fq^{1+N_1}/h_1, \ldots, Fq^{1+N_{s-1}}/h_{s-1}; q)_{|\mathbf{k}|}}{(Fq/d, Fq/h_1, \ldots, Fq/h_{s-1}; q)_{|\mathbf{k}|}} \left(\frac{a^{r+1} q^{1-(N_1+\cdots+N_{s-1})}}{BcdE} \right)^{|\mathbf{k}|}$$

$$\times \sum_{\substack{l_1,\ldots,l_r \geq 0 \\ 0 \leq |\mathbf{k}| \leq N_s}} \prod_{1 \leq i < j \leq r} \left(\frac{e_i z_i q^{l_i} - e_j z_j q^{l_j}}{e_i z_i - e_j z_j} \right) \prod_{i,j=1}^{r} \frac{(e_i f_j z_i q^{k_j}/a z_j; q)_{l_i}}{(q e_i z_i/e_j z_j; q)_{l_i}}$$

$$\times \prod_{i=1}^{r} \frac{(e z_i q^{-|\mathbf{k}|}/F; q)_{l_i}}{(e_i z_i q/h_s; q)_{l_i}} \cdot \frac{(q^{-N_s}; q)_{|\mathbf{l}|}}{(E h_s q^{-N_s}/a^r; q)_{|\mathbf{l}|}} q^{|\mathbf{l}|} \bigg).$$

Now, we evaluate the inner multiple sum by the $a_i \mapsto e_i f_i q^{k_i}/a$, $b \mapsto q^{-|\mathbf{k}|}/F$, $c \mapsto q/h_s$, $x_i \mapsto e_i z_i$, $i = 1, \ldots, r$, and $N \mapsto N_s$ case of the A_{r-1} q-Pfaff–Saalschütz summation in Theorem 3.3, and obtain

$$\prod_{j=1}^{s} \left[\frac{(Fq/h_j; q)_{N_j}}{(q/h_j; q)_{N_j}} \prod_{i=1}^{r} \frac{(a z_i q/f_i h_j; q)_{N_j}}{(a z_i q/h_j; q)_{N_j}} \right]$$

$$\times \frac{(Fq/d, aq/cF; q)_{\infty}}{(aq/c, q/d; q)_{\infty}} \prod_{i,j=1}^{r} \frac{(q z_i/z_j, a z_i q/e_j f_i z_j, f_j z_i q/b_i z_j; q)_{\infty}}{(q f_j z_i/f_i z_j, q z_i/b_i z_j, a z_i q/e_j z_j; q)_{\infty}}$$

$$\times \prod_{i=1}^{r} \frac{(a z_i q, q/a z_i, Fq/e_i z_i, a z_i q/b_i F, a z_i q/d f_i, f_i q/c z_i; q)_{\infty}}{(F f_i q/a z_i, a z_i q/F f_i, a z_i q/b_i, q/e_i z_i, q/c z_i, a z_i q/d; q)_{\infty}}$$

$$\times \sum_{k_1,\ldots,k_r = -\infty}^{\infty} \Bigg(\prod_{1 \leq i < j \leq r} \left(\frac{f_i q^{k_i}/z_i - f_j q^{k_j}/z_j}{f_i/z_i - f_j/z_j} \right) \prod_{i=1}^{r} \left(\frac{1 - F f_i q^{k_i + |\mathbf{k}|}/a z_i}{1 - F f_i/a z_i} \right)$$

$$\times \prod_{i,j=1}^{r} \frac{(e_j f_i z_j/a z_i; q)_{k_i}}{(f_i z_j q/b_j z_i; q)_{k_i}} \prod_{i=1}^{r} \frac{(b_i F/a z_i; q)_{|\mathbf{k}|}}{(Fq/e_i z_i; q)_{|\mathbf{k}|}}$$

$$\times \prod_{i=1}^{r} \frac{(d f_i/a z_i, f_i h_1/a z_i, \ldots, f_i h_s/a z_i; q)_{k_i}}{(f_i q/c z_i, f_i h_1 q^{-N_1}/a z_i, \ldots, f_i h_s q^{-N_s}/a z_i; q)_{k_i}}$$

$$\times \frac{(cF/a, Fq^{1+N_1}/h_1, \ldots, Fq^{1+N_s}/h_s; q)_{|\mathbf{k}|}}{(Fq/d, Fq/h_1, \ldots, Fq/h_s; q)_{|\mathbf{k}|}} \left(\frac{a^{r+1} q^{1-|N|}}{BcdE} \right)^{|\mathbf{k}|} \Bigg),$$

which is the right side of (4.3). $\square$

A special case of Theorem 4.2 immediately gives the following summation formula as a corollary. It is an A_{r-1} extension of an identity due to W. C. Chu [**6**, Theorem 2].

COROLLARY 4.3 (A multilateral very-well-poised A_{r-1} q-IPD type summation). *Let a, $b_1, \ldots, b_r$, d, $e_1, \ldots, e_r$, $z_1, \ldots, z_r$, and $h_1, \ldots, h_s$ be indeterminate, let $N_1, \ldots, N_s$ be nonnegative integers, let $|N| = \sum_{i=1}^{s} N_i$, $r \geq 1$, and suppose that none of the denominators in (4.4) vanishes. Then*

$$(4.4) \quad {}_{6+2s}\Psi_{6+2s}^{(r)} \bigg[a; b_1, \ldots, b_r; \frac{a}{B}, d; e_1, \ldots, e_r; z_1, \ldots, z_r;$$

$$\frac{aq^{1+N_1}}{h_1}, \ldots, \frac{aq^{1+N_s}}{h_s}; h_1, \ldots, h_s \bigg| q, \frac{a^r q^{1-|N|}}{dE} \bigg]$$

$$= \frac{(Bq/d, q; q)_{\infty}}{(Bq, q/d; q)_{\infty}} \prod_{i=1}^{r} \frac{(a z_i q, q/a z_i, Bq/e_i z_i, a z_i q/db_i; q)_{\infty}}{(a z_i q/b_i, q/e_i z_i, Bq/a z_i, a z_i q/d; q)_{\infty}}$$

$$\times \prod_{i,j=1}^{r} \frac{(qz_i/z_j, az_iq/e_jb_iz_j;q)_\infty}{(qz_i/b_iz_j, az_iq/e_jz_j;q)_\infty} \prod_{j=1}^{s} \left[\frac{(Bq/h_j;q)_{N_j}}{(q/h_j;q)_{N_j}} \prod_{i=1}^{r} \frac{(az_iq/b_ih_j;q)_{N_j}}{(az_iq/h_j;q)_{N_j}} \right],$$

provided $|a^r q^{1-|N|}/dE| < 1$.

PROOF. In (4.3), we let $c \to a/B$ and $f_i \to b_i$, for $i = 1, \ldots, r$. In this case the $_{6+2s}\Psi^{(r)}_{6+2s}$ series on the right side terminates from below, and from above, and evalutes to one. In particular, the appearance of the factor

$$\prod_{i,j=1}^{r} (b_iz_jq/b_jz_i;q)_{k_i}^{-1}$$

makes the terms in the series vanish unless $k_i \geq 0$, $i = 1, \ldots, r$. Similarly, the appearance of the factor

$$(1;q)_{|\mathbf{k}|}$$

ensures that if $|\mathbf{k}| > 0$, the terms of the series are zero. In total, only the term where $k_1 = \cdots = k_r = 0$ survives, and that term is just one. $\square$

A further specialization of Corollary 4.3, namely the case $e_i \to a$, $i = 1, \ldots, r$, yields an r-dimensional generalization of G. Gasper's [**11**, Eq. (5.13)] very-well-poised $_{6+2s}\phi_{5+2s}$ summation.

COROLLARY 4.4 (A very-well-poised A_{r-1} q-IPD type summation). *Let a, $b_1, \ldots, b_r$, d, $z_1, \ldots, z_r$, and $h_1, \ldots, h_s$ be indeterminate, let $N_1, \ldots, N_s$ be nonnegative integers, let $|N| = \sum_{i=1}^{s} N_i$, $r \geq 1$, and suppose that none of the denominators in (4.5) vanishes. Then*

$$(4.5) \quad \sum_{k_1,\ldots,k_r=0}^{\infty} \left(\prod_{1 \leq i < j \leq r} \left(\frac{z_iq^{k_i} - z_jq^{k_j}}{z_i - z_j} \right) \prod_{i=1}^{r} \left(\frac{1 - az_iq^{k_i+|\mathbf{k}|}}{1 - az_i} \right) \right.$$

$$\times \prod_{i,j=1}^{r} \frac{(b_jz_i/z_j;q)_{k_i}}{(qz_i/z_j;q)_{k_i}} \prod_{i=1}^{r} \frac{(az_i;q)_{|\mathbf{k}|}}{(az_iq/b_i;q)_{|\mathbf{k}|}}$$

$$\times \prod_{i=1}^{r} \frac{(az_i/B, az_iq^{1+N_1}/h_1, \ldots, az_iq^{1+N_s}/h_s;q)_{k_i}}{(az_iq/d, az_iq/h_1, \ldots, az_iq/h_s;q)_{k_i}}$$

$$\left. \times \frac{(d, h_1, \ldots, h_s;q)_{|\mathbf{k}|}}{(Bq, h_1q^{-N_1}, \ldots, h_sq^{-N_s};q)_{|\mathbf{k}|}} \left(\frac{q^{1-|N|}}{d} \right)^{|\mathbf{k}|} \right)$$

$$= \frac{(Bq/d, q;q)_\infty}{(Bq, q/d;q)_\infty} \prod_{i=1}^{r} \frac{(az_iq, az_iq/b_id;q)_\infty}{(az_iq/b_i, az_iq/d;q)_\infty}$$

$$\times \prod_{j=1}^{s} \left[\frac{(Bq/h_j;q)_{N_j}}{(q/h_j;q)_{N_j}} \prod_{i=1}^{r} \frac{(az_iq/b_ih_j;q)_{N_j}}{(az_iq/h_j;q)_{N_j}} \right],$$

provided $|q^{1-|N|}/d| < 1$.

To derive the multilateral q-IPD type transformation in Theorem 4.6 we need the following lemma, which is easily established by applying Theorem 3.4 twice.

LEMMA 4.5. *Let $a_1, \ldots, a_r$, $b_1, \ldots, b_r$, $c_1, \ldots, c_r$, $x_1, \ldots, x_r$, $y_1, \ldots, y_r$, and z be indeterminate, let $r \geq 1$, and suppose that none of the denominators in (4.6) vanishes. Then*

$$\text{(4.6)} \quad \sum_{k_1,\dots,k_r=-\infty}^{\infty} \prod_{1\leq i<j\leq r}\left(\frac{x_i q^{k_i} - x_j q^{k_j}}{x_i - x_j}\right) \prod_{i,j=1}^{r} \frac{(a_j x_i/x_j; q)_{k_i}}{(b_j x_i/x_j; q)_{k_i}} z^{|\mathbf{k}|}$$

$$= \frac{(Az, q/Az; q)_\infty}{(Azq^r/C, Cq^{1-r}/Az; q)_\infty} \prod_{i,j=1}^{r} \frac{(qx_i/x_j, b_j x_i/a_i x_j, c_i y_i/a_i y_j, b_j y_i q/c_j y_j; q)_\infty}{(qy_i/y_j, b_j c_i y_i/a_i c_j y_j, qx_i/a_i x_j, b_j x_i/x_j; q)_\infty}$$

$$\times \sum_{k_1,\dots,k_r=-\infty}^{\infty} \prod_{1\leq i<j\leq r}\left(\frac{y_i q^{k_i} - y_j q^{k_j}}{y_i - y_j}\right) \prod_{i,j=1}^{r} \frac{(a_j y_i q/c_j y_j; q)_{k_i}}{(b_j y_i q/c_j y_j; q)_{k_i}} z^{|\mathbf{k}|},$$

where $|Bq^{1-r}/A| < |z| < 1$.

We have

THEOREM 4.6 (A multilateral A_{r-1} q-IPD type transformation). *Let* $a_1, \dots, a_r$, $b_1, \dots, b_r$, $c_1, \dots, c_r$, $x_1, \dots, x_r$, $y_1, \dots, y_r$, $h_{11}, \dots, h_{rs}$, *and* z *be indeterminate, let* $N_{11}, \dots, N_{rs}$ *be nonnegative integers, let* $r \geq 1$, *and suppose that none of the denominators in* (4.7) *vanishes. Then*

$$\text{(4.7)} \quad \sum_{k_1,\dots,k_r=-\infty}^{\infty} \Bigg(\prod_{1\leq i<j\leq r}\left(\frac{x_i q^{k_i} - x_j q^{k_j}}{x_i - x_j}\right) \prod_{i,j=1}^{r} \frac{(a_j x_i/x_j; q)_{k_i}}{(b_j x_i/x_j; q)_{k_i}}$$

$$\times \prod_{i=1}^{r} \prod_{j=1}^{s} \frac{(h_{ij} q^{N_{ij}}; q)_{|\mathbf{k}|}}{(h_{ij}; q)_{|\mathbf{k}|}} z^{|\mathbf{k}|} \Bigg)$$

$$= \prod_{i=1}^{r} \prod_{j=1}^{s} \frac{(h_{ij} q^r/C; q)_{N_{ij}}}{(h_{ij}; q)_{N_{ij}}} \cdot \frac{(Az, q/Az; q)_\infty}{(Azq^r/C, Cq^{1-r}/Az; q)_\infty}$$

$$\times \prod_{i,j=1}^{r} \frac{(qx_i/x_j, b_j x_i/a_i x_j, c_i y_i/a_i y_j, b_j y_i q/c_j y_j; q)_\infty}{(qy_i/y_j, b_j c_i y_i/a_i c_j y_j, qx_i/a_i x_j, b_j x_i/x_j; q)_\infty}$$

$$\times \sum_{k_1,\dots,k_r=-\infty}^{\infty} \Bigg(\prod_{1\leq i<j\leq r}\left(\frac{y_i q^{k_i} - y_j q^{k_j}}{y_i - y_j}\right) \prod_{i,j=1}^{r} \frac{(a_j y_i q/c_j y_j; q)_{k_i}}{(b_j y_i q/c_j y_j; q)_{k_i}}$$

$$\times \prod_{i=1}^{r} \prod_{j=1}^{s} \frac{(h_{ij} q^{r+N_{ij}}/C; q)_{|\mathbf{k}|}}{(h_{ij} q^r/C; q)_{|\mathbf{k}|}} z^{|\mathbf{k}|} \Bigg),$$

provided $\left|Bq^{1-r-\sum_{i,j} N_{ij}}/A\right| < |z| < 1$.

PROOF. We proceed by induction on s. For $s = 0$ (4.7) is true by Lemma 4.5. So, suppose that the transformation is already shown for $s \mapsto s - 1$. Then (again using some elementary identities from [**13**, Appendix I]),

$$\sum_{k_1,\dots,k_r=-\infty}^{\infty} \Bigg(\prod_{1\leq i<j\leq r}\left(\frac{x_i q^{k_i} - x_j q^{k_j}}{x_i - x_j}\right) \prod_{i,j=1}^{r} \frac{(a_j x_i/x_j; q)_{k_i}}{(b_j x_i/x_j; q)_{k_i}}$$

$$\times \prod_{i=1}^{r} \prod_{j=1}^{s-1} \frac{(h_{ij} q^{N_{ij}}; q)_{|\mathbf{k}|}}{(h_{ij}; q)_{|\mathbf{k}|}} z^{|\mathbf{k}|} \cdot \prod_{i=1}^{r} \frac{(h_{is} q^{N_{is}}; q)_{|\mathbf{k}|}}{(h_{is}; q)_{|\mathbf{k}|}} \Bigg)$$

$$= \prod_{i=1}^{r} \frac{1}{(h_{is}; q)_{N_i}} \sum_{k_1,\dots,k_r=-\infty}^{\infty} \Bigg(\prod_{1\leq i<j\leq r}\left(\frac{x_i q^{k_i} - x_j q^{k_j}}{x_i - x_j}\right) \prod_{i,j=1}^{r} \frac{(a_j x_i/x_j; q)_{k_i}}{(b_j x_i/x_j; q)_{k_i}}$$

$$\times \prod_{i=1}^{r}\prod_{j=1}^{s-1}\frac{(h_{ij}q^{N_{ij}};q)_{|\mathbf{k}|}}{(h_{ij};q)_{|\mathbf{k}|}}z^{|\mathbf{k}|}\cdot\prod_{i=1}^{r}(h_{is}q^{|\mathbf{k}|};q)_{N_{is}}\Bigg).$$

Now we expand the last factors (those involving $(\cdot;q)_{N_{is}}$) by applying the $x_i \mapsto h_{is}$, $n_i \mapsto N_{is}$, $i=1,\ldots,r$, and $z \mapsto q^{|\mathbf{k}|+(N_{1s}+\cdots+N_{rs})}$ case of the A_{r-1} summation in Theorem 3.5. We obtain

$$\prod_{i=1}^{r}\frac{1}{(h_{is};q)_{N_i}}\sum_{k_1,\ldots,k_r=-\infty}^{\infty}\Bigg(\prod_{1\le i<j\le r}\left(\frac{x_iq^{k_i}-x_jq^{k_j}}{x_i-x_j}\right)\prod_{i,j=1}^{r}\frac{(a_jx_i/x_j;q)_{k_i}}{(b_jx_i/x_j;q)_{k_i}}$$

$$\times \prod_{i=1}^{r}\prod_{j=1}^{s-1}\frac{(h_{ij}q^{N_{ij}};q)_{|\mathbf{k}|}}{(h_{ij};q)_{|\mathbf{k}|}}z^{|\mathbf{k}|}$$

$$\times \sum_{\substack{0\le l_i\le N_{is}\\ i=1,\ldots,r}}\prod_{1\le i<j\le r}\left(\frac{h_{is}q^{l_i}-h_{js}q^{l_j}}{h_{is}-h_{js}}\right)\prod_{i,j=1}^{r}\frac{(q^{-N_{js}}h_{is}/h_{js};q)_{l_i}}{(qh_{is}/h_{js};q)_{l_i}}$$

$$\times \prod_{i=1}^{r}h_{is}^{l_i}\cdot q^{|\mathbf{k}||\mathbf{l}|+(N_{1s}+\cdots+N_{rs})|\mathbf{l}|-\binom{|\mathbf{l}|}{2}+\sum_{i=1}^{r}\binom{l_i}{2}}\Bigg)$$

$$=\prod_{i=1}^{r}\frac{1}{(h_{is};q)_{N_i}}\sum_{\substack{0\le l_i\le N_{is}\\ i=1,\ldots,r}}\Bigg(\prod_{1\le i<j\le r}\left(\frac{h_{is}q^{l_i}-h_{js}q^{l_j}}{h_{is}-h_{js}}\right)\prod_{i,j=1}^{r}\frac{(q^{-N_{js}}h_{is}/h_{js};q)_{l_i}}{(qh_{is}/h_{js};q)_{l_i}}$$

$$\times \prod_{i=1}^{r}h_{is}^{l_i}\cdot q^{(N_{1s}+\cdots+N_{rs})|\mathbf{l}|-\binom{|\mathbf{l}|}{2}+\sum_{i=1}^{r}\binom{l_i}{2}}$$

$$\times \sum_{k_1,\ldots,k_r=-\infty}^{\infty}\prod_{1\le i<j\le r}\left(\frac{x_iq^{k_i}-x_jq^{k_j}}{x_i-x_j}\right)\prod_{i,j=1}^{r}\frac{(a_jx_i/x_j;q)_{k_i}}{(b_jx_i/x_j;q)_{k_i}}$$

$$\times \prod_{i=1}^{r}\prod_{j=1}^{s-1}\frac{(h_{ij}q^{N_{ij}};q)_{|\mathbf{k}|}}{(h_{ij};q)_{|\mathbf{k}|}}\left(zq^{|\mathbf{l}|}\right)^{|\mathbf{k}|}\Bigg).$$

By the $z \mapsto zq^{|\mathbf{l}|}$ case of the inductive hypothesis we obtain

$$\prod_{i=1}^{r}\frac{1}{(h_{is};q)_{N_i}}\sum_{\substack{0\le l_i\le N_{is}\\ i=1,\ldots,r}}\Bigg(\prod_{1\le i<j\le r}\left(\frac{h_{is}q^{l_i}-h_{js}q^{l_j}}{h_{is}-h_{js}}\right)\prod_{i,j=1}^{r}\frac{(q^{-N_{js}}h_{is}/h_{js};q)_{l_i}}{(qh_{is}/h_{js};q)_{l_i}}$$

$$\times \prod_{i=1}^{r}h_{is}^{l_i}\cdot q^{(N_{1s}+\cdots+N_{rs})|\mathbf{l}|-\binom{|\mathbf{l}|}{2}+\sum_{i=1}^{r}\binom{l_i}{2}}$$

$$\times \prod_{i=1}^{r}\prod_{j=1}^{s-1}\frac{(h_{ij}q^r/C;q)_{N_{ij}}}{(h_{ij};q)_{N_{ij}}}\cdot\frac{(Azq^{|\mathbf{l}|},q^{1-|\mathbf{l}|}/Az;q)_{\infty}}{(Azq^{r+|\mathbf{l}|}/C,Cq^{1-r-|\mathbf{l}|}/Az;q)_{\infty}}$$

$$\times \prod_{i,j=1}^{r}\frac{(qx_i/x_j,b_jx_i/a_ix_j,c_iy_i/a_iy_j,b_jy_iq/c_jy_j;q)_{\infty}}{(qy_i/y_j,b_jc_iy_i/a_ic_jy_j,qx_i/a_ix_j,b_jx_i/x_j;q)_{\infty}}$$

$$\times \sum_{k_1,\ldots,k_r=-\infty}^{\infty}\prod_{1\le i<j\le r}\left(\frac{y_iq^{k_i}-y_jq^{k_j}}{y_i-y_j}\right)\prod_{i,j=1}^{r}\frac{(a_jy_iq/c_jy_j;q)_{k_i}}{(b_jy_iq/c_jy_j;q)_{k_i}}$$

$$\times \prod_{i=1}^{r} \prod_{j=1}^{s-1} \frac{(h_{ij}q^{r+N_{ij}}/C;q)_{|\mathbf{k}|}}{(h_{ij}q^{r}/C;q)_{|\mathbf{k}|}} \left(zq^{|\mathbf{l}|}\right)^{|\mathbf{k}|}\Bigg)$$

$$= \prod_{i=1}^{r} \frac{1}{(h_{is};q)_{N_i}} \prod_{i=1}^{r} \prod_{j=1}^{s-1} \frac{(h_{ij}q^{r}/C;q)_{N_{ij}}}{(h_{ij};q)_{N_{ij}}} \cdot \frac{(Az, q/Az; q)_\infty}{(Azq^r/C, Cq^{1-r}/Az; q)_\infty}$$

$$\times \prod_{i,j=1}^{r} \frac{(qx_i/x_j, b_j x_i/a_i x_j, c_i y_i/a_i y_j, b_j y_i q/c_j y_j; q)_\infty}{(qy_i/y_j, b_j c_i y_i/a_i c_j y_j, qx_i/a_i x_j, b_j x_i/x_j; q)_\infty}$$

$$\times \sum_{k_1,\ldots,k_r=-\infty}^{\infty} \left(\prod_{1\le i<j\le r} \left(\frac{y_i q^{k_i} - y_j q^{k_j}}{y_i - y_j}\right) \prod_{i,j=1}^{r} \frac{(a_j y_i q/c_j y_j; q)_{k_i}}{(b_j y_i q/c_j y_j; q)_{k_i}}\right.$$

$$\times \prod_{i=1}^{r} \prod_{j=1}^{s-1} \frac{(h_{ij}q^{r+N_{ij}}/C;q)_{|\mathbf{k}|}}{(h_{ij}q^{r}/C;q)_{|\mathbf{k}|}} z^{|\mathbf{k}|}$$

$$\times \sum_{\substack{0\le l_i \le N_{is}\\ i=1,\ldots,r}} \prod_{1\le i<j\le r} \left(\frac{h_{is}q^{l_i} - h_{js}q^{l_j}}{h_{is} - h_{js}}\right) \prod_{i,j=1}^{r} \frac{(q^{-N_{js}}h_{is}/h_{js}; q)_{l_i}}{(qh_{is}/h_{js}; q)_{l_i}}$$

$$\times \prod_{i=1}^{r} h_{is}^{l_i} \cdot \left(\frac{q^r}{C}\right)^{|\mathbf{l}|} q^{|\mathbf{k}||\mathbf{l}|+(N_{1s}+\cdots+N_{rs})|\mathbf{l}|-\binom{|\mathbf{l}|}{2}+\sum_{i=1}^{r}\binom{l_i}{2}}\Bigg).$$

Now, we evaluate the inner multiple sum by the $x_i \mapsto h_{is}$, $n_i \mapsto N_{is}$, $i = 1,\ldots,r$, and $z \mapsto q^{r+|\mathbf{k}|+(N_{1s}+\cdots+N_{rs})}/C$ case of the A_{r-1} summation in Theorem 3.5. We obtain

$$\prod_{i=1}^{r} \frac{1}{(h_{is};q)_{N_i}} \prod_{i=1}^{r} \prod_{j=1}^{s-1} \frac{(h_{ij}q^r/C;q)_{N_{ij}}}{(h_{ij};q)_{N_{ij}}} \cdot \frac{(Az, q/Az; q)_\infty}{(Azq^r/C, Cq^{1-r}/Az; q)_\infty}$$

$$\times \prod_{i,j=1}^{r} \frac{(qx_i/x_j, b_j x_i/a_i x_j, c_i y_i/a_i y_j, b_j y_i q/c_j y_j; q)_\infty}{(qy_i/y_j, b_j c_i y_i/a_i c_j y_j, qx_i/a_i x_j, b_j x_i/x_j; q)_\infty}$$

$$\times \sum_{k_1,\ldots,k_r=-\infty}^{\infty} \left(\prod_{1\le i<j\le r} \left(\frac{y_i q^{k_i} - y_j q^{k_j}}{y_i - y_j}\right) \prod_{i,j=1}^{r} \frac{(a_j y_i q/c_j y_j; q)_{k_i}}{(b_j y_i q/c_j y_j; q)_{k_i}}\right.$$

$$\left.\times \prod_{i=1}^{r} \prod_{j=1}^{s-1} \frac{(h_{ij}q^{r+N_{ij}}/C;q)_{|\mathbf{k}|}}{(h_{ij}q^r/C;q)_{|\mathbf{k}|}} z^{|\mathbf{k}|} \cdot \prod_{i=1}^{r}(h_i q^{r+|\mathbf{k}|}/C;q)_{N_{is}}\right),$$

which, after an elementary manipulation of q-shifted factorials, gives us the right side of (4.7), as desired. $\square$

Finally, we provide four more multilateral transformations of q-IPD type. Unfortunately, we were not able to find multiple extensions of Propositions 2.3 or 2.4 which are as deep as the identities in Theorems 4.2 and 4.6. The following theorems are obtained by combining Propositions 2.3 and 2.4 each with Lemmas 3.6 and 3.7, thus giving rise to four different multilateral transformations.

THEOREM 4.7 (A multilateral A_{r-1} q-IPD type transformation). *Let a, b, $c_1,\ldots,c_r$, d, $e_1,\ldots,e_r$, $x_1,\ldots,x_r$, $y_1,\ldots,y_r$, and $h_1,\ldots,h_s$ be indeterminate, let N be an integer, let $m_1,\ldots,m_s$ be nonnegative integers, let $|m|=\sum_{i=1}^{s} m_i$, $r\ge 1$, and suppose that none of the denominators in (4.8) vanishes. Then*

$$(4.8) \quad \sum_{k_1,\ldots,k_r=-\infty}^{\infty} \left(\prod_{1 \le i < j \le r} \left(\frac{x_i q^{k_i} - x_j q^{k_j}}{x_i - x_j} \right) \prod_{i,j=1}^{r} (c_j x_i/x_j; q)_{k_i}^{-1} \prod_{i=1}^{r} x_i^{rk_i - |\mathbf{k}|} \right.$$
$$\times (-1)^{(r-1)|\mathbf{k}|} q^{-\binom{|\mathbf{k}|}{2} + r \sum_{i=1}^{r} \binom{k_i}{2}}$$
$$\left. \times \frac{(a, b, h_1 q^{m_1}, \ldots, h_s q^{m_s}; q)_{|\mathbf{k}|}}{(d, h_1, \ldots, h_s; q)_{|\mathbf{k}|}} \left(\frac{Eq^{1-r-N}}{ab} \right)^{|\mathbf{k}|} \right)$$
$$= (Eq^{-r})^N \frac{(Eq^{1-r}/a, Eq^{1-r}/b, dq^r/E; q)_\infty}{(q/a, q/b, d; q)_\infty}$$
$$\times \prod_{i=1}^{s} \frac{(h_i q^r/E; q)_{m_i}}{(h_i; q)_{m_i}} \prod_{i,j=1}^{r} \frac{(q x_i/x_j, c_j y_i q/e_j y_j; q)_\infty}{(q y_i/y_j, c_j x_i/x_j; q)_\infty}$$
$$\times \sum_{k_1,\ldots,k_r=-\infty}^{\infty} \left(\prod_{1 \le i < j \le r} \left(\frac{y_i q^{k_i} - y_j q^{k_j}}{y_i - y_j} \right) \prod_{i,j=1}^{r} (c_j y_i q/e_j y_j; q)_{k_i}^{-1} \prod_{i=1}^{r} y_i^{rk_i - |\mathbf{k}|} \right.$$
$$\times (-1)^{(r-1)|\mathbf{k}|} q^{-\binom{|\mathbf{k}|}{2} + r \sum_{i=1}^{r} \binom{k_i}{2}}$$
$$\left. \times \frac{(aq^r/E, bq^r/E, h_1 q^{r+m_1}/E, \ldots, h_s q^{r+m_s}/E; q)_{|\mathbf{k}|}}{(dq^r/E, h_1 q^r/E, \ldots, h_s q^r/E; q)_{|\mathbf{k}|}} \left(\frac{Eq^{1-r-N}}{ab} \right)^{|\mathbf{k}|} \right),$$

provided $|Eq^{1-r}/ab| < |q^N| < |Eq^{|m|}/Cd|$.

PROOF. We have, for $|Eq^{1-r}/ab| < |q^N| < |Eq^{|m|}/Cd|$,

$$(4.9) \quad {}_{2+s}\psi_{2+s} \left[\begin{matrix} a, b, h_1 q^{m_1}, \ldots, h_s q^{m_s} \\ Cq^{1-r}, d, h_1, \ldots, h_s \end{matrix} ; q, \frac{Eq^{1-r-N}}{ab} \right]$$
$$= (Eq^{-r})^N \frac{(Eq^{1-r}/a, Eq^{1-r}/b, Cq/E, dq^r/E; q)_\infty}{(q/a, q/b, Cq^{1-r}, d; q)_\infty} \prod_{i=1}^{s} \frac{(h_i q^r/E; q)_{m_i}}{(h_i; q)_{m_i}}$$
$$\times {}_{2+s}\psi_{2+s} \left[\begin{matrix} aq^r/E, bq^r/E, h_1 q^{r+m_1}/E, \ldots, h_s q^{r+m_s}/E \\ Cq/E, dq^r/E, h_1 q^r/E, \ldots, h_s q^r/E \end{matrix} ; q, \frac{Eq^{1-r-N}}{ab} \right],$$

by the q-IPD type transformation in (2.11). Now we apply Lemma 3.6 to the ${}_{2+s}\psi_{2+s}$'s on the left and on the right side of this transformation. Specifically, we rewrite the ${}_{2+s}\psi_{2+s}$ on left side of (4.9) by the $b_i \mapsto c_i$, $i = 1, \ldots, r$, and

$$f(n) = \frac{(a, b, h_1 q^{m_1}, \ldots, h_s q^{m_s}; q)_n}{(d, h_1, \ldots, h_s; q)_n} \left(\frac{Eq^{1-r-N}}{ab} \right)^n$$

case of Lemma 3.6. The ${}_{2+s}\psi_{2+s}$ on the right side of (4.9) is rewritten by the $b_i \mapsto c_i q/e_i$, $x_i \mapsto y_i$, $i = 1, \ldots, r$, and

$$f(n) = \frac{(aq^r/E, bq^r/E, h_1 q^{r+m_1}/E, \ldots, h_s q^{r+m_s}/E; q)_n}{(dq^r/E, h_1 q^r/E, \ldots, h_s q^r/E; q)_n} \left(\frac{Eq^{1-r-N}}{ab} \right)^n$$

case of Lemma 3.6. Finally, we divide both sides of the resulting equation by

$$(4.10) \quad \frac{(q; q)_\infty}{(Cq^{1-r}; q)_\infty} \prod_{i,j=1}^{r} \frac{(c_j x_i/x_j; q)_\infty}{(q x_i/x_j; q)_\infty}$$

and simplify to obtain (4.8). $\square$

THEOREM 4.8 (A multilateral A_{r-1} q-IPD type transformation). Let $a_1, \ldots, a_r$, b, $c_1, \ldots, c_r$, d, $e_1, \ldots, e_r$, $x_1, \ldots, x_r$, $y_1, \ldots, y_r$, and $h_1, \ldots, h_s$ be indeterminate, let N be an integer, let $m_1, \ldots, m_s$ be nonnegative integers, let $|m| = \sum_{i=1}^{s} m_i$, $r \geq 1$, and suppose that none of the denominators in (4.11) vanishes. Then

$$(4.11) \quad \sum_{k_1,\ldots,k_r=-\infty}^{\infty} \left(\prod_{1 \leq i < j \leq r} \left(\frac{x_i q^{k_i} - x_j q^{k_j}}{x_i - x_j} \right) \prod_{i,j=1}^{r} \frac{(a_j x_i/x_j; q)_{k_i}}{(c_j x_i/x_j; q)_{k_i}} \right.$$
$$\left. \times \frac{(b, h_1 q^{m_1}, \ldots, h_s q^{m_s}; q)_{|\mathbf{k}|}}{(d, h_1, \ldots, h_s; q)_{|\mathbf{k}|}} \left(\frac{Eq^{1-r-N}}{Ab} \right)^{|\mathbf{k}|} \right)$$
$$= (Eq^{-r})^N \frac{(Eq^{1-r}/b, dq^r/E; q)_{\infty}}{(q/b, d; q)_{\infty}} \prod_{i=1}^{s} \frac{(h_i q^r/E; q)_{m_i}}{(h_i; q)_{m_i}}$$
$$\times \prod_{i,j=1}^{r} \frac{(qx_i/x_j, c_j x_i/a_i x_j, c_j y_i q/e_j y_j, e_i y_i/a_i y_j; q)_{\infty}}{(qy_i/y_j, c_j e_i y_i/a_i e_j y_j, c_j x_i/x_j, x_i q/a_i x_j; q)_{\infty}}$$
$$\times \sum_{k_1,\ldots,k_r=-\infty}^{\infty} \left(\prod_{1 \leq i < j \leq r} \left(\frac{y_i q^{k_i} - y_j q^{k_j}}{y_i - y_j} \right) \prod_{i,j=1}^{r} \frac{(a_j y_i q/e_j y_j; q)_{k_i}}{(c_j y_i q/e_j y_j; q)_{k_i}} \right.$$
$$\left. \times \frac{(bq^r/E, h_1 q^{r+m_1}/E, \ldots, h_s q^{r+m_s}/E; q)_{|\mathbf{k}|}}{(dq^r/E, h_1 q^r/E, \ldots, h_s q^r/E; q)_{|\mathbf{k}|}} \left(\frac{Eq^{1-r-N}}{Ab} \right)^{|\mathbf{k}|} \right),$$

provided $|Eq^{1-r}/Ab| < |q^N| < |Eq^{|m|}/Cd|$.

PROOF. We have, for $|Eq^{1-r}/Ab| < |q^N| < |Eq^{|m|}/Cd|$,

$$(4.12) \quad {}_{2+s}\psi_{2+s} \left[\begin{matrix} A, b, h_1 q^{m_1}, \ldots, h_s q^{m_s} \\ Cq^{1-r}, d, h_1, \ldots, h_s \end{matrix} ; q, \frac{Eq^{1-r-N}}{Ab} \right]$$
$$= (Eq^{-r})^N \frac{(Eq^{1-r}/A, Eq^{1-r}/b, Cq/E, dq^r/E; q)_{\infty}}{(q/A, q/b, Cq^{1-r}, d; q)_{\infty}} \prod_{i=1}^{s} \frac{(h_i q^r/E; q)_{m_i}}{(h_i; q)_{m_i}}$$
$$\times {}_{2+s}\psi_{2+s} \left[\begin{matrix} Aq^r/E, bq^r/E, h_1 q^{r+m_1}/E, \ldots, h_s q^{r+m_s}/E \\ Cq/E, dq^r/E, h_1 q^r/E, \ldots, h_s q^r/E \end{matrix} ; q, \frac{Eq^{1-r-N}}{Ab} \right],$$

by the q-IPD type transformation in (2.11). Now we apply Lemma 3.7 to the ${}_{2+s}\psi_{2+s}$'s on the left and on the right side of this transformation. Specifically, we rewrite the ${}_{2+s}\psi_{2+s}$ on left side of (4.12) by the $b_i \mapsto c_i$, $i = 1, \ldots, r$, and

$$g(n) = \frac{(b, h_1 q^{m_1}, \ldots, h_s q^{m_s}; q)_n}{(d, h_1, \ldots, h_s; q)_n} \left(\frac{Eq^{1-r-N}}{Ab} \right)^n$$

case of Lemma 3.7. The ${}_{2+s}\psi_{2+s}$ on the right side of (4.12) is rewritten by the $a_i \mapsto a_i q/e_i$, $b_i \mapsto c_i q/e_i$, $x_i \mapsto y_i$, $i = 1, \ldots, r$, and

$$g(n) = \frac{(bq^r/E, h_1 q^{r+m_1}/E, \ldots, h_s q^{r+m_s}/E; q)_n}{(dq^r/E, h_1 q^r/E, \ldots, h_s q^r/E; q)_n} \left(\frac{Eq^{1-r-N}}{Ab} \right)^n$$

case of Lemma 3.7. Finally, we divide both sides of the resulting equation by

$$(4.13) \quad \frac{(q, Cq^{1-r}/A; q)_{\infty}}{(Cq^{1-r}, q/A; q)_{\infty}} \prod_{i,j=1}^{r} \frac{(c_j x_i/x_j, x_i q/a_i x_j; q)_{\infty}}{(qx_i/x_j, c_j x_i/a_i x_j; q)_{\infty}}$$

and simplify to obtain (4.11). □

THEOREM 4.9 (A multilateral A_{r-1} q-IPD type transformation). Let $a_1, \ldots, a_r$, b, $c_1, \ldots, c_r$, d, $e_1, \ldots, e_r$, $x_1, \ldots, x_r$, $y_1, \ldots, y_r$, and $h_1, \ldots, h_s$ be indeterminate, let N be an integer, let $m_1, \ldots, m_s$ be nonnegative integers, let $|m| = \sum_{i=1}^{s} m_i$, $r \geq 1$, and suppose that none of the denominators in (4.14) vanishes. Then

$$(4.14) \quad \sum_{k_1,\ldots,k_r=-\infty}^{\infty} \left(\prod_{1 \leq i < j \leq r} \left(\frac{x_i q^{k_i} - x_j q^{k_j}}{x_i - x_j} \right) \prod_{i,j=1}^{r} (c_j x_i / x_j; q)_{k_i}^{-1} \prod_{i=1}^{r} x_i^{r k_i - |\mathbf{k}|} \right.$$

$$\times (-1)^{(r-1)|\mathbf{k}|} q^{-\binom{|\mathbf{k}|}{2} + r \sum_{i=1}^{r} \binom{k_i}{2}}$$

$$\left. \times \frac{(A, b, h_1 q^{m_1}, \ldots, h_s q^{m_s}; q)_{|\mathbf{k}|}}{(d, h_1, \ldots, h_s; q)_{|\mathbf{k}|}} \left(\frac{E q^{1-r-N}}{Ab} \right)^{|\mathbf{k}|} \right)$$

$$= (E q^{-r})^N \frac{(E q^{1-r}/b, Cq/E, dq^r/E; q)_{\infty}}{(q/A, q/b, d; q)_{\infty}}$$

$$\times \prod_{i=1}^{s} \frac{(h_i q^r / E; q)_{m_i}}{(h_i; q)_{m_i}} \prod_{i,j=1}^{r} \frac{(q x_i/x_j, e_j y_i / a_j y_j; q)_{\infty}}{(q y_i/y_j, c_j x_i / x_j; q)_{\infty}}$$

$$\times \sum_{k_1,\ldots,k_r=-\infty}^{\infty} \left(\prod_{1 \leq i < j \leq r} \left(\frac{y_i q^{k_i} - y_j q^{k_j}}{y_i - y_j} \right) \prod_{i,j=1}^{r} (e_j y_i / a_j y_j; q)_{k_i}^{-1} \prod_{i=1}^{r} y_i^{r k_i - |\mathbf{k}|} \right.$$

$$\times (-1)^{(r-1)|\mathbf{k}|} q^{-\binom{|\mathbf{k}|}{2} + r \sum_{i=1}^{r} \binom{k_i}{2}}$$

$$\left. \times \frac{(E/C, Eq^{1-r}/d, Eq^{1-r}/h_1, \ldots, Eq^{1-r}/h_s; q)_{|\mathbf{k}|}}{(Eq^{1-r}/b, Eq^{1-r-m_1}/h_1, \ldots, Eq^{1-r-m_s}/h_s; q)_{|\mathbf{k}|}} \left(\frac{Cd q^{N-|m|}}{E} \right)^{|\mathbf{k}|} \right),$$

provided $|Eq^{1-r}/Ab| < |q^N| < |Eq^{|m|}/Cd|$.

PROOF. We have, for $|Eq^{1-r}/Ab| < |q^N| < |Eq^{|m|}/Cd|$,

$$(4.15) \quad {}_{2+s}\psi_{2+s} \left[\begin{array}{c} A, b, h_1 q^{m_1}, \ldots, h_s q^{m_s} \\ Cq^{1-r}, d, h_1, \ldots, h_s \end{array} ; q, \frac{Eq^{1-r-N}}{Ab} \right]$$

$$= (Eq^{-r})^N \frac{(Eq^{1-r}/A, Eq^{1-r}/b, Cq/E, dq^r/E; q)_{\infty}}{(q/A, q/b, Cq^{1-r}, d; q)_{\infty}} \prod_{i=1}^{s} \frac{(h_i q^r/E; q)_{m_i}}{(h_i; q)_{m_i}}$$

$$\times {}_{2+s}\psi_{2+s} \left[\begin{array}{c} E/C, Eq^{1-r}/d, Eq^{1-r}/h_1, \ldots, Eq^{1-r}/h_s \\ Eq^{1-r}/A, Eq^{1-r}/b, Eq^{1-r-m_1}/h_1, \ldots, Eq^{1-r-m_s}/h_s \end{array} ; q, \frac{Cd q^{N-|m|}}{E} \right],$$

by the q-IPD type transformation in (2.12). Now we apply Lemma 3.6 to the ${}_{2+s}\psi_{2+s}$'s on the left and on the right side of this transformation. Specifically, we rewrite the ${}_{2+s}\psi_{2+s}$ on left side of (4.15) by the $b_i \mapsto c_i$, $i = 1, \ldots, r$, and

$$f(n) = \frac{(A, b, h_1 q^{m_1}, \ldots, h_s q^{m_s}; q)_n}{(d, h_1, \ldots, h_s; q)_n} \left(\frac{Eq^{1-r-N}}{Ab} \right)^n$$

case of Lemma 3.6. The ${}_{2+s}\psi_{2+s}$ on the right side of (4.15) is rewritten by the $b_i \mapsto c_i/a_i$, $x_i \mapsto y_i$, $i = 1, \ldots, r$, and

$$f(n) = \frac{(E/C, Eq^{1-r}/d, Eq^{1-r}/h_1, \ldots, Eq^{1-r}/h_s; q)_n}{(Eq^{1-r}/b, Eq^{1-r-m_1}/h_1, \ldots, Eq^{1-r-m_s}/h_s; q)_n} \left(\frac{Cd q^{N-|m|}}{E} \right)^n$$

case of Lemma 3.6. Finally, we divide both sides of the resulting equation by (4.10) and simplify to obtain (4.14). □

THEOREM 4.10 (A multilateral A_{r-1} q-IPD type transformation). Let $a_1, \ldots, a_r$, b, $c_1, \ldots, c_r$, d, $e_1, \ldots, e_r$, $x_1, \ldots, x_r$, $y_1, \ldots, y_r$, and $h_1, \ldots, h_s$ be indeterminate, let N be an integer, let $m_1, \ldots, m_s$ be nonnegative integers, let $|m| = \sum_{i=1}^{s} m_i$, $r \geq 1$, and suppose that none of the denominators in (4.16) vanishes. Then

$$(4.16) \quad \sum_{k_1,\ldots,k_r=-\infty}^{\infty} \left(\prod_{1 \leq i < j \leq r} \left(\frac{x_i q^{k_i} - x_j q^{k_j}}{x_i - x_j} \right) \prod_{i,j=1}^{r} \frac{(a_j x_i/x_j; q)_{k_i}}{(c_j x_i/x_j; q)_{k_i}} \right.$$

$$\left. \times \frac{(b, h_1 q^{m_1}, \ldots, h_s q^{m_s}; q)_{|\mathbf{k}|}}{(d, h_1, \ldots, h_s; q)_{|\mathbf{k}|}} \left(\frac{Eq^{1-r-N}}{Ab} \right)^{|\mathbf{k}|} \right)$$

$$= (Eq^{-r})^N \frac{(Eq^{1-r}/b, dq^r/E; q)_\infty}{(q/b, d; q)_\infty} \prod_{i=1}^{s} \frac{(h_i q^r/E; q)_{m_i}}{(h_i; q)_{m_i}}$$

$$\times \prod_{i,j=1}^{r} \frac{(qx_i/x_j, c_j x_i/a_i x_j, e_j y_i/a_j y_j, c_i y_i q/e_i y_j; q)_\infty}{(qy_i/y_j, c_i e_j y_i/a_j e_i y_j, c_j x_i/x_j, x_i q/a_i x_j; q)_\infty}$$

$$\times \sum_{k_1,\ldots,k_r=-\infty}^{\infty} \left(\prod_{1 \leq i < j \leq r} \left(\frac{y_i q^{k_i} - y_j q^{k_j}}{y_i - y_j} \right) \prod_{i,j=1}^{r} \frac{(e_j y_i/c_j y_j; q)_{k_i}}{(e_j y_i/a_j y_j; q)_{k_i}} \right.$$

$$\left. \times \frac{(Eq^{1-r}/d, Eq^{1-r}/h_1, \ldots, Eq^{1-r}/h_s; q)_{|\mathbf{k}|}}{(Eq^{1-r}/b, Eq^{1-r-m_1}/h_1, \ldots, Eq^{1-r-m_s}/h_s; q)_{|\mathbf{k}|}} \left(\frac{Cdq^{N-|m|}}{E} \right)^{|\mathbf{k}|} \right),$$

provided $|Eq^{1-r}/Ab| < |q^N| < |Eq^{|m|}/Cd|$.

PROOF. We have, for $|Eq^{1-r}/Ab| < |q^N| < |Eq^{|m|}/Cd|$, (4.15) by the q-IPD type transformation in (2.12). Now we apply Lemma 3.7 to the $_{2+s}\psi_{2+s}$'s on the left and on the right side of this transformation. Specifically, we rewrite the $_{2+s}\psi_{2+s}$ on left side of (4.15) by the $b_i \mapsto c_i$, $i = 1, \ldots, r$, and

$$g(n) = \frac{(b, h_1 q^{m_1}, \ldots, h_s q^{m_s}; q)_n}{(d, h_1, \ldots, h_s; q)_n} \left(\frac{Eq^{1-r-N}}{Ab} \right)^n$$

case of Lemma 3.7. The $_{2+s}\psi_{2+s}$ on the right side of (4.15) is rewritten by the $a_i \mapsto e_i/c_i$, $b_i \mapsto e_i/a_i$, $x_i \mapsto y_i$, $i = 1, \ldots, r$, and

$$g(n) = \frac{(Eq^{1-r}/d, Eq^{1-r}/h_1, \ldots, Eq^{1-r}/h_s; q)_n}{(Eq^{1-r}/b, Eq^{1-r-m_1}/h_1, \ldots, Eq^{1-r-m_s}/h_s; q)_n} \left(\frac{Cdq^{N-|m|}}{E} \right)^n$$

case of Lemma 3.7. Finally, we divide both sides of the resulting equation by (4.13) and simplify to obtain (4.16). □

The $e_i = a_i q$, $i = 1, \ldots, r$, cases of Theorems 4.9 and 4.10 yield two A_{r-1} extensions of W. C. Chu's [**5**, Eq. 15] bilateral transformation. If we specialize these identities further by setting $c_i = q$, $i = 1, \ldots, r$, we obtain two A_{r-1} extensions of G. Gasper's [**10**, Eq. (19)] q-IPD type transformation.

Acknowledgements

We wish to thank Professors George Gasper and Mourad Ismail for valuable suggestions which led to an improvement of the article.

References

[1] G. E. Andrews, *q-Series: Their development and application in analysis, number theory, combinatorics, physics and computer algebra*, CBMS Regional Conference Lectures Series **66** (Amer. Math. Soc., Providence, RI, 1986).

[2] W. N. Bailey, "Series of hypergeometric type which are infinite in both directions", *Quart. J. Math.* (Oxford) **7** (1936), 105–115.

[3] G. Bhatnagar, "D_n basic hypergeometric series", *The Ramanujan J.* **3** (1999), 175–203.

[4] G. Bhatnagar and S. C. Milne, "Generalized bibasic hypergeometric series and their $U(n)$ extensions", *Adv. Math.* **131** (1997), 188–252.

[5] W. C. Chu, "Partial fractions and bilateral summations", *J. Math. Phys.* **35** (1994), 2036–2042; erratum: *ibid* **36** (1995), 5198–5199.

[6] W. C. Chu, "Partial fraction expansions and well-poised bilateral series", *Acta Sci. Math.* (Szeged) **64** (1998), 495–513.

[7] J. L. Fields and M. E. H. Ismail, "Polynomial expansions", *Math. Comp.* **29** (1975), 894–902.

[8] J. L. Fields and J. Wimp, "Expansions of hypergeometric functions in hypergeometric functions", *Math. Comp.* **15** (1961), 894–902.

[9] C. Fox, "The expression of hypergeometric series in terms of similar series", *Proc. London Math. Soc.* (2) **26** (1927), 201–210.

[10] G. Gasper, "Summation formulas for basic hypergeometric series", *SIAM J. Math. Anal.* **12** (1981), 196–200.

[11] G. Gasper, "Elementary derivations of summation and transformation formulas for q-series", in Special Functions, q-Series and Related Topics (M. E. H. Ismail, D. R. Masson and M. Rahman, eds.), Amer. Math. Soc., Providence, R. I., Fields Institute Communications **14** (1997), 55–70.

[12] G. Gasper and M. Rahman, "An indefinite bibasic summation formula and some quadratic, cubic, and quartic summation and transformation formulas", *Canad. J. Math.* **42** (1990), 1–27.

[13] G. Gasper and M. Rahman, *Basic hypergeometric series*, Encyclopedia of Mathematics And Its Applications 35, Cambridge University Press, Cambridge (1990).

[14] I. M. Gessel and C. Krattenthaler, "Cylindric Partitions", *Trans. Amer. Math. Soc.* **349** (1997), 429–479.

[15] R. A. Gustafson, "Multilateral summation theorems for ordinary and basic hypergeometric series in $U(n)$", *SIAM J. Math. Anal.* **18** (1987), 1576–1596.

[16] R. A. Gustafson, "The Macdonald identities for affine root systems of classical type and hypergeometric series very well-poised on semi-simple Lie algebras", *in* Ramanujan International Symposium on Analysis (Dec. 26th to 28th, 1987, Pune, India), N. K. Thakare (ed.) (1989), 187–224.

[17] J. Haglund, "Rook theory and hypergeometric series", *Adv. Appl. Math.* **17** (1996), 408–459.

[18] M. E. H. Ismail, private communication, February 2001.

[19] P. W. Karlsson, "Hypergeometric functions with integral parameter differences", *J. Math. Phys.* **12** (1971), 270–271.

[20] S. C. Milne, "An elementary proof of the Macdonald identities for $A_\ell^{(1)}$", *Adv. Math.* **57** (1985), 34–70.

[21] S. C. Milne, "The multidimensional $_1\Psi_1$ sum and Macdonald identities for $A_l^{(1)}$", in Theta Functions Bowdoin 1987 (L. Ehrenpreis and R. C. Gunning, eds.), Proc. Sympos. Pure Math. **49** (1989), 323–359.

[22] S. C. Milne, "Balanced $_3\phi_2$ summation theorems for $U(n)$ basic hypergeometric series", *Adv. Math.* **131** (1997), 93–187.

[23] S. C. Milne and G. M. Lilly, "Consequences of the A_l and C_l Bailey transform and Bailey lemma", *Discrete Math.* **139** (1995), 319–346.

[24] S. C. Milne and J. W. Newcomb, "$U(n)$ very-well-poised $_{10}\phi_9$ transformations", *J. Comput. Appl. Math.* **68** (1996), 239–285.

[25] S. C. Milne and M. Schlosser, "A new A_n extension of Ramanujan's $_1\psi_1$ summation with applications to multilateral A_n series", preprint arXiv:math.CA/0010162, to appear in *Rocky Mount. J. Math.*, spec. vol. "Special Functions 2000" (Tempe, AZ, May 29 – June 9, 2000).

[26] B. M. Minton, "Generalized hypergeometric function of unit argument", *J. Math. Phys.* **11** (1970), 1375–1376.

[27] M. Schlosser, "Multidimensional matrix inversions and A_r and D_r basic hypergeometric series", *The Ramanujan J.* **1** (1997), 243–274.

[28] M. Schlosser, "Some new applications of matrix inversions in A_r", *The Ramanujan J.* **3** (1999), 405–461.

[29] M. Schlosser, "Summation theorems for multidimensional basic hypergeometric series by determinant evaluations", *Discrete Math.* **210** (2000), 151–169.

[30] M. Schlosser, "A simple proof of Bailey's very-well-poised $_6\psi_6$ summation", preprint arXiv:math.CA/0007046, to appear in *Proc. Amer. Math. Soc.*

[31] M. Schlosser, "Elementary derivations of identities for bilateral basic hypergeometric series", preprint arXiv:math.CA/0010161, submitted.

[32] L. J. Slater, "General transformations of bilateral series", *Quart. J. Math.* (Oxford) **3** (1952), 73–80.

[33] L. J. Slater, *Generalized hypergeometric functions*, Cambridge Univ. Press, London/New York, 1966.

[34] D. Stanton, "An elementary approach to the Macdonald identities", *in q-Series and partitions* (D. Stanton, ed.), The IMA volumes in mathematics and its applications, vol. 18, Springer-Verlag, 1989, 139–150.

[35] A. Verma, "Some transformations of series with arbitrary terms", *Instituto Lombardo* (Rend. Sc.) A **106**, 342–353.

DEPARTMENT OF MATHEMATICS, THE OHIO STATE UNIVERSITY, 231 WEST 18TH AVENUE, COLUMBUS, OHIO 43210, USA

Current address: Institut für Mathematik der Universität Wien, Strudlhofgasse 4, A-1090 Wien, Austria

E-mail address: schlosse@ap.univie.ac.at

URL: http://www.mat.univie.ac.at/People/mschloss

Completeness of Basic Trigonometric System in $\mathcal{L}^p$

Sergei K. Suslov

ABSTRACT. We discuss several results on the completeness of the basic trigonometric system in $\mathcal{L}^p$-spaces in the framework of a general approach to the basic Fourier series.

1. Introduction

In this paper we discuss the completeness of the q-trigonometric system as a part of a program on the detailed investigation of the basic Fourier series introduced recently by Bustoz and Suslov [5]. One can see references [5], [8], [27], [28], [29], [30] for an introduction to the theory of q-Fourier series and references [6], [13], [14], [15], [16], [25], [26] regarding to the corresponding basic exponential function on a q-quadratic grid [21]. The case of the basic Fourier series on a q-linear grid is under investigation in [4]. In the current paper we establish several results on the completeness of the basic trigonometric system $\{\mathcal{E}_q(x; i\omega_n)\}_{n=-\infty}^{\infty}$ in the weighted $\mathcal{L}_\rho^p$-spaces for $1 \leq p < \infty$, using methods of the theory of entire functions [3], [5], [18] and [19].

The basic exponential function on a q-quadratic grid can be introduced as

$$(1.1) \quad \mathcal{E}_q(x; \alpha) = \frac{(\alpha^2; q^2)_\infty}{(q\alpha^2; q^2)_\infty}$$

$$\times \sum_{n=0}^{\infty} \frac{q^{n^2/4} \alpha^n}{(q;q)_n} (-i)^n \left(-iq^{(1-n)/2}e^{i\theta}, -iq^{(1-n)/2}e^{-i\theta}; q\right)_n,$$

where $x = \cos\theta$ and $|\alpha| < 1$; see [5], [14], [15], [16], [25], [26], and a review paper [29] for more details. We assume that $0 < q < 1$ and use the standard notations [7] for the basic hypergeometric series and for the q-shifted factorials throughout the paper. Analytic continuation in a larger domain can be given with the help of the following generating function (see, for example, [16] and [25])

$$(1.2) \quad (q\alpha^2; q^2)_\infty \mathcal{E}_q(x; \alpha) = \sum_{n=0}^{\infty} \frac{q^{n^2/4}}{(q;q)_n} \alpha^n H_n(x|q),$$

1991 *Mathematics Subject Classification.* Primary 33D45, 42C10; Secondary 33D15.
Key words and phrases. Basic trigonometric functions, basic Fourier series, completeness of the basic trigonometric system.
The author was supported by NSF grant # DMS 9803443.

where $H_n(x|q)$ are the continuous q-Hermite polynomials

$$(1.3) \qquad H_n(\cos\theta|q) = \sum_{k=0}^{n} \frac{(q;q)_n}{(q;q)_k (q;q)_{n-k}} e^{i(n-2k)\theta}.$$

The right hand side of (1.2) is an entire function in both variables x and α. Following [5] we introduce

$$(1.4) \qquad e(x,\alpha) := \sum_{n=0}^{\infty} \frac{q^{n^2/4}}{(q;q)_n} \alpha^n H_n(x|q),$$

which is an entire function of order zero in α for all finite values of x by Lemma 1 of [5].

Completeness of the trigonometric system $\{e^{inx}\}_{n=-\infty}^{\infty}$ on the interval $(-\pi,\pi)$ is one of the fundamental facts in the theory of trigonometric series (see, for example, [1], [2], [17], [18], [19], [20], [31], and [32]). Bustoz and Suslov [5] proved that the basic trigonometric system $\{\mathcal{E}_q(x;i\omega_n)\}_{n=-\infty}^{\infty}$ is complete in the weighted $\mathcal{L}^2_{\rho_1}(-1,1)$ space where $\rho_1(x)$ is the weight function for the continuous q-ultraspherical polynomials $C_m(x;\beta|q)$ with $\beta = q^{1/2}$ when $\omega_n = 0, \pm\omega_1, \pm\omega_2, \pm\omega_3, \ldots$ and $\omega_0 = 0 < \omega_1 < \omega_2 < \omega_3 < \ldots$ are nonnegative zeros of certain basic sine function

$$(1.5) \qquad S_q(\eta;\omega) = \frac{(-i\omega;q^{1/2})_\infty - (i\omega;q^{1/2})_\infty}{2i(-q\omega^2;q^2)_\infty},$$

which can be viewed as an analog of $\sin \pi\omega$. See [5] for the details and [30] for an independent proof.

These results admit an interesting generalization. Levinson [19] and Levin [18] had established the completeness of systems of functions $\{e^{i\lambda_n x}\}$ on the interval $(-\pi,\pi)$ where $\{\lambda_n\}$ is a sequence of complex numbers. Many different criteria for the completeness of these systems are known. We shall mention only two classical results here; see Chapter I of [19] and Appendix III of [18] for more details.

THEOREM 1. *If $|\lambda_n| \leq |n| + 1/2p$, $n = 0, \pm 1, \pm 2, \ldots$, then the system $\{e^{i\lambda_n x}\}$ is complete in $\mathcal{L}^p(-\pi,\pi)$, $1 < p < \infty$.*

THEOREM 2. *If $\{\lambda_n\}$ is the set of zeros of an entire function $h(\lambda)$ of exponential type and*

$$(1.6) \qquad \lim_{|\gamma|\to\infty} |h(i\gamma)| e^{-\pi|\gamma|} > 0,$$

then the system $\{e^{i\lambda_n x}\}$ is complete in $\mathcal{L}^p(-\pi,\pi)$, $1 \leq p < \infty$.

In the current paper we present theorems similar to Theorems 1 and 2 for the basic trigonometric system $\{e(x,i\omega_n)\}$ that extends the completeness results of [5] and [30]. The paper is organized as follows. In Section 2 we discuss analogs of Theorem 2 and then consider several examples including an analog of Theorem 1 in Section 3.

2. Completeness Theorems

A set $\mathcal{B}$ of vectors of the linear normalized space V is said to be complete in V if every vector $\chi \in V$ can be approximated to any degree of accuracy by means of an expression of the form

$$(2.1) \qquad \alpha_1 e_1 + \alpha_2 e_2 + \ldots + \alpha_n e_n,$$

where $e_k \in \mathcal{B}$. We shall use the Criterion for the Completeness of a Set of Vectors in Linear Normalized Spaces [1].

CRITERION 1. *The necessary and sufficient condition that a set $\mathcal{B}$ in V be a complete set is, that every linear functional φ in V which vanishes for any vector $e_k \in \mathcal{B}$ be identically zero.*

The completeness of the basic trigonometric system when the corresponding eigenvalues are the zeros of Jackson's q-Bessel function [10], [11],

$$(2.2) \qquad J_\nu^{(2)}(x;q) = \frac{(q^{\nu+1};q)_\infty}{(q;q)_\infty} \sum_{n=0}^{\infty} q^{(\nu+n)n} \frac{(-1)^n (x/2)^{\nu+2n}}{(q;q)_n (q^{\nu+1};q)_n},$$

has been established in [29]. The following theorem holds.

THEOREM 3. *Let $j_{\mu,k}(q)$ be the zeros of Jackson's q-Bessel function $J_\mu^{(2)}(x;q)$. The q-trigonometric system $\{\mathcal{E}_q(x;i\omega_k)\}_{k=-\infty}^{\infty}$ is complete in the weighted $\mathcal{L}_\rho^2(-1,1)$ space, where $\rho(x)$ is the weight function of the continuous q-ultraspherical polynomials $C_m(x;q^\nu|q)$ for $0 < \nu < 1/2$, if: (a) $\omega_k = \frac{1}{2} j_{\nu-1,k}(q)$, $k = \pm 1, \pm 2, \ldots$; (b) $\omega_k = \frac{1}{2} j_{\nu,k}(q)$, $k = 0, \pm 1, \pm 2, \ldots$.*

The proof of this theorem in [29] uses an independent method based on Ismail and Zhang's basic analog of the expansion formula of the plane wave in terms of the spherical harmonics [16] and the q-Lommel polynomials introduced in [11]. The most important practical case $\nu = 1/2$, when the corresponding systems are orthogonal, is considered in [30]. In this paper we establish more general results using the theory of entire functions.

The study of completeness of the classical and basic trigonometric systems amounts to the study of zeros of certain entire function. We shall first extend the following result by Levinson [19], [18].

THEOREM 4. *For the system $\{e^{i\lambda_n x}\}$, in which the λ_n, $n = 1, 2, 3, \ldots$ are complex numbers, to be incomplete in $\mathcal{L}^p(-\pi, \pi)$, $1 \leq p < \infty$, it is necessary and sufficient that there exist an entire function $f(\lambda)$, zero at all λ_n, $n = 1, 2, 3, \ldots$, that can be represented in the form*

$$(2.3) \qquad f(\lambda) = \int_{-\pi}^{\pi} e^{i\lambda x} \phi(x) \, dx,$$

where $\phi(x) \in \mathcal{L}^r(-\pi, \pi)$ and $1/p + 1/r = 1$ (for $p = 1$ the $\phi(x)$ is a bounded measurable function).

Let $\rho(x)$ be an integrable function,

$$(2.4) \qquad \int_{-1}^{1} \rho(x) \, dx < \infty,$$

positive almost everywhere, $\rho(x) > 0$, on $(-1, 1)$. We shall assume that this condition holds throughout the paper and consider the $\mathcal{L}_\rho^p(-1,1)$-spaces with respect to this weight function if it is not stated otherwise. The case of a more general finite measure can be discussed in a similar fashion. For the basic trigonometric system a somewhat similar theorem holds.

THEOREM 5. *The basic trigonometric system $\{e(x, i\omega_n)\}$, where $\{\omega_n\}$ is an infinite sequence of complex numbers, is incomplete in the weighted space $\mathcal{L}_\rho^p(-1,1)$,*

$1 \leq p < \infty$ if and only if there exist an entire function $f(\omega)$ of order zero, such that $f(\omega_n) = 0$ for all ω_n, that can be represented in the form

$$(2.5) \qquad f(\omega) = \int_{-1}^{1} e(x, i\omega) \ \phi(x) \ \rho(x) \ dx,$$

where $\phi(x) \in \mathcal{L}_\rho^r(-1, 1)$ not equivalent to zero and $1/p + 1/r = 1$ (for $p = 1$ the $\phi(x)$ is a bounded measurable function).

PROOF. We can use arguments similar to the proof of Theorem 4 in [19] and [18]. If the system $\{e(x, i\omega_n)\}$ is incomplete in $\mathcal{L}_\rho^p(-1, 1)$, then by the Criterion for the Completeness of a Set of Vectors in Linear Normalized Spaces [1] there is a linear functional φ in $\mathcal{L}_\rho^p(-1, 1)$, not identically zero, which vanishes for all the function $e(x, i\omega_n)$: $\varphi(e(x, i\omega_n)) = 0$. By the Riesz Representation Theorem [1], [20], [23] the general form of a linear functional in $\mathcal{L}_\rho^p(-1, 1)$, $1 \leq p < \infty$ is

$$(2.6) \qquad \varphi(\chi) = \int_{-1}^{1} \chi(x) \ \phi(x) \ \rho(x) \ dx,$$

where $\phi(x) \in \mathcal{L}_\rho^r(-1, 1)$ and $1/p + 1/r = 1$ (for $p = 1$ the $\phi(x)$ is a bounded measurable function). Thus,

$$(2.7) \qquad \varphi(e(x, i\omega_n)) = \int_{-1}^{1} e(x, i\omega_n) \ \phi(x) \ \rho(x) \ dx = 0$$

and the function $f(\omega)$ given by (2.5) is zero at all the points ω_n.

Therefore the study of completeness of the basic trigonometric systems amounts to the study of zeros of the function $f(\omega)$. By the Hölder inequality

$$\int_{-1}^{1} |\phi(x)| \ \rho(x) \ dx \leq \left(\int_{-1}^{1} |\phi(x)|^r \ \rho(x) \ dx \right)^{1/r} \left(\int_{-1}^{1} \rho(x) \ dx \right)^{1/p},$$

thus, $\phi(x)$ is integrable on $(-1, 1)$ for $1 \leq p < \infty$,

$$(2.8) \qquad \int_{-1}^{1} |\phi(x)| \ \rho(x) \ dx = A < \infty.$$

Let us show that the $f(\omega)$ given by (2.5) is an entire function. The series in (1.4) converges absolutely and uniformly in x on $[-1, 1]$ and integrating with respect to the finite measure $d\mu(x) = |\phi(x)| \rho(x) dx$ we obtain by the Lebesgue Dominated Convergence Theorem

$$f(\omega) = \int_{-1}^{1} \phi(x) \left(\sum_{k=0}^{\infty} \frac{q^{k^2/4}}{(q;q)_k} (i\omega)^k H_k(x|q) \right) \rho(x) dx$$

$$(2.9) \qquad = \sum_{k=0}^{\infty} \frac{q^{k^2/4}}{(q;q)_k} (i\omega)^k \int_{-1}^{1} \phi(x) H_k(x|q) \rho(x) dx,$$

or

$$(2.10) \qquad f(\omega) = \sum_{k=0}^{\infty} c_k \ \omega^k$$

with

$$(2.11) \qquad c_k = i^k \frac{q^{k^2/4}}{(q;q)_k} \int_{-1}^{1} \phi(x) H_k(x|q) \rho(x) dx.$$

Using the uniform upper bound for the continuous q-Hermite polynomials

$$|H_k(x|q)| < q^{-k\varepsilon} \frac{(q;q)_k}{(q,q^{2\varepsilon},q^{1-2\varepsilon};q)_\infty}, \tag{2.12}$$

when $|x| < x(\varepsilon) = (q^\varepsilon + q^{-\varepsilon})/2$ and $0 < \varepsilon < 1/2$ (see (12.5) of [27]), one gets

$$|c_k| \leq B q^{k^2/4 - \varepsilon k}, \tag{2.13}$$

where B is a positive constant. The radius of convergence of series (2.9) is

$$\frac{1}{R} = \lim_{k\to\infty} (|c_k|)^{1/k} \leq \lim_{k\to\infty} \left(B^{1/k} q^{k/4-\varepsilon} \right) = 0. \tag{2.14}$$

Therefore the $f(\omega)$ is an entire function. The order of this entire function is [18]

$$\lim_{k\to\infty} \left(\frac{k \log k}{-\log |c_k|} \right) \leq \lim_{k\to\infty} \left(\frac{k \log k}{-\log \left(B q^{k^2/4 - \varepsilon k} \right)} \right)$$

$$= \lim_{k\to\infty} \left(\frac{k \log k}{-(k^2/4 - \varepsilon k)\log q - \log B} \right) = 0. \tag{2.15}$$

This completes the proof. $\square$

REMARK 1. As in the classical case [18], Theorem 5 is also valid in $\mathcal{C}(-1,1)$. The corresponding entire function of order zero can be represented as

$$f(\omega) = \int_{-1}^{1} e(x, i\omega) \, d\sigma(x), \tag{2.16}$$

where $\sigma(x)$ is a function of bounded variation on the interval $(-1,1)$. This result follows directly from the general form of the linear functional in $\mathcal{C}(-1,1)$ [1], [17], [20], [23].

We can now prove the following analog of Theorem 2.

THEOREM 6. Let $\{\omega_n\}$ be the set of zeros of an entire function $h(\omega)$ of order zero and suppose that all these zeros ω_n are simple. If on the imaginary axis

$$\lim_{|\omega|\to\infty} \frac{e(x(\varepsilon), |\omega|)}{|h(\omega)|} = 0, \tag{2.17}$$

where $x(\varepsilon) = (q^\varepsilon + q^{-\varepsilon})/2$ and $0 \leq \varepsilon < 1/2$, then the system $\{e(x, i\omega_n)\}$ is complete in $\mathcal{L}^p_\rho(-1,1)$, $1 \leq p < \infty$.

PROOF. In view of Theorem 5 the basic trigonometric system $\{e(x, i\omega_n)\}$ is incomplete in $\mathcal{L}^p_\rho(-1,1)$ if and only if there is an entire function of order zero $f(\omega)$ given by (2.5) where $\phi(x) \in \mathcal{L}^r_\rho(-1,1)$ is not zero almost everywhere on $[-1,1]$. In view of (2.8),

$$|f(\omega)| \leq \int_{-1}^{1} |\phi(x) \, e(x, i\omega)| \, \rho(x) \, dx$$

$$\leq e(x(\varepsilon), |\omega|) \int_{-1}^{1} |\phi(x)| \, \rho(x) \, dx$$

$$= A \, e(x(\varepsilon), |\omega|), \tag{2.18}$$

where we have used the uniform bound from Lemma 2 of [5],

$$|e(\cos\theta, i\omega)| \leq e(x(\varepsilon), |\omega|), \quad \varepsilon > 0. \tag{2.19}$$

Our next step is to show that under the hypotheses of the theorem the function $f(\omega)$ is identically zero. Consider the quotient

$$(2.20) \qquad g(\omega) = \frac{f(\omega)}{h(\omega)}$$

of two entire functions. The functions $f(\omega)$ and $h(\omega)$ have the same zeros, so $g(\omega)$ is an entire function since $h(\omega)$ has only simple zeros. The order of this entire function is zero because both $f(\omega)$ and $h(\omega)$ are of order zero (see [**18**], Corollary of Theorem 12 on p. 24). Moreover, this function $g(\omega)$ is bounded on the imaginary axis. Indeed, by (2.18)

$$(2.21) \qquad |g(\omega)| \leq \frac{Ae\left(x\left(\varepsilon\right), |\omega|\right)}{|h(\omega)|}$$

and the limit $|\omega| \to \infty$ of the right hand side exists by (2.17). But an entire function of order zero bounded on a line must be a constant (see Theorems 21–22 and Corollary on pp. 49–51 of [**18**]). Therefore

$$(2.22) \qquad f(\omega) = C\ h(\omega)$$

and

$$(2.23) \qquad |C| \leq \frac{Ae\left(x\left(\varepsilon\right), |\omega|\right)}{|h(\omega)|} \to 0$$

as $|\omega| \to \infty$ by (2.17). Thus, $f(\omega)$ is identically zero and by (2.9)

$$(2.24) \qquad \int_{-1}^{1} H_k\left(x|q\right) \phi\left(x\right) \rho\left(x\right) dx = 0$$

for all $k = 0, 1, 2, \ldots$. But the system $\{H_k(x|q)\}_{k=0}^{\infty}$ is complete in $\mathcal{L}_{\rho}^{r}(-1,1)$, $1 \leq r < \infty$ (see, for example, [**1**], [**24**], or [**9**]) and, therefore, $\phi(x) = 0$ almost everywhere on $(-1, 1)$ due to the positivity of $\rho(x)$. This contradiction proves that the system $\{e(x, i\omega_n)\}$ is complete in $\mathcal{L}_{\rho}^{p}(-1,1)$, $1 \leq p < \infty$. $\square$

In order to verify the limiting condition (2.17) one can use the following uniform upper bounds for the q-exponential function found in [**27**].

LEMMA 1. *Let* $-x(\varepsilon) < x < x(\varepsilon)$, *where* $x(\varepsilon) = (q^{\varepsilon} + q^{-\varepsilon})/2$, $0 < q < 1$, *and* $0 < \varepsilon < 1/2$. *Then*

$$(2.25) \qquad |e(x, i\omega)| < \frac{\left(-q^{1/4-\varepsilon} |\omega|; q^{1/2}\right)_{\infty}}{(q, q^{2\varepsilon}, q^{1-2\varepsilon}; q)_{\infty}}$$

and

$$(2.26) \qquad |e(x, i\omega)| < \frac{\left(q^{1/2}, -q^{1/4-\varepsilon} |\omega|, -q^{1/4+\varepsilon}/|\omega|; q^{1/2}\right)_{\infty}}{(q, q^{2\varepsilon}, q^{1-2\varepsilon}; q)_{\infty}}$$

for all real values of the ω.

(See Appendix of [**27**] for the proof of this Lemma.)

The proof of Theorem 6 is based on the upper bound (2.18) for the entire function $f(\omega)$. In the case of the $\mathcal{L}^2$-space the Cauchy–Schwarz inequality leads to another estimate for this function.

LEMMA 2. Let $f(\omega)$ be defined by (2.5) where $\phi(x)$ is in $\mathcal{L}^2_\rho(-1,1)$ with $\rho(x) = \rho_1(x) w(x)$, $w(x) > 0$ is a continuous function on $[-1,1]$ and $\rho_1(x)$ is the weight function of the continuous q-ultraspherical polynomials $C_m(x; q^{1/2}|q)$. Then

$$|f(\omega)| \leq \sqrt{\pi} \frac{(q^{1/2};q)_\infty}{(q;q)_\infty} \left\|\phi(x)\sqrt{w(x)}\right\|_\rho \tag{2.27}$$

$$\times \left(\frac{\left|(i\omega; q^{1/2})_\infty\right|^2 - \left|(-i\omega; q^{1/2})_\infty\right|^2}{2 \operatorname{Im} \omega}\right)^{1/2},$$

where

$$\left\|\phi(x)\sqrt{w(x)}\right\|_\rho = \left(\int_{-1}^1 \left|\phi(x)\sqrt{w(x)}\right|^2 \rho(x)\, dx\right)^{1/2}. \tag{2.28}$$

PROOF. By the Cauchy–Schwarz inequality

$$|f(\omega)| \leq \int_{-1}^1 |e(x, i\omega)|\, |\phi(x)|\, \rho(x)\, dx \tag{2.29}$$

$$\leq \left(\int_{-1}^1 |e(x, i\omega)|^2 \rho_1(x)\, dx\right)^{1/2}$$

$$\times \left(\int_{-1}^1 \left|\phi(x)\sqrt{w(x)}\right|^2 \rho(x)\, dx\right)^{1/2}.$$

The integral involving the q-exponential functions in the right hand side can be evaluated as the following special case

$$\int_0^\pi e(\cos\theta; \alpha)\, e(\cos\theta; \beta)\, (e^{2i\theta}, e^{-2i\theta}; q)_{1/2}\, d\theta \tag{2.30}$$

$$= \pi \frac{(q^{1/2};q)^2_\infty}{(q;q)^2_\infty} \frac{(-\alpha, -\beta; q^{1/2})_\infty - (\alpha, \beta; q^{1/2})_\infty}{(\alpha + \beta)}$$

of an integral evaluated by Ismail and Stanton [15]. By the hypothesis of the lemma function $w(x)$ is bounded on $[-1,1]$ and the integral in (2.28) converges. This completes the proof. □

REMARK 2. It is worth mentioning that the special case of the Ismail and Stanton integral (2.30) gives rise to an independent proof of the orthogonality property of the basic trigonometric system $\{\mathcal{E}_q(x; i\omega_n)\}_{n=-\infty}^\infty$ on the zeros of the q-sine function $S_q(\eta; \omega)$ given by (1.5); see [29] for the details.

The last inequality leads to the following theorem.

THEOREM 7. Let $\{\omega_n\}$ be the set of zeros of an entire function $h(\omega)$ of order zero and suppose that all these zeros ω_n are simple. If

$$\lim_{|\gamma| \to \infty} \frac{(-\gamma; q^{1/2})^2_\infty - (\gamma; q^{1/2})^2_\infty}{2\gamma |h(i\gamma)|^2} = 0, \tag{2.31}$$

then the system $\{e(x, i\omega_n)\}$ is complete in $\mathcal{L}^2_\rho(-1, 1)$, where $\rho(x) = \rho_1(x) w(x)$, $w(x) > 0$ is a continuous function on $[-1, 1]$ and $\rho_1(x)$ is the weight function for the continuous q-ultraspherical polynomials $C_m(x; q^{1/2}|q)$.

PROOF. Suppose that the system $\{e(x, i\omega_n)\}$ is incomplete in $\mathcal{L}^2_\rho(-1,1)$. Then by Theorem 5 there is an entire function of order zero $f(\omega)$ given by (2.5), where the corresponding function $\phi(x)$, not equivalent to zero, is in $\mathcal{L}^2_\rho(-1,1)$, such that $f(\omega_n) = 0$ at all the points ω_n. Moreover, inequality (2.27) holds. In the same manner as in the proof of Theorem 6, we can see that function $g(\omega)$ given by (2.20) is an entire function of order zero and on the imaginary axis

$$(2.32) \qquad |g(\omega)| = \left|\frac{f(\omega)}{h(\omega)}\right| \leq B \left(\frac{(-\gamma; q^{1/2})_\infty^2 - (\gamma; q^{1/2})_\infty^2}{2\gamma |h(i\gamma)|^2}\right)^{1/2} \to 0$$

as $|\gamma| \to \infty$ by (2.31) (B is a constant). Thus, $g(\omega)$ is bounded on the imaginary axis and, therefore, must be a constant. The value of this constant is zero due to (2.32). Thus, $f(\omega)$ is identically zero and by (2.9) equations (2.24) hold for all $k = 0, 1, 2, \ldots$. But the system $\{H_k(x|q)\}_{k=0}^\infty$ is closed and $\phi(x) = 0$ almost everywhere on $(-1,1)$. Therefore the system $\{e(x, i\omega_n)\}$ is complete in $\mathcal{L}^2_\rho(-1,1)$. □

REMARK 3. In view of

$$(2.33) \qquad \left(-|\gamma|; q^{1/2}\right)_\infty^2 - \left(|\gamma|; q^{1/2}\right)_\infty^2 \leq \left(-|\gamma|; q^{1/2}\right)_\infty^2,$$

relation (2.31) can be replaced by a weaker condition

$$(2.34) \qquad \lim_{|\gamma| \to \infty} \frac{\left(-|\gamma|; q^{1/2}\right)_\infty}{|\gamma|^{1/2} |h(i\gamma)|} = 0.$$

3. Examples

Let us discuss several examples of the complete basic trigonometric systems $\{e(x, i\omega_n)\}$ using Theorems 6 and 7.

3.1. Some infinite products.
Consider the entire function of order zero

$$(3.1) \qquad h(\omega) = (\omega^2/\alpha^2; q^2)_\infty = \prod_{k=0}^\infty \left(1 - \frac{\omega^2}{\alpha^2} q^{2k}\right),$$

where the zeros are given explicitly as

$$(3.2) \qquad \omega_n = \alpha q^{-n}, \qquad \omega_{-n} = -\omega_n, \qquad n = 0, 1, 2, \ldots.$$

By Theorem 6 the system $\{e(x, i\omega_n)\}_{n=-\infty}^\infty$ is complete in $\mathcal{L}^p_\rho(-1,1)$ if the condition (2.17) is satisfied. In view of (2.25),

$$(3.3) \qquad \left|\frac{e(x, i\omega)}{h(\omega)}\right| < C \frac{\left(-q^{1/4-\varepsilon} |\omega|; q^{1/2}\right)_\infty}{\left(-|\omega|^2/\alpha^2; q^2\right)_\infty}$$

on the imaginary axis $\omega = i\gamma$. Consider an arbitrary sequence $\{\gamma_k\}$ of real numbers such that $|\gamma_k| \to \infty$ and present these numbers in an 'exponential form',

$$(3.4) \qquad \gamma_k = \beta_k q^{-n_k}, \qquad 1 < |\beta_k| < q^{-1}.$$

Then, by (I.7) of [7]

$$(3.5) \qquad \frac{\left(-q^{1/4-\varepsilon} |\gamma_k|; q^{1/2}\right)_\infty}{\left(-\gamma_k^2/\alpha^2; q^2\right)_\infty}$$

$$= \frac{\left(-q^{1/4-\varepsilon}\left|\gamma_k\right|;q^{1/2}\right)_{2n_k}}{\left(-\gamma_k^2/\alpha^2;q^2\right)_{n_k}} \frac{\left(-q^{1/4-\varepsilon+n_k}\left|\gamma_k\right|;q^{1/2}\right)_\infty}{\left(-q^{2n_k}\gamma_k^2/\alpha^2;q^2\right)_\infty}$$

$$= \left(q^{1-2\varepsilon}\alpha^2\right)^{n_k} \frac{\left(-q^{1/4+\varepsilon-n_k}/\left|\gamma_k\right|;q^{1/2}\right)_{2n_k}}{\left(-q^{2-2n_k}\alpha^2/\gamma_k^2;q^2\right)_{n_k}}$$

$$\times \frac{\left(-q^{1/4-\varepsilon+n_k}\left|\gamma_k\right|;q^{1/2}\right)_\infty}{\left(-q^{2n_k}\gamma_k^2/\alpha^2;q^2\right)_\infty}$$

$$< \left(q^{1-2\varepsilon}\alpha^2\right)^{n_k} \frac{\left(-q^{1/4+\varepsilon};q^{1/2}\right)_{2n_k}}{\left(-q^4\alpha^2;q^2\right)_{n_k}} \frac{\left(-q^{-3/4-\varepsilon};q^{1/2}\right)_\infty}{\left(-1/\alpha^2;q^2\right)_\infty} \to 0$$

as $k \to \infty$ when $q^{1-2\varepsilon}\alpha^2 < 1$. Thus, the system $\{e(x, \pm i\alpha q^{-n})\}_{n=0}^\infty$ is complete in $\mathcal{L}_\rho^p(-1,1)$, $1 \le p < \infty$, provided that $0 < \alpha^2 < q^{-1}$.

Choosing the function

$$(3.6) \qquad h(\omega) = \omega\left(\omega^2/\alpha^2;q^2\right)_\infty = \omega \prod_{k=0}^\infty \left(1 - \frac{\omega^2}{\alpha^2}q^{2k}\right),$$

one gets that the system $\{1, e(x, \pm i\alpha q^{-n})\}_{n=0}^\infty$ is complete in $\mathcal{L}_\rho^p(-1,1)$, $1 \le p < \infty$, if $0 < \alpha^2 < q^{-2}$. Condition (2.34) and Theorem 7 give the same results but for $\mathcal{L}_\rho^2(-1,1)$.

In a similar fashion, we can establish the following result somewhat similar to Theorem 1.

THEOREM 8. *The system $\{e(x, \pm i\omega_n)\}_{n=1}^\infty$, where $\omega_n = \alpha_n q^{1-n}$, is complete in $\mathcal{L}_\rho^p(-1,1)$, $p \ge 1$, if $q < \alpha_n < 1$. The similar system $\{e(x, i\omega_n)\}_{n=-\infty}^\infty$, where $\omega_n = \alpha_n q^{1-n}$, and $\omega_{-n} = -\omega_n$ is complete in $\mathcal{L}_\rho^p(-1,1)$, $1 \le p < \infty$, if $1 < \alpha_n < q^{-1}$.*

PROOF. Consider the first case. By the definition the convergence exponent of the sequence $\{\omega_n\}$ is the greatest lower bound of numbers λ for which the series

$$(3.7) \qquad \sum_{n=1}^\infty \frac{1}{\left|\omega_n\right|^\lambda}$$

converges. For the given sequence $\{\omega_n\}$ this series converges for every $\lambda > 0$ and the convergence exponent is zero. By Borel's Theorem [**18**], the order of the corresponding canonical product does not exceed the convergence exponent of the sequence of zeros. Thus, the infinite product

$$(3.8) \qquad h(\omega) = \prod_{n=1}^\infty \left(1 - \frac{\omega^2}{\alpha_n^2}q^{2n-2}\right)$$

is an entire function of order zero. One can repeat the previous consideration to verify the validity of the limiting condition (2.17). Indeed, for a sequence of real numbers $\{\gamma_k\}$ of the form (3.4) we can write

$$(3.9) \qquad h(i\gamma_k) = \prod_{n=1}^{n_k}\left(1 + \frac{\gamma_k^2}{\alpha_n^2}q^{2n-2}\right) \prod_{n=n_k+1}^\infty \left(1 + \frac{\gamma_k^2}{\alpha_n^2}q^{2n-2}\right)$$

$$= \frac{\gamma_k^{2n_k}q^{n_k(n_k-1)}}{\prod_{n=1}^{n_k}\alpha_n^2} \prod_{n=1}^{n_k}\left(1 + \frac{\alpha_n^2}{\gamma_k^2}q^{2-2n}\right) \prod_{n=n_k+1}^\infty \left(1 + \frac{\gamma_k^2}{\alpha_n^2}q^{2n-2}\right)$$

and

$$\text{(3.10)} \quad \frac{\left(-q^{1/4-\varepsilon}\left|\gamma_k\right|;q^{1/2}\right)_\infty}{\left|h\left(i\gamma_k\right)\right|}$$

$$= \left(q^{1-2\varepsilon}\right)^{n_k} \left(\prod_{n=1}^{n_k} \alpha_n^2\right) \frac{\left(-q^{1/4+\varepsilon-n_k}/\left|\gamma_k\right|;q^{1/2}\right)_{2n_k}}{\prod_{n=1}^{n_k}\left(1+q^{2-2n}\alpha_n^2/\gamma_k^2\right)}$$

$$\times \frac{\left(-q^{1/4-\varepsilon+n_k}\left|\gamma_k\right|;q^{1/2}\right)_\infty}{\prod_{n=n_k+1}^{\infty}\left(1+q^{2n-2}\gamma_k^2/\alpha_n^2\right)}$$

$$< \left(q^{1-2\varepsilon}\right)^{n_k} \frac{\left(-q^{1/4+\varepsilon};q^{1/2}\right)_{2n_k}}{\left(-q^2;q^2\right)_{n_k}} \frac{\left(-q^{-3/4-\varepsilon};q^{1/2}\right)_\infty}{\left(-1;q^2\right)_\infty} \to 0$$

as $k \to \infty$ when $0 < \varepsilon < 1/2$. Thus, the system $\{e(x,\pm i\alpha_n q^{-n})\}_{n=0}^{\infty}$ is complete in $\mathcal{L}_\rho^p(-1,1)$, $1 \leq p < \infty$.

In the second case we repeat the same consideration for the infinite product of the form

$$\text{(3.11)} \quad h(\omega) = \omega \prod_{n=1}^{\infty} \left(1 - \frac{\omega^2}{\alpha_n^2} q^{2n-2}\right).$$

This completes the proof. $\square$

3.2. Basic sine and cosine functions. The q-sine and cosine functions under consideration can be introduced as

$$\text{(3.12)} \quad S_q(\eta;\omega) = \frac{(-i\omega;q^{1/2})_\infty - (i\omega;q^{1/2})_\infty}{2i(-q\omega^2;q^2)_\infty}$$

$$= \frac{1}{(-q\omega^2;q^2)_\infty} \sum_{k=0}^{\infty} (-1)^k \frac{q^{k(k+1/2)}}{(q^{1/2};q^{1/2})_{2k+1}} \omega^{2k+1}$$

and

$$\text{(3.13)} \quad C_q(\eta;\omega) = \frac{(-i\omega;q^{1/2})_\infty + (i\omega;q^{1/2})_\infty}{2(-q\omega^2;q^2)_\infty}$$

$$= \frac{1}{(-q\omega^2;q^2)_\infty} \sum_{k=0}^{\infty} (-1)^k \frac{q^{k(k-1/2)}}{(q^{1/2};q^{1/2})_{2k}} \omega^{2k}.$$

Functions $S_q(\eta;\omega)$ and $C_q(\eta;\omega)$ are special cases of the q-sine $S_q(x;\omega)$ and q-cosine $C_q(x;\omega)$ functions in two independent variables x and ω when $x = \eta = \left(q^{1/4}+q^{-1/4}\right)/2$; see [5] for more details.

The large ω-asymptotics of $S_q(\eta;\omega)$ and $C_q(\eta;\omega)$ can be investigated on the basis of the following expressions

$$\text{(3.14)} \quad C_q(\eta;\omega) = \frac{(q^{1/2}\omega^2, q^{3/2}/\omega^2; q^2)_\infty}{(q^{1/2};q)_\infty (q, -q\omega^2, -q/\omega^2; q^2)_\infty} C_q(\eta;\frac{q^{1/2}}{\omega})$$

$$- \omega \frac{(q^{3/2}\omega^2, q^{1/2}/\omega^2; q^2)_\infty}{(q^{1/2};q)_\infty (q, -q\omega^2, -q/\omega^2; q^2)_\infty} S_q(\eta;\frac{q^{1/2}}{\omega})$$

and

$$\text{(3.15)} \quad S_q(\eta;\omega) = \omega \frac{(q^{3/2}\omega^2, q^{1/2}/\omega^2; q^2)_\infty}{(q^{1/2};q)_\infty (q, -q\omega^2, -q/\omega^2; q^2)_\infty} C_q(\eta;\frac{q^{1/2}}{\omega})$$

$$+ \frac{(q^{1/2}\omega^2, q^{3/2}/\omega^2; q^2)_\infty}{(q^{1/2}; q)_\infty (q, -q\omega^2, -q/\omega^2; q^2)_\infty} S_q(\eta; \frac{q^{1/2}}{\omega}).$$

These formulas follow directly from (4.3) and (4.4) of [**8**]; see also (5.16) and (5.17) of [**5**]; when we substitute $e^{i\theta} = q^{1/4}$ [**28**]. One can easily see that equations (3.12)–(3.13) and (3.14)–(3.15) determine the asymptotic behavior of the basic trigonometric functions $S_q(\eta; \omega)$ and $C_q(\eta; \omega)$ for the large values of the variable ω.

Let $0 = \omega_0 < \omega_1 < \omega_2 < \omega_3 < ...$ be positive zeros of $S_q(\eta; \omega)$ and let $\varpi_1 < \varpi_2 < \varpi_3 < ...$ be positive zeros of $C_q(\eta; \omega)$. We remind the reader that when $0 < q < 1$ all zeros of $S_q(\eta; \omega)$ and $C_q(\eta; \omega)$ are real. Also these zeros are simple, the positive zeros of the basic sine function $S_q(\eta; \omega)$ are interlaced with those of the basic cosine function $C_q(\eta; \omega)$; see Theorems 1–4 of [**5**]. The asymptotic behavior of the large zeros of these q-trigonometric functions has been discussed in Theorems 5 and 6 of [**5**]; see also [**8**] for numerical investigation of these zeros and [**28**] for a rigorous proof of the asymptotic formulas numerically found in [**8**].

The completeness of the basic trigonometric system $\{\mathcal{E}_q(x; i\omega_n)\}_{n=-\infty}^{\infty}$ has been established in [**5**]. Here we extend this result to $\mathcal{L}_\rho^p(-1, 1)$, $1 \le p < \infty$.

THEOREM 9. *The systems of basic trigonometric function $\{\mathcal{E}_q(x; i\omega_n)\}_{n=-\infty}^{\infty}$, where $0 = \omega_0 < \omega_1 < \omega_2 < \omega_3 < ...$ are the positive zeros of $S_q(\eta; \omega)$, $\omega_{-n} = -\omega_n$, and $\{\mathcal{E}_q(x; i\varpi_n)\}_{n=-\infty}^{\infty}$, where $\varpi_1 < \varpi_2 < \varpi_3 < ...$ are the positive zeros of $C_q(\eta; \omega)$, $\varpi_{-n} = -\varpi_n$, are complete in $\mathcal{L}_\rho^p(-1, 1)$, $1 \le p < \infty$.*

PROOF. In the case of the q-sine function consider the entire function of order zero

$$(3.16) \qquad h(\omega) = (-q\omega^2; q^2)_\infty S_q(\eta; \omega).$$

We need to verify the limiting condition (2.17), say, using the sequences of the real numbers $\{\gamma_k\}$ of the form (3.4). The large asymptotic of this function follows from (3.15). The leading term is $\omega \left(q^{3/2}\omega^2; q^2\right)_\infty$, which corresponds to $\alpha^2 = q^{-3/2} < q^{-2}$ in (3.6) and condition (2.17) is satisfied. The case of the basic cosine function can be considered in a similar manner. We leave the details to the reader. □

The systems of the basic trigonometric functions in Theorem 9 are orthogonal. See [**5**] for more details.

3.3. Jackson's q-Bessel functions. Let us consider an extension of Theorem 3. The Jackson q-Bessel functions given by (2.2) can be transformed with the help of a quadratic transformation formula [**13**], [**22**] as

$$(3.17) \qquad J_\nu^{(2)}(x; q) = \frac{(q^{\nu+1}; q)_\infty}{(q; q)_\infty} \left(\frac{x}{2}\right)^\nu (-x^2/4; q^2)_\infty$$

$$\times {}_2\varphi_1\left(\begin{array}{c} -q^{\nu+1}, -q^{\nu+2} \\ q^{2\nu+2} \end{array}; q^2, -x^2/4\right)$$

$$(3.18) \qquad = \frac{(q^{\nu+1}; q)_\infty}{(q; q)_\infty} \left(\frac{x}{2}\right)^\nu$$

$$\times \frac{(-q^{\nu+1}, -q^{\nu+2}, x^2 q^{\nu+1}/4, 4q^{1-\nu}/x^2; q^2)_\infty}{(q, q^{2\nu+2}, -4q^2/x^2; q^2)_\infty}$$

$$\times {}_2\varphi_1\left(\begin{array}{c} -q^{1+\nu}, -q^{1-\nu} \\ q \end{array}; q^2, -\frac{4q}{x^2}\right)$$

$$+ \frac{(q^{\nu+1}; q)_\infty}{(q; q)_\infty} \left(\frac{x}{2}\right)^\nu$$

$$\times \frac{\left(-q^\nu, -q^{\nu+1}, x^2 q^{\nu+2}/4, 4q^{-\nu}/x^2; q^2\right)_\infty}{(q^{-1}, q^{2\nu+2}, -4q^2/x^2; q^2)_\infty}$$

$$\times \ {}_2\varphi_1\left(\begin{array}{c} -q^{2+\nu}, -q^{2-\nu} \\ q^3 \end{array}; q^2, -\frac{4q}{x^2}\right),$$

where we have used (III.32) of [7] in order to transform (3.17) to (3.18). Let us consider the entire function

(3.19) $$h(\omega) = \omega^{1-\nu} J^{(2)}_{\nu-1}(2\omega; q).$$

Expression (3.18) shows that the leading term of the $h(\omega)$ as $|\omega| \to \infty$ is $(q^\nu \omega^2; q^2)_\infty$, which corresponds to the case $\alpha^2 = q^{-\nu} < q^{-1}$ in (3.1) and, therefore, condition (2.17) is satisfied when $0 < \nu < 1$.

In a similar fashion, consider the entire function

(3.20) $$h_1(\omega) = \omega^{1-\nu} J^{(2)}_\nu(2\omega; q)$$

with the leading term $\omega \left(q^{\nu+1}\omega^2; q^2\right)_\infty$ as $|\omega| \to \infty$. This corresponds to $\alpha^2 = q^{-\nu-1} < q^{-2}$ in (3.6) and condition (2.17) is satisfied when $0 < \nu < 1$. As a result we obtain the following extension of Theorem 3.

THEOREM 10. *Let $j_{\mu,k}(q)$ be the zeros of Jackson's q-Bessel function $J^{(2)}_\mu(x; q)$. The q-trigonometric system $\{\mathcal{E}_q(x; i\omega_k)\}_{k=-\infty}^\infty$ is complete in $\mathcal{L}^p_\rho(-1, 1)$, $1 \le p < \infty$, for $0 < \nu < 1$, if: (a) $\omega_k = \frac{1}{2}j_{\nu-1,k}(q)$, $k = \pm 1, \pm 2, \ldots$; (b) $\omega_k = \frac{1}{2}j_{\nu,k}(q)$, $k = 0, \pm 1, \pm 2, \ldots$.*

REMARK 4. Theorem 9 is, obviously, the special case $\nu = 1/2$ of Theorem 10 due to the relations (5.34)–(5.35) of [5] between the q-Bessel and q-trigonometric functions, but the systems in Theorem 10 are not orthogonal if $\nu \ne 1/2$. There is an independent proof of Theorem 3 in [29] where $0 < \nu < 1/2$; the case $\nu = 1/2$ is discussed in [30].

In a similar manner, one can prove the completeness of the basic trigonometric system on the zeros of the q-Bessel function on a q-quadratic grid considered in [12]. We leave this and other examples to the reader.

The author thanks Joaquin Bustoz and John McDonald for valuable discussions.

References

[1] N. I. Akhiezer, *Theory of Approximation*, Frederick Ungar Publishing Co., New York, 1956.
[2] N. K. Bary, *A Treatise on Trigonometric Series*, in two volumes, Macmillan, New York, 1964.
[3] R. P. Boas, *Entire Functions*, Academic Press, New York, 1954.
[4] J. Bustoz and J. L. Cordoso, *Basic analog of Fourier series on a q-linear grid*, submitted to J. Approx. Th.
[5] J. Bustoz and S. K. Suslov, *Basic analog of Fourier series on a q-quadratic grid*, Methods and Applications of Analysis **5** (1998), 1–38.
[6] R. Floreanini, J. LeTourneux, and L. Vinet, *Symmetries and continuous q-orthogonal polynomials*, in: "Algebraic Methods and q-Special Functions", eds. J. F. van Diejen and L. Vinet, CRM Proceedings & Lecture Notes, Vol. 22, American Mathematical Society, 1999, pp. 135–144.
[7] G. Gasper and M. Rahman, *Basic Hypergeometric Series*, Cambridge University Press, Cambridge, 1990.

[8] R. W. Gosper, Jr. and S. K. Suslov, *Numerical investigation of basic Fourier series*, in "*q-Series from a Contemporary Perspective*", eds. M. E. H. Ismail and D. R. Stanton, Contemporary Mathematics, Vol. 254, American Mathematical Society, 2000, pp. 199–227.

[9] J. R. Higgins, *Completeness and Basic Properties of Sets of Special Functions*, Cambridge University Press, Cambridge, 1977.

[10] M. E. H. Ismail, *The basic Bessel functions and polynomials*, SIAM J. Math. Anal. **12** (1981), 454–468.

[11] M. E. H. Ismail, *The zeros of basic Bessel functions, the functions $J_{\nu+ax}(x)$, and associated orthogonal polynomials*, J. Math. Anal. Appl. **86** (1982), 1–19.

[12] M. E. H. Ismail, D. R. Masson, and S. K. Suslov, *The q-Bessel functions on a q-quadratic grid*, in: "*Algebraic Methods and q-Special Functions*", eds. J. F. van Diejen and L. Vinet, CRM Proceedings & Lecture Notes, Vol. 22, American Mathematical Society, 1999, pp. 183–200.

[13] M. E. H. Ismail, M. Rahman, and D. Stanton, *Quadratic q-exponentials and connection coefficient problems*, Proc. Amer. Math. Soc. **127** (1999) #10, 2931–2941.

[14] M. E. H. Ismail, M. Rahman, and R. Zhang, *Diagonalization of certain integral operators II*, J. Comp. Appl. Math. **68** (1996), 163–196.

[15] M. E. H. Ismail and D. Stanton, *Addition theorems for the q-exponential functions*, in "*q-Series from a Contemporary Perspective*", eds. M. E. H. Ismail and D. R. Stanton, Contemporary Mathematics, Vol. 254, American Mathematical Society, 2000, pp. 235–245.

[16] M. E. H. Ismail and R. Zhang, *Diagonalization of certain integral operators*, Advances in Math. **108** (1994), 1–33.

[17] A. N. Kolmogorov and S. V. Fomin, *Introductory Real Analysis*, Dover, New York, 1970.

[18] B. Ya. Levin, *Distribution of Zeros of Entire Functions*, Translations of Mathematical Monographs, Vol. 5, Amer. Math. Soc., Providence, Rhode Island, 1980.

[19] N. Levinson, *Gap and Density Theorems*, Amer. Math. Soc. Colloq. Publ., Vol. 36, New York, 1940.

[20] J. N. McDonald and N. A. Weiss, *A Course in Real Analysis*, Academic Press, New York, 1999.

[21] A. F. Nikiforov, S. K. Suslov, and V. B. Uvarov, *Classical Orthogonal Polynomials of a Discrete Variable*, Nauka, Moscow, 1985 [in Russian]; English translation, Springer–Verlag, Berlin, 1991.

[22] M. Rahman, *An integral representation and some transformation properties of q-Bessel functions*, J. Math. Anal. Appl. **125** (1987), 58–71.

[23] F. Riesz and B. Sz.-Nagy, *Functional Analysis*, Dover, New York, 1990.

[24] G. Szegö, *Orthogonal Polynomials*, Amer. Math. Soc. Colloq. Publ., Vol. 23, Rhode Island, 1939.

[25] S. K. Suslov, *"Addition" theorems for some q-exponential and q-trigonometric functions*, Methods and Applications of Analysis **4** (1997), 11–32.

[26] S. K. Suslov, *Another addition theorem for the q-exponential function*, J. Phys. A: Math. Gen. **33** (2000), L375–L380.

[27] S. K. Suslov, *Some expansions in basic Fourier series and related topics*, J. Approx. Th., to appear.

[28] S. K. Suslov, *Asymptotics of zeros of basic sine and cosine functions*, submitted.

[29] S. K. Suslov, *Basic exponential functions on a q-quadratic grid*, in: "*Special Functions 2000: Current Perspective and Future Directions*", eds. J. Bustoz, M. E. H. Ismail and S. K. Suslov, NATO Science Series II: Mathematics, Physics and Chemistry, Vol. 30, Kluwer Academic Publishers, Dordrecht – Boston – London, 2001, pp. 411–456.

[30] S. K. Suslov, *A note on completeness of basic trigonometric system in $\mathcal{L}^2$*, submitted to Rocky Mountain J. Math.

[31] G. P. Tolstov, *Fourier Series*, Dover, New York, 1962.

[32] A. Zygmund, *Trigonometric Series*, second edition, Cambridge University Press, Cambridge, 1968.

DEPARTMENT OF MATHEMATICS, ARIZONA STATE UNIVERSITY, TEMPE, AZ 85287-1804, U.S.A.
E-mail address: sks@asu.edu
URL: http://hahn.la.asu.edu/~suslov/index.html

The generalized Borwein conjecture. I. The Burge transform

S. Ole Warnaar

ABSTRACT. Given an arbitrary ordered pair of coprime integers (a,b) we obtain a pair of identities of the Rogers–Ramanujan type. These identities have the same product side as the (first) Andrews–Gordon identity for modulus $2ab \pm 1$, but an altogether different sum side, based on the representation of $(a/b - 1)^{\pm 1}$ as a continued fraction. Our proof, which relies on the Burge transform, first establishes a binary tree of polynomial identities. Each identity in this Burge tree settles a special case of Bressoud's generalized Borwein conjecture.

1. Introduction

Several years ago P. Borwein communicated the following observation to G. E. Andrews [4].

CONJECTURE 1.1 (First Borwein conjecture). *The polynomials $A_n(q)$, $B_n(q)$ and $C_n(q)$ defined by*

$$(1.1) \qquad \prod_{k=1}^{n}(1-q^{3k-2})(1-q^{3k-1}) = A_n(q^3) - qB_n(q^3) - q^2 C_n(q^3)$$

have nonnegative coefficients.

As so often in mathematics, the simplicity of the above claim is rather deceptive, and the conjecture still lacks proof. Following the motto 'if you can't prove it, generalize it', Andrews [4], Bressoud [10], and Ismail, Kim and Stanton [14] have extended the Borwein conjecture in several directions. This is the first of a series of papers devoted to Bressoud's generalization. To see how Bressoud's conjecture arises most naturally from Conjecture 1.1 we follow Andrews [4] and rewrite A_n, B_n and C_n as a sum over q-binomial coefficients. First we need the usual definitions of the q-shifted factorial

$$(a;q)_n = (a)_n = \prod_{k=0}^{n-1}(1-aq^k) \quad \text{for } n \in \mathbb{Z}_+$$

2000 *Mathematics Subject Classification.* Primary 05A15, 05A19; Secondary 33D15.
Work supported by the Australian Research Council.

and q-binomial coefficient

$$
(1.2) \qquad \begin{bmatrix} n \\ m \end{bmatrix}_q = \begin{bmatrix} n \\ m \end{bmatrix} = \begin{cases} \dfrac{(q)_n}{(q)_m (q)_{n-m}} & \text{for } m, n-m \in \mathbb{Z}_+ \\ 0 & \text{otherwise.} \end{cases}
$$

Important later will be that for m and $n-m$ nonnegative integers, the q-binomial coefficient is a polynomial in q with only positive coefficients. Also introducing the notation $(a_1, \ldots, a_k; q)_n = (a_1; q)_n \cdots (a_k; q)_n$, we can apply the q-binomial theorem [13, Eq. (II.4)] to expand the left-hand side of (1.1) as

$$
\begin{aligned}
(q, q^2; q^3)_n &= \sum_{j=-n}^{n} (-1)^j q^{j(3j-1)/2} \begin{bmatrix} 2n \\ n-j \end{bmatrix}_{q^3} \\
&= \sum_{\mu=-1}^{1} (-1)^\mu q^{\mu(3\mu-1)/2} \sum_{j \in \mathbb{Z}} (-1)^j q^{3j(9j+6\mu-1)/2} \begin{bmatrix} 2n \\ n - 3j - \mu \end{bmatrix}_{q^3}.
\end{aligned}
$$

From this one can read off

$$
(1.3) \qquad A_n(q) = \sum_{j \in \mathbb{Z}} (-1)^j q^{j(9j-1)/2} \begin{bmatrix} 2n \\ n - 3j \end{bmatrix}
$$

and similar expressions for B_n and C_n. To pass from this to Bressoud's conjecture we need to recall an important result from partition theory.

Let λ be a partition and λ' its conjugate. The (i,j)th node of λ is the node in the ith row and jth column of the Ferrers diagram of λ. The dth diagonal of λ is formed by the nodes with coordinates $(i, i-d)$. The hook difference at node (i,j) is defined as $\lambda_i - \lambda'_j$. In [5, Thm. 1] Andrews et al. prove the following theorem using recurrences.

THEOREM 1.1. *The generating function $D_{K,i}(N, M; \alpha, \beta)$ of partitions with at most M parts, largest part not exceeding N, and hook differences on the $(1-\beta)$th diagonal at least $\beta - i + 1$ and on the $(\alpha - 1)$th diagonal at most $K - \alpha - i - 1$ is given by*

$$
(1.4) \quad D_{K,i}(N, M; \alpha, \beta; q) = D_{K,i}(N, M; \alpha, \beta)
$$
$$
= \sum_{j \in \mathbb{Z}} \left\{ q^{j((\alpha+\beta)Kj + K\beta - (\alpha+\beta)i)} \begin{bmatrix} M+N \\ M - Kj \end{bmatrix} - q^{((\alpha+\beta)j+\beta)(Kj+i)} \begin{bmatrix} M+N \\ M - Kj - i \end{bmatrix} \right\}.
$$

Here the following conditions apply: $\alpha, \beta \geq 0$ and $\beta - i \leq N - M \leq K - \alpha - i$ with the added restrictions that the largest part exceeds $M - i$ if $\beta = 0$ and the number of parts exceeds $N - K + i$ if $\alpha = 0$.

If we follow [10] and define $G(N, M; \alpha, \beta, K) = D_{2K,K}(N, M; \alpha, \beta)$, then

$$
(1.5) \qquad G(N, M; \alpha, \beta, K) = \sum_{j \in \mathbb{Z}} (-1)^j q^{\frac{1}{2} Kj((\alpha+\beta)j + \alpha - \beta)} \begin{bmatrix} M+N \\ N - Kj \end{bmatrix}.
$$

Now observe that the expression (1.3) for $A_n(q)$ is exactly of this type. Explicitly,

$$
\begin{aligned}
A_n(q) &= G(n, n; 4/3, 5/3, 3) \\
B_n(q) &= G(n+1, n-1; 2/3, 7/3, 3) \\
C_n(q) &= G(n+1, n-1; 1/3, 8/3, 3).
\end{aligned}
$$

Unfortunately there is the complication that α and β have assumed noninteger values so that Theorem 1.1 cannot be applied to interpret A_n–C_n as generating functions. Although no progress has been made in proving Conjecture 1.1, the following generalization is clearly suggested [10].

CONJECTURE 1.2 (Bressoud's generalized Borwein conjecture). *Let K be a positive integer and $N, M, \alpha K, \beta K$ be nonnegative integers such that $1 \leq \alpha + \beta \leq 2K - 1$ (strict inequalities when $K = 2$) and $\beta - K \leq N - M \leq K - \alpha$. Then $G(N, M; \alpha, \beta, K; q)$ is a polynomial in q with nonnegative coefficients.*

Of course, when both α and β are integers the conjecture becomes a special case of Theorem 1.1. When $M + N$ is even and $\alpha = (K - N + M \pm 1)/2$, $\beta = (K + N - M \mp 1)/2$ the conjecture was proven by Ismail *et al.* [14, Thm. 5]. We should also remark that not all of the polynomials $G(N, M; \alpha, \beta, K; q)$ are independent. Besides the obvious symmetry

$$G(N, M; \alpha, \beta, K) = G(M, N; \beta, \alpha, K)$$

there also holds

(1.6) $G(N, M; \alpha, \beta, K; 1/q)$
$$= q^{-MN} G(N, M; K - \alpha - N + M, K - \beta + N - M, K; q)$$

and

$$G(N, M; \alpha, \beta, K) = G(N, M - 1; \alpha, \beta, K) + q^M G(N - 1, M; \alpha + 1, \beta - 1, K)$$
$$= G(N - 1, M; \alpha, \beta, K) + q^N G(N, M - 1; \alpha - 1, \beta + 1, K)$$

as follows from

(1.7) $$\begin{bmatrix} n \\ m \end{bmatrix}_{1/q} = q^{m(m-n)} \begin{bmatrix} n \\ m \end{bmatrix}_q$$

and

(1.8) $$\begin{bmatrix} n \\ m \end{bmatrix} = \begin{bmatrix} n-1 \\ m \end{bmatrix} + q^{n-m} \begin{bmatrix} n-1 \\ m-1 \end{bmatrix} = \begin{bmatrix} n-1 \\ m-1 \end{bmatrix} + q^m \begin{bmatrix} n-1 \\ m \end{bmatrix}.$$

These symmetries, not all of which are independent, are consistent with the bounds imposed on Conjecture 1.2.

This paper is the first in a series in which we apply known (the present paper) and new (subsequent papers) transformation formulas for simple q-polynomials to obtain identities for $G(N, M; \alpha, \beta, K)$ that prove its positivity of coefficients. A few simple examples of such identities can already be found in the literature and we quote

(1.9) $$G(L, L; 1/2, 1, 2) = \sum_{n=0}^{L} q^{nL} \begin{bmatrix} L \\ n \end{bmatrix}$$

(1.10) $$G(L, L; 1, 3/2, 2) = \sum_{n=0}^{L} q^{n^2} \begin{bmatrix} L \\ n \end{bmatrix}$$

$$G(L, L; 1/2, 3/2, 2) = (1 + q^L)(-q^2; q^2)_{L-1}.$$

The first two, which are dual in the sense of (1.6), were found by Bressoud [9, Eqs. (8) and (9)] and are bounded analogues of the Euler- and first Rogers–Ramanujan identity. The third identity is due to Ismail *et al.* [14, Prop. 2 (3)]. Although we will add infinitely many new identities to the above three, we failed to obtain identities that would settle the original Borwein conjecture. What our results do suggest however is that this is perhaps more a practical than fundamental problem. Often it is necessary to first guess identities before proving them and it seems that when it comes to the (generalized) Borwein conjecture the guessing is by far the hardest part of the game. For example, almost anyone familiar with q-series will have little trouble proving the following identities for $G(L, L; 1, 2/3, 3)$ and $G(L+1, L-1; 1/3, 4/3, 3)$:

$$(1.11) \qquad \sum_{j \in \mathbb{Z}} (-1)^j q^{\frac{1}{2}j(5j+1)} \begin{bmatrix} 2L \\ L-3j \end{bmatrix} = \sum_{n,i \geq 0} q^{n^2+i(L+n)} \begin{bmatrix} 2L-2i-n \\ n \end{bmatrix} \begin{bmatrix} L-i-n \\ i \end{bmatrix}$$

and

$$(1.12) \quad \sum_{j \in \mathbb{Z}} (-1)^j q^{\frac{1}{2}j(5j+3)} \begin{bmatrix} 2L \\ L-3j-1 \end{bmatrix} = \sum_{n,i \geq 0} q^{n(n+1)+i(L+n+1)} \begin{bmatrix} 2L-2i-n-1 \\ n \end{bmatrix} \begin{bmatrix} L-i-n-1 \\ i \end{bmatrix},$$

but to guess these results is quite a bit harder. In particular we note that when L tends to infinity only the terms with $i = 0$ contribute to the sums on the right. Also using Jacobi's triple product identity [2, Thm. 2.8] thus yields for $|q| < 1$

$$\sum_{n=0}^{\infty} \frac{q^{n^2}}{(q)_n} = \frac{(q^2, q^3, q^5; q^5)}{(q;q)_\infty} \quad \text{and} \quad \sum_{n=0}^{\infty} \frac{q^{n(n+1)}}{(q)_n} = \frac{(q, q^4, q^5; q^5)}{(q;q)_\infty},$$

which are the celebrated Rogers–Ramanujan identities [16]. Admittedly, we did not guess (1.11) and (1.12) but obtained them by transforming simpler identities, but since we were not so lucky with the original Borwein conjecture, a good guess is exactly what is needed. (See also the discussion in section 8.)

As a bonus of our identities for $G(M, N; \alpha, \beta, K)$ we obtain many new identities of the Rogers–Ramanujan type. Some nice examples worth stating in the introduction are the following four series of identities featuring the Fibonacci numbers F_k defined recursively as $F_0 = 0$, $F_1 = 1$ and $F_k = F_{k-1} + F_{k-2}$.

THEOREM 1.2. *For $|q| < 1$ and $k \geq 4$ there holds*

$$\sum_{m_1,\ldots,m_{k-2} \geq 0} \frac{q^{m_1^2 + \cdots + m_{k-2}^2}}{(q)_{2m_1}} \left(\prod_{j=2}^{k-3} \begin{bmatrix} m_{j-1}+m_j-m_{j+1} \\ 2m_j \end{bmatrix} \right) \begin{bmatrix} m_{k-3} \\ m_{k-2} \end{bmatrix}$$
$$= \frac{(q^{F_k F_{k-1}}, q^{F_k F_{k-1}+1}, q^{2F_k F_{k-1}+1}; q^{2F_k F_{k-1}+1})_\infty}{(q;q)_\infty}.$$

Taking $\sum_{m \geq 0} q^{m^2}/(q)_m$ as the left-hand side when $k = 3$ includes the first Rogers–Ramanujan identity in this series of identities. For $k = 4$ Theorem 1.2 is [3, Eq. (5.8); $k = 1$].

THEOREM 1.3. *For $|q| < 1$ and $k \geq 4$ there holds*

$$\sum_{m_1,\ldots,m_{k-2} \geq 0} \frac{q^{m_1^2 + \cdots + m_{k-3}^2 + m_{k-3}m_{k-2}}}{(q)_{2m_1}} \left(\prod_{j=2}^{k-3} \begin{bmatrix} m_{j-1}+m_j-m_{j+1} \\ 2m_j \end{bmatrix} \right) \begin{bmatrix} m_{k-3} \\ m_{k-2} \end{bmatrix}$$
$$= \frac{(q^{F_k F_{k-1}-1}, q^{F_k F_{k-1}}, q^{2F_k F_{k-1}-1}; q^{2F_k F_{k-1}-1})_\infty}{(q;q)_\infty}.$$

THEOREM 1.4. *For $|q| < 1$ and $k \geq 5$ there holds*

$$\sum_{m_1,\ldots,m_{k-2}\geq 0} \frac{q^{(m_1+m_2)^2+m_2^2+\cdots+m_{k-2}^2}}{(q)_{m_1}(q)_{2m_2}} \left(\prod_{j=3}^{k-3} \begin{bmatrix} m_{j-1}+m_j-m_{j+1} \\ 2m_j \end{bmatrix}\right) \begin{bmatrix} m_{k-3} \\ m_{k-2} \end{bmatrix}$$
$$= \frac{(q^{F_k F_{k-2}}, q^{F_k F_{k-2}+1}, q^{2F_k F_{k-2}+1}; q^{2F_k F_{k-2}+1})_\infty}{(q;q)_\infty}.$$

The cases $k = 3$ and $k = 4$ may be included by taking $\sum_{m\geq 0} q^{m^2}/(q)_m$ and $\sum_{m_1,m_2\geq 0} q^{(m_1+m_2)^2+m_2^2}/(q)_{m_1}(q)_{m_2}$ as respective left-hand sides.

THEOREM 1.5. *For $|q| < 1$ and $k \geq 5$ there holds*

$$\sum_{m_1,\ldots,m_{k-2}\geq 0} \frac{q^{(m_1+m_2)^2+m_2^2+\cdots+m_{k-3}^2+m_{k-3}m_{k-2}}}{(q)_{m_1}(q)_{2m_2}} \left(\prod_{j=3}^{k-3} \begin{bmatrix} m_{j-1}+m_j-m_{j+1} \\ 2m_j \end{bmatrix}\right) \begin{bmatrix} m_{k-3} \\ m_{k-2} \end{bmatrix}$$
$$= \frac{(q^{F_k F_{k-2}-1}, q^{F_k F_{k-2}}, q^{2F_k F_{k-2}-1}; q^{2F_k F_{k-2}-1})_\infty}{(q;q)_\infty}.$$

2. The Burge transform

The Burge transform, which is a generalization of a special case of the Bailey lemma, provides an iterative method for proving polynomial analogues of q-series identities [**11**]. Before we state the actual transform let us define the polynomial $\mathcal{B}(L, M, a, b; q)$ as

(2.1) $\quad \mathcal{B}(L, M, a, b; q) = \mathcal{B}(L, M, a, b) = \begin{bmatrix} L+M+a-b \\ L+a \end{bmatrix} \begin{bmatrix} L+M-a+b \\ L-a \end{bmatrix}.$

Note that $\mathcal{B}(L, M, a, b)$ is nonzero iff $L + a$ and $M + b$ are integers such that $0 \leq |b| \leq M$ and $0 \leq |a| \leq L$. Moreover, it satisfies the symmetries

$$\mathcal{B}(L, M, -a, -b) = \mathcal{B}(L, M, a, b)$$

(2.2) $\quad \mathcal{B}(L, M, a, b) = \mathcal{B}(M, L, b, a)$

(2.3) $\quad \mathcal{B}(L, M, a, b; 1/q) = q^{2ab-2LM} \mathcal{B}(L, M, a, b; q)$

and becomes proportional to the q-binomial coefficient when either L or M tends to infinity

(2.4) $\quad \lim_{M\to\infty} \mathcal{B}(L, M, a, b) = \begin{bmatrix} 2L \\ L-a \end{bmatrix}/(q)_{2L}.$

The most interesting properties of $\mathcal{B}$ are however the following two transformations.

THEOREM 2.1. *For L, M, a, b integers such that not $-L + a \leq -b \leq L + a < b \leq M$ or $-L - a \leq b \leq L - a < -b \leq M$,*

(2.5) $\quad \sum_{i\geq 0} q^{i^2} \begin{bmatrix} 2L+M-i \\ 2L \end{bmatrix} \mathcal{B}(L-i, i, a, b) = q^{b^2} \mathcal{B}(L, M, a+b, b)$

and

(2.6) $\quad \sum_{i\geq 0} q^{i^2} \begin{bmatrix} 2L+M-i \\ 2L \end{bmatrix} \mathcal{B}(i, L-i, b, a) = q^{b^2} \mathcal{B}(L, M, a+b, b).$

These two results are known as the Burge transform [**11, 12, 19**], and are an immediate consequence of the q-Saalschütz sum [**13**, Eq. (II.12)], or, equivalently, of [**2**, Eq. (3.3.11)]. The second transform of course follows from the first by the symmetry relation (2.2). The conditions imposed on the parameters are due to the fact that the left side may be zero when the right side is not. They perhaps appear cumbersome but are in fact quite innocent. In all our applications of the Burge transform a and b will have the same signature. It is not hard to see that the conditions then become void. For example, assuming $0 < b \leq M$ we need to inspect the first condition only. But $a \geq 0$ clearly avoids the occurrence of $-b \leq L + a < b \leq L - a$. A similar reasoning applies for negative b.

3. The generalized Borwein conjecture

To derive manifestly positive representations for $G(N, M; \alpha, \beta, K)$ using the Burge transform the following lemma will be crucial.

LEMMA 3.1. *For $L, M \geq 0$ there holds*

$$(3.1) \qquad \sum_{j \in \mathbb{Z}} (-1)^j q^{\frac{1}{2}j(3j+1)} \mathcal{B}(L, M, j, j) = \begin{bmatrix} L + M \\ M \end{bmatrix}.$$

PROOF. Take Rogers' q-Dougall sum [**13**, (II.21)]

$$\sum_{k=0}^{n} \frac{1 - aq^{2k}}{1 - a} \frac{(a, b, c, q^{-n}; q)_k}{(q, aq/b, aq/c, aq^{n+1}; q)_k} \left(\frac{aq^{n+1}}{bc}\right)^k = \frac{(aq, aq/bc; q)_n}{(aq/b, aq/c; q)_n}$$

and let $a \to 1$, $b \to q^{-M}$, $c \to \infty$ and $n \to L$. After some simplifications this gives (3.1). $\square$

When either L or M tends to infinity (3.1) simplifies to Rogers' polynomial analogue of the Euler identity [**17**, §1]

$$\sum_{j \in \mathbb{Z}} (-1)^j q^{\frac{1}{2}j(3j+1)} \begin{bmatrix} 2M \\ M - j \end{bmatrix} = \frac{(q)_{2M}}{(q)_M}.$$

In the following we will iterate the two Burge transformations (2.5) and (2.6) to transform (3.1) into a binary tree of polynomial identities. First, by application of either (2.5) or (2.6) one finds

$$(3.2) \qquad \sum_{j \in \mathbb{Z}} (-1)^j q^{\frac{1}{2}j(5j+1)} \mathcal{B}(L, M, 2j, j) = \sum_{n \geq 0} q^{n^2} \begin{bmatrix} 2L + M - n \\ 2L \end{bmatrix} \begin{bmatrix} L \\ n \end{bmatrix}.$$

This is a doubly-bounded analogue of the first Rogers–Ramanujan identity. When M goes to infinity we recover Bressoud's identity (1.10) and when L goes to infinity we find the following well-known specialization of Watson's ${}_8\phi_7$ transform (see e.g. [**9**, Eq. (1.11)] or [**15**, Eq. (39)])

$$\sum_{j \in \mathbb{Z}} (-1)^j q^{\frac{1}{2}j(5j+1)} \begin{bmatrix} 2M \\ M - j \end{bmatrix} = \frac{(q)_{2M}}{(q)_M} \sum_{n \geq 0} q^{n^2} \begin{bmatrix} M \\ n \end{bmatrix}.$$

Without too much effort one can utilize (3.2) to also obtain a doubly-bounded analogue of the second Rogers–Ramanujan identity.

LEMMA 3.2. *For $L, M \geq 0$ there holds*

$$(3.3) \quad \sum_{j \in \mathbb{Z}} (-1)^j q^{\frac{1}{2}j(5j+3)} \begin{bmatrix} L+M+j \\ M-j-1 \end{bmatrix} \begin{bmatrix} L+M-j \\ M+j \end{bmatrix} = \sum_{n \geq 0} q^{n(n+1)} \begin{bmatrix} 2L+M-n \\ 2L+1 \end{bmatrix} \begin{bmatrix} L \\ n \end{bmatrix}.$$

PROOF. Suppressing their L-dependence, we denote the left sides of (3.3) and (3.2) by f_M and g_M, respectively. By application of the recurrence (1.8)

$$f_M = \sum_{j \in \mathbb{Z}} (-1)^j q^{\frac{1}{2}j(5j+3)} \begin{bmatrix} L+M+j \\ M-j-1 \end{bmatrix} \begin{bmatrix} L+M-j-1 \\ M+j-1 \end{bmatrix}$$

$$+ q^M \sum_{j \in \mathbb{Z}} (-1)^j q^{\frac{5}{2}j(j+1)} \begin{bmatrix} L+M+j \\ M-j-1 \end{bmatrix} \begin{bmatrix} L+M-j-1 \\ M+j \end{bmatrix}.$$

Since the second term on the right changes sign after the variable change $j \to -j-1$ it vanishes. The first term is again split using (1.8) leading to

$$(3.4) \quad f_M = f_{M-1} + q^{M-1} g_{M-1}.$$

Next we let f_M and g_M denote the right sides of (3.3) and (3.2). A single application of (1.8) show that (3.4) again holds. Since equation (3.3) trivializes to $0 = 0$ for $M = 0$ this settles the lemma. $\square$

It is easy to see that by computing $(3.3) + q^M (3.2)$ we also get

$$(3.5) \quad \sum_{j \in \mathbb{Z}} (-1)^j q^{\frac{1}{2}j(5j+3)} \begin{bmatrix} L+M+j+1 \\ M-j \end{bmatrix} \begin{bmatrix} L+M-j \\ M+j \end{bmatrix} = \sum_{n \geq 0} q^{n(n+1)} \begin{bmatrix} 2L+M-n+1 \\ 2L+1 \end{bmatrix} \begin{bmatrix} L \\ n \end{bmatrix}.$$

Although the identities (3.3) and (3.5) cannot be written in terms of the polynomials $\mathcal{B}(L, M, a, b)$ they can be iterated by a simple modification of the Burge transform. We will not pursue this here and only take the large M limit to arrive at the isolated positivity result

$$G(L+1, L; 1/2, 2, 2) = \sum_{j \in \mathbb{Z}} (-1)^j q^{\frac{1}{2}j(5j+3)} \begin{bmatrix} 2L+1 \\ L-2j \end{bmatrix} = \sum_{n \geq 0} q^{n(n+1)} \begin{bmatrix} L \\ n \end{bmatrix}.$$

After this intermezzo we continue to iterate (3.2). To state the general result some more notation is needed. Assume that (a, b) is a pair of coprime integers such that $1 \leq b < a$, and define a nonnegative integer n and positive integers $a_0, \ldots, a_n$ as the order and partial quotients of the continued fraction representation of $(a/b - 1)^{\text{sign}(a-2b)}$ ($\text{sign}(0) = 0$), i.e.,

$$(3.6) \quad \left(\frac{a}{b} - 1\right)^{\text{sign}(a-2b)} = [a_0, \ldots, a_n] = a_0 + \cfrac{1}{a_1 + \cfrac{1}{\cdots + \cfrac{1}{a_n}}}.$$

We denote the continued fraction corresponding to (a, b) by $\text{cf}(a, b)$, and note the obvious symmetry $\text{cf}(a, b) = \text{cf}(a, a - b)$. By abuse of notation we sometimes write $\text{cf}(a, b) = (a/b - 1)^{\text{sign}(a-2b)}$.

Before we continue, we make the following important remark concerning continued fractions. For any admissible (a, b) such that $(a, b) \neq (2, 1)$ the continued fraction $\text{cf}(a, b)$ is not unique. Indeed, for $c_n \geq 2$ the continued fractions $[c_0, \ldots, c_n]$ and $[c_0, \ldots, c_n - 1, 1]$ represent the same rational number. This means in particular that the order and partial quotients of $\text{cf}(a, b)$ defined in (3.6) are not unique. In the following we will use these to define several other quantities, which are therefore not unique either. Later we will show however, that whichever choice is made, the final objects of interest are independent of the choice of representation. This allows use to choose at our convenience the representation with either $a_n = 1$ or $a_n \geq 2$. For $(a,b) = (2,1)$ we of course only have $\mathrm{cf}(2,1) = [1]$.

We now continue by further defining the partial sums $t_j = \sum_{k=0}^{j-1} a_k$ for $j = 1, \ldots, n+1$. We also introduce $t_0 = 0$ and $d(a,b) = t_{n+1}$. Note that $d(a,b)$ is insensitive to the choice of representation of $\mathrm{cf}(a,b)$. Finally we define a $d(a,b) \times d(a,b)$ matrix $\mathcal{I}(a,b)$ with entries

$$\mathcal{I}(a,b)_{j,k} = \begin{cases} \delta_{j,k+1} + \delta_{j,k-1} & \text{for } j \neq t_i \\ \delta_{j,k+1} + \delta_{j,k} - \delta_{j,k-1} & \text{for } j = t_i \end{cases}$$

and a corresponding Cartan-type matrix $C(a,b) = 2I - \mathcal{I}(a,b)$ where I is the $d(a,b) \times d(a,b)$ identity matrix. Note that the matrix $\mathcal{I}(a,b)$ has the following block-structure:

$$\mathcal{I}(a,b) = \begin{pmatrix} T_{a_0} & \begin{matrix} \\ -1 \end{matrix} & & \\ \begin{matrix} & 1 \end{matrix} & \ddots & \begin{matrix} \\ -1 \end{matrix} \\ & \begin{matrix} & 1 \end{matrix} & T_{a_n} \end{pmatrix}$$

where T_i is the incidence matrix of the tadpole graph with i vertices, i.e., $(T_i)_{j,k} = \delta_{j,k+1} + \delta_{j,k-1} + \delta_{j,k}\delta_{j,i}$. A change from the representation of $\mathrm{cf}(a,b)$ with $a_n \geq 2$ to the representation with $a_n = 1$ corresponds to the transformation

$$(3.7) \qquad T_{a_n} \to \begin{pmatrix} T_{a_n - 1} & \begin{matrix} \\ -1 \end{matrix} \\ \begin{matrix} 1 \end{matrix} & 1 \end{pmatrix}$$

which leaves all but two matrix elements unchanged. (Given that $\mathcal{I}(a,b)$ and $C(a,b)$ depend not only on (a,b) but also on the choice of representation of $\mathrm{cf}(a,b)$ the fastidious reader might prefer the notation $\mathcal{I}(\mathrm{cf}(a,b))$ and $C(\mathrm{cf}(a,b))$.)

With the above notation we define a polynomial for each ordered pair of positive, coprime integers (a,b) by

$$(3.8) \qquad F_{a,b}(L,M) = \sum_{m \in \mathbb{Z}_+^{d(a,b)}} q^{mC(a,b)m} \begin{bmatrix} 2L+M-m_1 \\ 2L \end{bmatrix} \prod_{j=1}^{d(a,b)} \begin{bmatrix} \tau_j m_j + n_j \\ \tau_j m_j \end{bmatrix}$$

for $a \leq 2b$, and

$$(3.9) \qquad F_{a,b}(L,M) = \sum_{m \in \mathbb{Z}_+^{d(a,b)}} q^{L(L-2m_1)+mC(a,b)m} \begin{bmatrix} L+M+m_1 \\ 2L \end{bmatrix} \prod_{j=1}^{d(a,b)} \begin{bmatrix} \tau_j m_j + n_j \\ \tau_j m_j \end{bmatrix}$$

for $a \geq 2b$. Here

$$(3.10) \quad mC(a,b)m = \sum_{j,k=1}^{d(a,b)} m_j C(a,b)_{j,k} m_k = \sum_{j=0}^{n} \Big(m_{t_j+1}^2 + \sum_{k=t_j+1}^{t_{j+1}-1} (m_k - m_{k+1})^2\Big)$$

and $\tau_1 = \cdots = \tau_{d(a,b)-1} = 2$, $\tau_{d(a,b)} = 1$. The auxiliary variables n_j in the summand are integers defined by the (m,n)-system

$$(3.11) \quad n_j = L\delta_{j,1} - \sum_{k=1}^{d(a,b)} C(a,b)_{j,k} m_k \quad \text{for } j = 1, \ldots, d(a,b).$$

We remark that the polynomials $F_{a,b}(L,M)$ bear close resemblance to the much-studied fermionic polynomials arising in the quasiparticle description of $c < 1$ conformal field theory [**6, 7, 12, 18**].

For each admissible (a,b) the polynomial $F_{a,b}(L,M)$ is defined twice. To show that these definitions are consistent we first assume that $(a,b) \neq (2,1)$ and that we have chosen cf(a,b) such that $a_n \geq 2$. Making the variable change $m_{d(a,b)} \leftrightarrow n_{d(a,b)}$ then yields the representation with $a_n = 1$. The easiest way to see this is perhaps to first eliminate the n_j, then to make the variable change $m_{d(a,b)} \to m_{d(a,b)-1} - m_{d(a,b)}$ and to rewrite this again in the form using the n_j. Recalling (3.7) and the definition of T_i and $C(a,b)$ this gives the desired result. When $(a,b) = (2,1)$ there is only one continued fraction representation, but $F_{2,1}(L,M)$ is defined in both (3.8) and (3.9). Since cf$(2,1) = [1]$, one finds $d(2,1) = 1$ and $\mathcal{I}(2,1) = C(2,1) = (1)$. From (3.11) we further find $n_1 = L - m_1$. Hence, according to (3.8)

$$F_{2,1}(L,M) = \sum_{m_1 \geq 0} q^{m_1^2} \begin{bmatrix} 2L + M - m_1 \\ 2L \end{bmatrix} \begin{bmatrix} L \\ m_1 \end{bmatrix}$$

and according to (3.9)

$$F_{2,1}(L,M) = \sum_{m_1 \geq 0} q^{(L-m_1)^2} \begin{bmatrix} L + M + m_1 \\ 2L \end{bmatrix} \begin{bmatrix} L \\ m_1 \end{bmatrix}.$$

The variable change $m_1 \to L - m_1$ shows these two expressions are consistent.

We are now ready to state the identities obtained by applying the Burge transform to (3.2).

THEOREM 3.1. *For L,M nonnegative integers and (a,b) a pair of coprime integers such that $1 \leq b < a$ there holds*

$$(3.12) \quad \sum_{j \in \mathbb{Z}} (-1)^j q^{\frac{1}{2}j((2ab+1)j+1)} \mathcal{B}(L,M,aj,bj) = F_{a,b}(L,M).$$

We postpone the proof till the next section and instead continue with examples and corollaries.

In our first example we take $(a,b) = (7,5)$ and explicitly calculate $F_{7,5}(L,M)$ and $F_{7,2}(L,M)$. As continued fraction we choose the representation cf$(7,5) = [2,1,1]$. Hence $n = 2$, $a_0 = 2$, $a_1 = a_2 = 1$, $t_1 = 2$, $t_2 = 3$ and $t_3 = d(7,5) = 4$. The matrices $\mathcal{I}(7,5)$ and $C(7,5)$ are thus found to be

$$\mathcal{I}(7,5) = \left(\begin{array}{cc|c|c} 0 & 1 & 0 & 0 \\ 1 & 1 & -1 & 0 \\ \hline 0 & 1 & 1 & -1 \\ \hline 0 & 0 & 1 & 1 \end{array}\right), \quad C(7,5) = \left(\begin{array}{cc|c|c} 2 & -1 & 0 & 0 \\ -1 & 1 & 1 & 0 \\ \hline 0 & -1 & 1 & 1 \\ \hline 0 & 0 & -1 & 1 \end{array}\right).$$

Inserting this in (3.8) and (3.9) (using the symmetry cf(7,5) = cf(7,2)) yields

$$F_{7,5}(L,M) = \sum_{m_1,\ldots,m_4 \geq 0} q^{m_1^2+(m_1-m_2)^2+m_3^2+m_4^2} \begin{bmatrix} 2L+M-m_1 \\ 2L \end{bmatrix} \begin{bmatrix} L+m_2 \\ 2m_1 \end{bmatrix}$$
$$\times \begin{bmatrix} m_1+m_2-m_3 \\ 2m_2 \end{bmatrix} \begin{bmatrix} m_2+m_3-m_4 \\ 2m_3 \end{bmatrix} \begin{bmatrix} m_3 \\ m_4 \end{bmatrix}$$

and

$$F_{7,2}(L,M) = \sum_{m_1,\ldots,m_4 \geq 0} q^{(L-m_1)^2+(m_1-m_2)^2+m_3^2+m_4^2} \begin{bmatrix} L+M+m_1 \\ 2L \end{bmatrix} \begin{bmatrix} L+m_2 \\ 2m_1 \end{bmatrix}$$
$$\times \begin{bmatrix} m_1+m_2-m_3 \\ 2m_2 \end{bmatrix} \begin{bmatrix} m_2+m_3-m_4 \\ 2m_3 \end{bmatrix} \begin{bmatrix} m_3 \\ m_4 \end{bmatrix}.$$

Next we consider Theorem 3.1 in the large M limit, and with the same notation as above we define $F_{a,b}(L) = (q)_{2L} \lim_{M \to \infty} F_{a,b}(L,M)$. Explicitly this yields

$$(3.13) \qquad F_{a,b}(L) = \sum_{m \in \mathbb{Z}_+^{d(a,b)}} q^{mC(a,b)m} \prod_{j=1}^{d(a,b)} \begin{bmatrix} \tau_j m_j + n_j \\ \tau_j m_j \end{bmatrix}$$

for $a \leq 2b$, and

$$(3.14) \qquad F_{a,b}(L) = \sum_{m \in \mathbb{Z}_+^{d(a,b)}} q^{L(L-2m_1)+mC(a,b)m} \prod_{j=1}^{d(a,b)} \begin{bmatrix} \tau_j m_j + n_j \\ \tau_j m_j \end{bmatrix}$$

for $a \geq 2b$. Then, using (2.4), the following result arises after letting M tend to infinity in (3.12).

COROLLARY 3.1. *For L a nonnegative integer and (a,b) a pair of coprime integers such that $1 \leq b < a$ there holds*

$$(3.15) \qquad G(L,L;b,b+1/a,a) = \sum_{j \in \mathbb{Z}} (-1)^j q^{\frac{1}{2}j((2ab+1)j+1)} \begin{bmatrix} 2L \\ L-aj \end{bmatrix} = F_{a,b}(L).$$

Obviously, $F_{a,b}(L)$ is a polynomial with only nonnegative coefficients. This leads to our next corollary.

COROLLARY 3.2. *For (a,b) a pair of coprime integers such that $1 \leq b < a$ the polynomial $G(L,L;b,b+1/a,a)$ has nonnegative coefficients.*

Taking the large L limit of Theorem 3.1 is more intricate due to the L-dependent term in the exponent of q in (3.9). To overcome this complication one first has to make a change of variables expressing the kernel of $F_{a,b}(L,M)$ (with $a \geq 2b$) in terms of the variables $n_1, \ldots, n_{a_0}, m_{a_0+1}, \ldots, m_{d(a,b)}$ instead of $m_1, \ldots, m_{d(a,b)}$. To achieve this, first note that (3.11) implies

$$(3.16) \qquad m_j = L - jm_{a_0+1} - \sum_{k=1}^{a_0} \min(j,k) n_k \quad \text{for } j = 1,\ldots,a_0,$$

and hence $m_j - m_{j+1} = m_{a_0+1} + n_{j+1} + \cdots + n_{a_0}$. (If for $(a,b) = (a,1)$ we take the representation cf$(a,1) = [a-1]$ then $a_0 = d(a,1) = a-1$ in which case $m_{a_0+1} := 0$.)

Inserting this in (3.9), using (3.10) and defining $N_j = n_j + \cdots + n_{a_0}$, gives

$$F_{a,b}(L,M) = \sum_{\substack{n_1,\ldots,n_{a_0} \geq 0 \\ m_{a_0+1},\ldots,m_{d(a,b)} \geq 0}} q^{(N_1+m_{a_0+1})^2 + \cdots + (N_{a_0}+m_{a_0+1})^2}$$

$$\times q^{\sum_{j,k=a_0+1}^{d(a,b)} m_j C(a,b)_{j,k} m_k} \begin{bmatrix} L+M+m_1 \\ 2L \end{bmatrix} \prod_{j=1}^{a_0} \begin{bmatrix} \tau_j m_j + n_j \\ n_j \end{bmatrix} \prod_{j=a_0+1}^{d(a,b)} \begin{bmatrix} \tau_j m_j + n_j \\ \tau_j m_j \end{bmatrix},$$

with $a \geq 2b$. Here it is to be understood that the auxiliary variables $m_1,\ldots,m_{a_0}$ and $n_{a_0+2},\ldots,n_{d(a,b)}$ follow from (3.16) and (3.11), respectively. The auxiliary variable n_{a_0+1} is special in that it needs both equations for its computation,

$$n_{a_0+1} = L - a_0 m_{a_0+1} - \sum_{k=1}^{a_0} k n_k - \sum_{k=a_0+1}^{d(a,b)} C(a,b)_{a_0+1,k} m_k.$$

After these preliminaries we are prepared to take the large L limit of Theorem 3.1. All that remains to be done is to define the polynomials $\tilde{F}_{a,b}(M) = (q)_{2M} \lim_{L \to \infty} F_{a,b}(L,M)$, i.e.,

$$(3.17) \qquad \tilde{F}_{a,b}(M) = \sum_{m \in \mathbb{Z}_+^{d(a,b)}} \frac{(q)_{2M} \, q^{mC(a,b)m}}{(q)_{M-m_1}(q)_{\tau_1 m_1}} \prod_{j=2}^{d(a,b)} \begin{bmatrix} \tau_j m_j + n_j \\ \tau_j m_j \end{bmatrix}$$

for $a \leq 2b$, and

$$(3.18) \quad \tilde{F}_{a,b}(M) = \sum_{\substack{n_1,\ldots,n_{a_0} \geq 0 \\ m_{a_0+1},\ldots,m_{d(a,b)} \geq 0}} \frac{(q)_{2M} \, q^{(N_1+m_{a_0+1})^2 + \cdots + (N_{a_0}+m_{a_0+1})^2}}{(q)_{M-N_1-m_{a_0+1}}(q)_{n_1} \cdots (q)_{n_{a_0}}(q)_{\tau_{a_0+1} m_{a_0+1}}}$$

$$\times q^{\sum_{j,k=a_0+1}^{d(a,b)} m_j C(a,b)_{j,k} m_k} \prod_{j=a_0+2}^{d(a,b)} \begin{bmatrix} \tau_j m_j + n_j \\ \tau_j m_j \end{bmatrix}$$

for $a \geq 2b$. Of course τ_{a_0+1} is always 2 except when $(a,b) = (a,1)$ with the choice $\text{cf}(a,1) = [a-2,1]$.

Our next corollary can at last be stated as follows.

COROLLARY 3.3. *For M a nonnegative integer and (a,b) a pair of coprime integers such that $1 \leq b < a$ there holds*

$$G(M,M;a,a+1/b,b) = \sum_{j \in \mathbb{Z}} (-1)^j q^{\frac{1}{2}j((2ab+1)j+1)} \begin{bmatrix} 2M \\ M-bj \end{bmatrix} = \tilde{F}_{a,b}(M).$$

Continuing our previous example for $(a,b) = (7,5)$ we find

$$\tilde{F}_{7,5}(M) = (q)_{2M} \sum_{m_1,\ldots,m_4 \geq 0} \frac{q^{m_1^2 + (m_1-m_2)^2 + m_3^2 + m_4^2}}{(q)_{M-m_1}(q)_{2m_1}} \begin{bmatrix} m_1+m_2-m_3 \\ 2m_2 \end{bmatrix} \begin{bmatrix} m_2+m_3-m_4 \\ 2m_3 \end{bmatrix} \begin{bmatrix} m_3 \\ m_4 \end{bmatrix}$$

and

$$\tilde{F}_{7,2}(M) = (q)_{2M} \sum_{n_1,n_2,m_3,m_4 \geq 0} \frac{q^{(n_1+n_2+m_3)^2 + (n_2+m_3)^2 + m_3^2 + m_4^2}}{(q)_{M-m_3-n_1-n_2}(q)_{n_1}(q)_{n_2}(q)_{2m_3}} \begin{bmatrix} m_3 \\ m_4 \end{bmatrix}.$$

Next we return to Corollary 3.3 and take M to infinity. By the Jacobi triple product identity [**2**, Thm. 2.8] we then obtain the following two results.

THEOREM 3.2. *For $|q| < 1$ and (a,b) a pair of coprime integers such that $1 \leq b < a$ there holds*

$$\sum_{m \in \mathbb{Z}_+^{d(a,b)}} \frac{q^{mC(a,b)m}}{(q)_{\tau_1 m_1}} \prod_{j=2}^{d(a,b)} \begin{bmatrix} \tau_j m_j + n_j \\ \tau_j m_j \end{bmatrix} = \frac{(q^{ab}, q^{ab+1}, q^{2ab+1}; q^{2ab+1})_\infty}{(q;q)_\infty}$$

for $a \leq 2b$, and

$$\sum_{\substack{n_1,\ldots,n_{a_0} \geq 0 \\ m_{a_0+1},\ldots,m_{d(a,b)} \geq 0}} \frac{q^{(N_1+m_{a_0+1})^2+\cdots+(N_{a_0}+m_{a_0+1})^2}}{(q)_{n_1}\cdots(q)_{n_{a_0}}(q)_{\tau_{a_0+1}m_{a_0+1}}}$$

$$\times q^{\sum_{j,k=a_0+1}^{d(a,b)} m_j C(a,b)_{j,k} m_k} \prod_{j=a_0+2}^{d(a,b)} \begin{bmatrix} \tau_j m_j + n_j \\ \tau_j m_j \end{bmatrix} = \frac{(q^{ab}, q^{ab+1}, q^{2ab+1}; q^{2ab+1})_\infty}{(q;q)_\infty}$$

for $a \geq 2b$.

The simplest case of the theorem is $(a,b) = (k,1)$ or, by symmetry, $(a,b) = (k,k-1)$. Choosing cf$(k,1) = [k-1]$, one finds $n = 0$, $a_0 = k-1$, $d(k,1) = k-1$, and $\mathcal{I}(k,1) = T_{k-1}$. Hence

$$(3.19) \qquad \sum_{n_1,\ldots,n_{k-1} \geq 0} \frac{q^{N_1^2+\cdots+N_{k-1}^2}}{(q)_{n_1}\cdots(q)_{n_{k-1}}} = \frac{(q^k, q^{k+1}, q^{2k+1}; q^{2k+1})_\infty}{(q;q)_\infty}$$

for $k \geq 2$ where $N_j = n_j + \cdots + n_{k-1}$, and

$$\sum_{m_1,\ldots,m_{k-1} \geq 0} \frac{q^{M_1^2+\cdots+M_{k-1}^2}}{(q)_{2m_1}} \left(\prod_{j=2}^{k-2} \begin{bmatrix} m_{j-1}+m_{j+1} \\ 2m_j \end{bmatrix} \right) \begin{bmatrix} m_{k-2} \\ m_{k-1} \end{bmatrix}$$

$$= \frac{(q^{k(k-1)}, q^{k(k-1)+1}, q^{2k(k-1)+1}; q^{2k(k-1)+1})_\infty}{(q;q)_\infty}$$

for $k \geq 3$ with $M_j = m_j - m_{j-1}$ ($m_0 = 0$). The first of these two results is of course the (first) Andrews–Gordon identity for modulus $2k+1$ [1] which we have now embedded in a much larger family of Rogers–Ramanujan-type identities.

Another simple case of Theorem 3.2 occurs when $(a,b) = (F_k, F_{k-1})$ or (F_k, F_{k-2}) where $F_k = ((1+\sqrt{5})^k - (1-\sqrt{5})^k)/(2^k\sqrt{5})$ is the kth Fibonacci number; $F_0 = 0$, $F_1 = 1$ and $F_k = F_{k-1} + F_{k-2}$. Using the recurrence for the Fibonacci numbers and $F_2/F_1 = 1$ this yields

$$\text{cf}(F_k, F_{k-1}) = \left(\frac{F_k}{F_{k-1}} - 1 \right)^{-1} = \frac{F_{k-1}}{F_{k-2}} = 1 + \frac{1}{\frac{F_{k-2}}{F_{k-3}}} = \cdots = \underbrace{[1,\ldots,1]}_{k-2}$$

for $k \geq 3$. Hence $n = k-3$, $a_0 = \cdots = a_{k-3} = 1$ and $d(F_k, F_{k-1}) = k-2$. Inserting this in Theorem 3.2 and renaming n_1 as m_1 in the second equation yields the Theorems 1.2 and 1.4.

As third and final example we take $(a,b) = (2k-1, 2)$. Then cf$(a,b) = [k-2,1,1]$ leading to [**3**, Eq. (5.8)]

$$(3.20) \qquad \sum_{n_1,\ldots,n_k \geq 0} \frac{q^{N_1^2+\cdots+N_k^2}}{(q)_{n_1}\cdots(q)_{n_k}(q^{N_{k-1}+1})_{N_{k-1}}} = \frac{(q^{4k-2}, q^{4k-1}, q^{8k-3}; q^{8k-3})_\infty}{(q;q)_\infty}$$

for $k \geq 2$ with $N_j = n_j + \cdots + n_k$. (For $k = 2$ this follows from the first and for $k \geq 3$ from the second equation of Theorem 3.2, where we have changed $m_{k-1} \to n_{k-1}+n_k$ and $m_k \to n_k$).

Before we continue to explore the consequences of Theorem 3.1 we should perhaps remark that in the Andrews–Gordon identity (3.19) we can freely choose k. This means that we can always tune its right-hand side to coincide with the right-hand side of the first or second identity of Theorem 3.2. So what we have actually obtained are new representations for the sum side of the Andrews–Gordon identities. Specifically, if we consider the Andrews–Gordon identity for modulus $2k+1$ then we have found new sum sides for each ordered pair of coprime integers (a,b) such that $ab = k$. The most spectacular aspect of this is perhaps the fact that these new sums sides, though being considerably more cumbersome to write down, are much more efficient. In particular, the Andrews–Gordon identity for modulus $2k+1$ has a $(k-1)$-fold sum on the left-hand side whereas for (a,b) such that $ab = k$ (and $b > 1$) we obtain a sum side consisting of a $d(a,b)$-fold sum, where $d(a,b)$ is the sum of the partial quotients in the continued fraction of $(a/b-1)^{\mathrm{sign}(a-2b)}$. For instance, from our on-going example we find

$$\sum_{m_1,\ldots,m_4 \geq 0} \frac{q^{m_1^2+(m_1-m_2)^2+m_3^2+m_4^2}}{(q)_{2m_1}} \begin{bmatrix} m_1+m_2-m_3 \\ 2m_2 \end{bmatrix} \begin{bmatrix} m_2+m_3-m_4 \\ 2m_3 \end{bmatrix} \begin{bmatrix} m_3 \\ m_4 \end{bmatrix}$$
$$= \frac{(q^{35}, q^{36}, q^{71}; q^{71})_\infty}{(q;q)_\infty}$$

and the modulus 29 identity corresponding (3.20) with $k = 4$, which are to be compared with

$$\sum_{n_1,\ldots,n_{34} \geq 0} \frac{q^{N_1^2+\cdots+N_{34}^2}}{(q)_{n_1}\cdots(q)_{n_{34}}} = \frac{(q^{35}, q^{36}, q^{71}; q^{71})_\infty}{(q;q)_\infty}$$

and

$$\sum_{n_1,\ldots,n_{14} \geq 0} \frac{q^{N_1^2+\cdots+N_{14}^2}}{(q)_{n_1}\cdots(q)_{n_{14}}} = \frac{(q^{14}, q^{15}, q^{29}; q^{29})_\infty}{(q;q)_\infty}.$$

The most efficient sum sides of course occur in the identities involving the Fibonacci numbers, with $\log(2ab+1)/d(a,b) = \log(2F_k F_{k-1}+1)/(k-2) \to 2\log((1+\sqrt{5})/2)$ when k tends to infinity.

We have not yet come to the end of our list of corollaries to Theorem 3.1 and next we are going to exploit the fact that the polynomials in (3.12) are not reciprocal. If we define the polynomials $f_{a,b}(L,M)$ exactly as in (3.8) and (3.9) but change the term $mC(a,b)m$ to $\bar{m}C(a,b)m$ with $\bar{m} = (m_1,\ldots,m_{d(a,b)-1},0)$, i.e., $\bar{m}C(a,b)m = mC(a,b)m + m_{d(a,b)}(m_{d(a,b)-1} - m_{d(a,b)})$ and define the special case $f_{2,1}(L,M)$ as

$$f_{2,1}(L,M) = \sum_{m \geq 0} q^{Lm} \begin{bmatrix} 2L+M-m \\ 2L \end{bmatrix} \begin{bmatrix} L \\ m \end{bmatrix},$$

then the $q \to 1/q$ version of Theorem 3.1 can be states as follows.

COROLLARY 3.4. *For L, M nonnegative integers and (a,b) a pair of coprime integers such that $1 \leq b < a$ there holds*

$$\sum_{j \in \mathbb{Z}} (-1)^j q^{\frac{1}{2}j((2ab-1)j+1)} \mathcal{B}(L,M,aj,bj) = f_{a,b}(L,M).$$

The proof is obvious and merely requires (1.7) and (2.3) and the observation that $\bar{m}C(a,b)m = \sum_{j,k=1}^{d(a,b)}(\tau_j - 1)m_j C(a,b)_{j,k} m_k$.

The case $(a,b) = (3,1)$ deserves special attention. Taking $\mathrm{cf}(3,1) = [1,1]$ and replacing $m_1 \to L - i - n$ and $m_2 \to i$ gives

$$\sum_{j \in \mathbb{Z}} (-1)^j q^{\frac{1}{2}j(5j+1)} \mathcal{B}(L,M,3j,j) = \sum_{n,i \geq 0} q^{n^2 + i(L+n)} \begin{bmatrix} 2L+M-n-i \\ 2L \end{bmatrix} \begin{bmatrix} 2L-2i-n \\ n \end{bmatrix} \begin{bmatrix} L-i-n \\ i \end{bmatrix}.$$

This is our second doubly-bounded analogue of the first Rogers–Ramanujan identity, reducing to (1.11) in the large M limit. More generally, when $(a,b) = (k+1,1)$ we obtain new doubly-bounded analogues of the Andrews–Gordon identity (3.19)

$$\sum_{j \in \mathbb{Z}} (-1)^j q^{\frac{1}{2}j((2k+1)j+1)} \mathcal{B}(L,M,(k+1)j,j) = \sum_{n_1,\ldots,n_{k-1},i \geq 0} q^{N_1^2 + \cdots + N_{k-1}^2 + i(L+\tilde{N}_k)}$$

$$\times \begin{bmatrix} 2L+M-N_1-i \\ 2L \end{bmatrix} \begin{bmatrix} L-i(k-1)-\tilde{N}_k \\ i \end{bmatrix} \prod_{j=1}^{k-1} \begin{bmatrix} 2L-2ij-N_j-N_{j+1}-2\tilde{N}_j \\ n_j \end{bmatrix},$$

with $N_k = 0$ and $\tilde{N}_j = N_1 + \cdots + N_{j-1} = \sum_{l=1}^{k-1} \min(j-1,l) n_l$.

We now proceed exactly as before. First, defining $f_{a,b}(L)$ as in (3.13) and (3.14) but with $mC(a,b)m$ replaced by $\bar{m}C(a,b)m$ (and $f_{2,1}(L)$ as the right side of (1.9)) we find the following large M limit result.

COROLLARY 3.5. *For L a nonnegative integer and (a,b) a pair of coprime integers such that $1 \leq b < a$ there holds*

(3.21) $$G(L,L;b-1/a,b,a) = \sum_{j \in \mathbb{Z}} (-1)^j q^{\frac{1}{2}j((2ab-1)j+1)} \begin{bmatrix} 2L \\ L-aj \end{bmatrix} = f_{a,b}(L).$$

A different route to this corollary is to take (3.15), replace q by $1/q$ using (1.6) and (1.7), and to then replace b by $a - b$. From this remark it is clear that (3.21) does not yield new positivity results independent of (3.2).

To take the large L limit in Corollary 3.4 we define $\tilde{f}_{a,b}(M)$ as in (3.18) and (3.17) but with the usual $m_{d(a,b)}(m_{d(a,b)-1} - m_{d(a,b)})$ added to the exponent of q. There is one exception, namely, $\tilde{f}_{a,1}(M) = \tilde{F}_{a-1,1}(M)$ (with $\tilde{f}_{2,1}(M) = (q)_{2M}/(q)_M$).

COROLLARY 3.6. *For M a nonnegative integer and (a,b) a pair of coprime integers such that $1 \leq b < a$ there holds*

$$G(M,M;a-1/b,a,b) = \sum_{j \in \mathbb{Z}} (-1)^j q^{\frac{1}{2}j((2ab-1)j+1)} \begin{bmatrix} 2M \\ M-bj \end{bmatrix} = \tilde{f}_{a,b}(M).$$

Letting M tend to infinity we have reached our last theorem of this section.

THEOREM 3.3. *For $|q| < 1$ and (a,b) a pair of coprime integers such that $1 \leq b < a$ there holds*

$$\sum_{m \in \mathbb{Z}_+^{d(a,b)}} \frac{q^{\bar{m}C(a,b)m}}{(q)_{2m_1}} \prod_{j=2}^{d(a,b)} \begin{bmatrix} \tau_j m_j + n_j \\ \tau_j m_j \end{bmatrix} = \frac{(q^{ab-1}, q^{ab}, q^{2ab-1}; q^{2ab-1})_\infty}{(q;q)_\infty}.$$

for $a < 2b$, and

$$\sum_{\substack{n_1,\ldots,n_{a_0}\geq 0\\ m_{a_0+1},\ldots,m_{d(a,b)}\geq 0}} \frac{q^{(N_1+m_{a_0+1})^2+\cdots+(N_{a_0}+m_{a_0+1})^2}}{(q)_{n_1}\cdots(q)_{n_{a_0}}(q)_{2m_{a_0+1}}}$$

$$\times q^{\sum_{j,k=a_0+1}^{d(a,b)}\bar{m}_j C(a,b)_{j,k}m_k} \prod_{j=a_0+2}^{d(a,b)} \begin{bmatrix} \tau_j m_j + n_j \\ \tau_j m_j \end{bmatrix} = \frac{(q^{ab-1},q^{ab},q^{2ab-1};q^{2ab-1})_\infty}{(q;q)_\infty}$$

for $a > 2b \neq 2$ with $\bar{m}_j = m_j(1 - \delta_{j,d(a,b)})$.

As examples one finds Theorem 1.3 when $(a,b) = (F_k, F_{k-1})$, Theorem 1.5 when $(a,b) = (F_k, F_{k-2})$,

$$\sum_{m_1,\ldots,m_{k-1}\geq 0} \frac{q^{M_1^2+\cdots+M_{k-2}^2-m_{k-2}M_{k-1}}}{(q)_{2m_1}} \left(\prod_{j=2}^{k-2} \begin{bmatrix} m_{j-1}+m_{j+1} \\ 2m_j \end{bmatrix}\right) \begin{bmatrix} m_{k-2} \\ m_{k-1} \end{bmatrix}$$

$$= \frac{(q^{k(k-1)-1},q^{k(k-1)},q^{2k(k-1)-1};q^{2k(k-1)-1})_\infty}{(q;q)_\infty}$$

with $M_j = m_j - m_{j-1}$ $(m_0 = 0)$ when $(a,b) = (k, k-1)$ $(k \geq 3)$,

$$\sum_{n_1,\ldots,n_k\geq 0} \frac{q^{N_1^2+\cdots+N_{k-1}^2+N_{k-1}N_k}}{(q)_{n_1}\cdots(q)_{n_k}(q^{N_{k-1}+1})_{N_{k-1}}} = \frac{(q^{4k-3},q^{4k-2},q^{8k-5};q^{8k-5})_\infty}{(q;q)_\infty}$$

with $N_j = n_j + \cdots + n_k$ when $(a,b) = (2k-1, 2)$ $(k \geq 2)$, and

$$\sum_{m_1,\ldots,m_4\geq 0} \frac{q^{m_1^2+(m_1-m_2)^2+m_3^2+m_3m_4}}{(q)_{2m_1}} \begin{bmatrix} m_1+m_2-m_3 \\ 2m_2 \end{bmatrix} \begin{bmatrix} m_2+m_3-m_4 \\ 2m_3 \end{bmatrix} \begin{bmatrix} m_3 \\ m_4 \end{bmatrix}$$

$$= \frac{(q^{34},q^{35},q^{69};q^{69})_\infty}{(q;q)_\infty}$$

when $(a,b) = (7,5)$.

4. Proof of Theorem 3.1

Throughout we assume that (a,b) is a pair of positive, coprime integers such that $a > b$. When $(a,b) \neq (2,1)$ we will, for definiteness, choose the representation of cf(a,b) with $a_n \geq 2$. Now recall the definition of $d(a,b)$ as the sum of the partial quotients of cf(a,b) and note that $d(2,1) = 1$ and $d(a,b) > 1$ if $(a,b) \neq (2,1)$. Also note that for $(a,b) = (2,1)$ the theorem is nothing but the identity (3.2). (To see this recall that we already calculated $F_{2,1}(L,M)$ and that it coincides with the right side of (3.2).) We may use these facts to set up a proof by induction on $d(a,b)$.

Let us now assume that the theorem is valid for all (a',b') such that $d(a',b') = d$ and use this to prove its validity for all (a,b) such that $d(a,b) = d+1$. There are four different cases to be considered depending on the relative values of a and b. Since we already dealt with $(a,b) = (2,1)$ we may assume that $a \neq 2b$. For brevity let us denote the left-hand side of (3.12) by $B_{a,b}(L,M)$.

4.1. Proof for $a < 2b$. Let (a,b) be a pair such that $a < 2b$ and $d(a,b) = d+1$. Since $a < 2b$ it follows from (2.6) that $B_{a,b}(L,M)$ satisfies the recurrence

$$(4.1) \qquad B_{a,b}(L,M) = \sum_{i \geq 0} q^{i^2} \begin{bmatrix} 2L + M - i \\ 2L \end{bmatrix} B_{b,a-b}(i, L-i).$$

4.1.1. *Proof for* $\frac{3}{2}b < a < 2b$. If we write $a' = b$ and $b' = a - b$ then the condition $3b < 2a$ translates into $a' < 2b'$ and hence $\mathrm{cf}(a',b') = b'/(a'-b')$ which we denote by $[a'_0, \dots, a'_{n'}]$. Since $\mathrm{cf}(a,b) = b/(a-b) = 1 + 1/\mathrm{cf}(a',b')$ we conclude that $\mathrm{cf}(a,b) = [1, a'_0, \dots, a'_{n'}]$ and $d(a',b') = d(a,b) - 1 = d$. (Note that by excluding $(a,b) = (3,2)$, which gives $(a',b') = (2,1)$, we avoid complications due to the fact that, by our choice of representation, $\mathrm{cf}(3,2) = [2] = [1 + a'_0]$ and not $\mathrm{cf}(3,2) = [1,1] = [1, a'_0]$.) Consequently we may use our induction hypothesis to replace $B_{a',b'}$ in (4.1) by $F_{a',b'}$ defined in (3.8). Abbreviating $\mathcal{I}(a',b')$ and $C(a',b')$ by $\mathcal{I}'$ and C' this yields

$$B_{a,b}(L,M) = \sum_{i \geq 0} q^{i^2} \begin{bmatrix} 2L+M-i \\ 2L \end{bmatrix} \sum_{m' \in \mathbb{Z}_+^d} q^{m'C'm'} \begin{bmatrix} L+i-m'_1 \\ 2i \end{bmatrix} \prod_{j=1}^{d} \begin{bmatrix} \tau_j m'_j + n'_j \\ \tau_j m'_j \end{bmatrix},$$

with

$$(4.2) \qquad n'_j = i\delta_{j,1} - \sum_{k=1}^{d} C'_{j,k} m'_k \quad \text{for } j = 1, \dots, d.$$

Next rename i as m_1 and m'_j as m_{j+1}, and define

$$(4.3) \qquad \mathcal{I} = \left(\begin{array}{c|c} \begin{matrix} 1 & -1 \\ 1 & \end{matrix} & \\ \hline & \mathcal{I}' \end{array} \right)$$

and $C = 2I - \mathcal{I}$ with I the $(d+1) \times (d+1)$ identity matrix. Observing that $\mathcal{I} = \mathcal{I}(a,b)$ and $C = C(a,b)$ thanks to $\mathrm{cf}(a,b) = 1 + 1/\mathrm{cf}[a',b'] = [1, a'_0, \dots, a'_{n'}]$, this gives

$$(4.4) \qquad B_{a,b}(L,M) = \sum_{m \in \mathbb{Z}_+^{d+1}} q^{mC(a,b)m} \begin{bmatrix} 2L+M-m_1 \\ 2L \end{bmatrix} \prod_{j=1}^{d+1} \begin{bmatrix} \tau_j m_j + n_j \\ \tau_j m_j \end{bmatrix},$$

with

$$(4.5) \qquad n_j = L\delta_{j,1} - \sum_{k=1}^{d+1} C_{j,k}(a,b) m_k \quad \text{for } j = 1, \dots, d+1.$$

Since the right-hand side is exactly expression (3.8) for $F_{a,b}(L,M)$ with $d(a,b) = d+1$, this establishes (3.12) for $3b/2 < a < 2b$ and $d(a,b) = d+1$.

4.1.2. *Proof for* $a \leq \frac{3}{2}b$. Again we write $a' = b$ and $b' = a - b$ but this time the condition $2a \leq 3b$ yields $a' \geq 2b'$. Hence $\mathrm{cf}(a',b') = (a'-b')/b'$ which is denoted by $[a'_0, \dots, a'_{n'}]$. Since $\mathrm{cf}(a,b) = b/(a-b) = 1 + \mathrm{cf}(a',b')$ we conclude that $\mathrm{cf}(a,b) = [a'_0 + 1, \dots, a'_{n'}]$ and $d(a',b') = d(a,b) - 1 = d$. We may thus use the induction hypothesis to replace $B_{a',b'}$ in (4.1) by $F_{a',b'}$ defined in (3.9).

Abbreviating $\mathcal{I}(a',b')$ and $C(a',b')$ by $\mathcal{I}'$ and C' this yields

$$B_{a,b}(L,M) = \sum_{i\geq 0} q^{i^2} \begin{bmatrix} 2L+M-i \\ 2L \end{bmatrix} \sum_{m'\in\mathbb{Z}_+^d} q^{i(i-2m_1)+m'C'm'} \begin{bmatrix} L+m_1' \\ 2i \end{bmatrix} \prod_{j=1}^d \begin{bmatrix} \tau_j m_j' + n_j' \\ \tau_j m_j' \end{bmatrix},$$

with n_j' again given by (4.2). Once more we change $i \to m_1$ and $m_j' \to m_{j+1}$, and define

(4.6) $$\mathcal{I} = \left(\begin{array}{c|c} 0 & 1 \\ \hline 1 & \\ & \mathcal{I}' \end{array}\right)$$

and $C = 2I - \mathcal{I}$. It follows from $\mathrm{cf}(a,b) = 1 + \mathrm{cf}[a',b'] = [a_0'+1,\ldots,a_{n'}']$ that $\mathcal{I} = \mathcal{I}(a,b)$ and $C = C(a,b)$, again leading to (4.4) and (4.5), thus establishing (3.12) for $a \leq 3b/2$ and $d(a,b) = d+1$.

4.2. Proof for $a > 2b$. Let (a,b) be an admissible pair such that $a > 2b$ and $d(a,b) = d+1$. Since $a > 2b$ it follows from (2.5) that $B_{a,b}(L,M)$ satisfies the recurrence

(4.7) $$B_{a,b}(L,M) = \sum_{i\geq 0} q^{i^2} \begin{bmatrix} 2L+M-i \\ 2L \end{bmatrix} B_{a-b,b}(L-i,i).$$

4.2.1. Proof for $2b < a < 3b$. If we write $a' = a-b$ and $b' = b$ then the condition $a < 3b$ implies that $a' < 2b'$, and we may copy the first part of the first paragraph of section 4.1.1 and replace $B_{a',b'}$ in (4.7) by $F_{a',b'}$ defined in (3.8). With the same notation as before this leads to

$$B_{a,b}(L,M) = \sum_{i\geq 0} q^{i^2} \begin{bmatrix} 2L+M-i \\ 2L \end{bmatrix} \sum_{m'\in\mathbb{Z}_+^d} q^{m'C'm'} \begin{bmatrix} 2L-i-m_1' \\ 2L-2i \end{bmatrix} \prod_{j=1}^d \begin{bmatrix} \tau_j m_j' + n_j' \\ \tau_j m_j' \end{bmatrix},$$

with

(4.8) $$n_j' = (L-i)\delta_{j,1} - \sum_{k=1}^d C_{j,k}' m_k' \quad \text{for } j = 1,\ldots,d.$$

This time we relabel i as $L - m_1$ and m_j' as m_{j+1}, and define $\mathcal{I}$ as in (4.3) and $C = 2I - \mathcal{I}$. Since $\mathcal{I} = \mathcal{I}(a,b)$ and $C = C(a,b)$ thanks to $\mathrm{cf}(a,b) = 1 + 1/\mathrm{cf}[a',b']$ this gives

(4.9) $$B_{a,b}(L,M) = \sum_{m\in\mathbb{Z}_+^{d+1}} q^{L(L-2m_1)+mC(a,b)m} \begin{bmatrix} L+M+m_1 \\ 2L \end{bmatrix} \prod_{j=1}^{d+1} \begin{bmatrix} \tau_j m_j + n_j \\ \tau_j m_j \end{bmatrix},$$

with n_j given by (4.5). Since the right-hand side is exactly expression (3.9) for $d(a,b) = d+1$, this results in (3.12) for $2b < a < 3b$ and $d(a,b) = d+1$.

4.2.2. Proof for $a \geq 3b$. Writing $a' = a - b$ and $b' = b$ the condition $a \geq 3b$ becomes $a' \geq 2b'$, and we may copy the first part of the first paragraph of section 4.1.2 and replace $B_{a',b'}$ in (4.7) by $F_{a',b'}$ defined in (3.9). With the same notation as before this leads to

$$B_{a,b}(L,M) = \sum_{i\geq 0} q^{i^2} \begin{bmatrix} 2L+M-i \\ 2L \end{bmatrix} \sum_{m'\in\mathbb{Z}_+^d} q^{m'C'm'} \begin{bmatrix} L+m_1' \\ 2L-2i \end{bmatrix} \prod_{j=1}^d \begin{bmatrix} \tau_j m_j' + n_j' \\ \tau_j m_j' \end{bmatrix},$$

with n'_j given by (4.8). Renaming $i \to L - m_1$ and $m'_j \to m_{j+1}$, defining $\mathcal{I}$ as in (4.3) and $C = 2I - \mathcal{I}$, and observing that $\mathcal{I} = \mathcal{I}(a,b)$ and $C = C(a,b)$ because $\mathrm{cf}(a,b) = 1 + \mathrm{cf}[a',b']$, this again gives (4.9) and (4.5) establishing (3.12) for $a \geq 3b$ and $d(a,b) = d+1$ and completing the proof.

5. Further positivity results

So far we have no identities for $G(N, M; \alpha, \beta, K)$ with both α and β noninteger. To obtain such results our starting point will be yet another doubly-bounded analogue of the first Rogers–Ramanujan identity.

LEMMA 5.1. *For $L, M \geq 0$ there holds*

$$(5.1) \qquad \sum_{j \in \mathbb{Z}} (-1)^j q^{\frac{1}{2}j(5j+1)} \mathcal{B}(L, M, 2j+1, j) = \sum_{n \geq 0} q^{n^2} \begin{bmatrix} 2L+M-n-1 \\ 2L-1 \end{bmatrix} \begin{bmatrix} L-1 \\ n \end{bmatrix}.$$

PROOF. As a first step we add zero in the form

$$q^{L+1} \sum_{j \in \mathbb{Z}} (-1)^j q^{\frac{5}{2}j(j+1)} \begin{bmatrix} L+M+j \\ M-j-1 \end{bmatrix} \begin{bmatrix} L+M-j-1 \\ M+j \end{bmatrix}$$

to the left side of (5.1). By the recurrence (1.8) we are then to prove

$$(5.2) \qquad \sum_{j \in \mathbb{Z}} (-1)^j q^{\frac{1}{2}j(5j+1)} \begin{bmatrix} L+M+j \\ M-j \end{bmatrix} \begin{bmatrix} L+M-j-1 \\ M+j \end{bmatrix} = \sum_{n \geq 0} q^{n^2} \begin{bmatrix} 2L+M-n-1 \\ 2L-1 \end{bmatrix} \begin{bmatrix} L-1 \\ n \end{bmatrix}.$$

Suppressing their L-dependence, we denote the left sides of (5.2) and (3.2) (with $L \to L-1$ in the latter) by f_M and g_M, respectively. Using (1.8) we then get

$$f_M = g_M + q^L \sum_{j \in \mathbb{Z}} (-1)^j q^{\frac{5}{2}j(j+1)} \begin{bmatrix} L+M+j-1 \\ M-j-1 \end{bmatrix} \begin{bmatrix} L+M-j-1 \\ M+j \end{bmatrix}$$

$$= g_M + q^{2L-1} f_{M-1} + q^L \sum_{j \in \mathbb{Z}} (-1)^j q^{\frac{5}{2}j(j+1)} \begin{bmatrix} L+M+j-1 \\ M-j-1 \end{bmatrix} \begin{bmatrix} L+M-j-2 \\ M+j \end{bmatrix}$$

$$= g_M + q^{2L-1} f_{M-1}.$$

Next we let f_M and g_M denote the right sides of (5.2) and (3.2) (with, of course, $L \to L-1$ in the latter). One application of (1.8) shows that the same recurrence again holds. Since (3.2) holds and (5.2) is true for $M = 0$, we are done. □

Now that (5.1) has been proven we closely follow the work of the previous two sections. The present situation is, however, notationally more involved and further definitions related to continued fractions are needed. Let (a, b) be the usual ordered pair of coprime integers with associated continued fraction $\mathrm{cf}(a,b) = (a/b - 1)^{\mathrm{sign}(a-2b)} = [a_0, \ldots, a_n]$. In principle we could still allow for both representations of $\mathrm{cf}(a,b)$ but many of the equations below are sensitive to the chosen representation and to avoid unnecessary complications we demand that $a_n \geq 2$ for $(a,b) \neq (2,1)$. Given (a,b) we define a second pair $(\bar{a}, \bar{b})$ of positive, coprime integers as follows

$$(5.3) \qquad \frac{\bar{a}}{\bar{b}} = \begin{cases} [1, a_0, \ldots, a_{n-1}] = 1 + 1/[a_0, \ldots, a_{n-1}] & \text{for } a < 2b \\ [a_0+1, a_1, \ldots, a_{n-1}] = 1 + [a_0, \ldots, a_{n-1}] & \text{for } a > 2b, \end{cases}$$

with special cases $(\bar{a}, \bar{b}) = (1, 0)$ for $(a, b) = (a, 1)$ ($a \geq 2$) and $(\bar{a}, \bar{b}) = (1, 1)$ for $(a, b) = (a, a-1)$ ($a > 2$). Since $1/[1, c_0, \ldots, c_n] + 1/[c_0 + 1, \ldots, c_n] = 1$ it readily follows that for (a_1, b_1) and (a_2, b_2) such that $a_1 = a_2$ and $b_2 = a_1 - b_1$ there holds $\bar{a}_1 = \bar{a}_2 = \bar{b}_1 + \bar{b}_2$. It is also easy to see that if $a < 2b$ then $\bar{a} \leq 2\bar{b}$

and if $a \geq 2b$ then $\bar{a} \geq 2\bar{b}$. We also observe that $[1, a_0, \ldots, a_n] = a/b$ ($a \geq 2b$) and $[a_0 + 1, a_1, \ldots, a_n] = a/b$ ($a \leq 2b$). In the language of continued fractions this means that $\bar{a}/\bar{b}$ is the nth convergent of the continued fraction of a/b (which itself is of order $n + 1$). Care should however be taken with the anomalous case $(a, b) = (a, 1)$ for which $a/b = a = [a]$ and $(\bar{a}, \bar{b}) = (1, 0)$. As an example of the above definitions let $(a, b) = (19, 12)$. Then cf$(19, 12) = [1, 1, 2, 2]$, $\bar{a}/\bar{b} = [1, 1, 1, 2] = 8/5$ and $a/b = [1, 1, 1, 2, 2]$. Similarly, if $(a, b) = (19, 7)$, then cf$(19, 7) = [1, 1, 2, 2]$, $\bar{a}/\bar{b} = [2, 1, 2] = 8/3$ and $a/b = [2, 1, 2, 2]$.

We further need to define the analogues of the polynomials $F_{a,b}(L, M)$, which will be denoted by $H_{a,b}(L, M)$. For $(a, b) = (2, 1)$ we take it to be the right-hand side of (5.1) and for all other (a, b)

$$(5.4) \quad H_{a,b}(L, M) = \sum_{m \in \mathbb{Z}_+^{d(a,b)}} q^{mC(a,b)m + A_m} \begin{bmatrix} 2L + M - m_1 \\ 2L \end{bmatrix} \prod_{j=1}^{d(a,b)} \begin{bmatrix} \tau_j m_j + n_j - \delta_{j,d(a,b)} \\ \tau_j m_j - \delta_{j,d(a,b)-1} \end{bmatrix}$$

for $a < 2b$, and
$$(5.5)$$
$$H_{a,b}(L, M) = \sum_{m \in \mathbb{Z}_+^{d(a,b)}} q^{L(L-2m_1) + mC(a,b)m + A_m} \begin{bmatrix} L + M + m_1 \\ 2L \end{bmatrix} \prod_{j=1}^{d(a,b)} \begin{bmatrix} \tau_j m_j + n_j - \delta_{j,d(a,b)} \\ \tau_j m_j - \delta_{j,d(a,b)-1} \end{bmatrix}$$

for $a > 2b$. Here $mC(a,b)m$ and n_j are still given by (3.10) and (3.11). The term A_m denotes the linear term $2m_{d(a,b)} - 2m_{d(a,b)-1} + 1$. Since $a_n \geq 2$ we thus have

$$mC(a,b)m + A_m = \sum_{j=0}^{n} \Big(m_{t_j+1}^2 + \sum_{k=t_j+1}^{t_{j+1}-1} (m_k - m_{k+1} - \delta_{k,d(a,b)-1})^2 \Big).$$

The analogue of theorem 3.1 now breaks up into two separate statements.

THEOREM 5.1. *For L, M nonnegative integers and (a, b) a pair of coprime integers such that $1 \leq b < a$ there holds (i)*

$$(5.6) \quad \sum_{j \in \mathbb{Z}} (-1)^j q^{\frac{1}{2}j((2ab+1)j + 4\bar{a}b + 1) + \bar{a}\bar{b}} \mathcal{B}(L, M, aj + \bar{a}, bj + \bar{b}) = H_{a,b}(L, M)$$

for $a < 2b$ and cf(a, b) a continued fraction of even order (i.e., n even), or $a > 2b$ and cf(a, b) a continued fraction of odd order, (ii)

$$(5.7) \quad \sum_{j \in \mathbb{Z}} (-1)^j q^{\frac{1}{2}j((2ab+1)j + 4a\bar{b} + 1) + \bar{a}\bar{b}} \mathcal{B}(L, M, aj + \bar{a}, bj + \bar{b}) = H_{a,b}(L, M)$$

for $a < 2b$ and cf(a, b) of odd order, or $a \geq 2b$ and cf(a, b) of even order.

Instead of boring the reader with a complete list of analogues of the results of section 3, we restrict ourselves to the analogue of Corollary 3.2. This follows after letting M tend to infinity in Theorem 5.1 and using (2.4).

COROLLARY 5.1. *Let (a, b) be a pair of coprime integers such that $1 \leq b < a$, with $(a, b) = (2, 1)$ excluded. Then (i) $G(L + \bar{a}, L - \bar{a}; b - 2\bar{a}b/a, b + 1/a + 2\bar{a}b/a, a)$ is a polynomial with nonnegative coefficients if $a < 2b$ and cf(a, b) has even order, or $a > 2b$ and cf(a, b) has odd order, (ii) $G(L + \bar{a}, L - \bar{a}; b - 2\bar{b}, b + 1/a + 2\bar{b}, a)$ is a polynomial with nonnegative coefficients if $a < 2b$ and cf(a, b) has odd order, or $a > 2b$ and cf(a, b) has even order.*

No further positivity results arise from Theorem 5.1. If we replace q by $1/q$ and then let M tend to infinity we get results which are equivalent to those obtained by exploiting the duality (1.6) in Corollary 5.1. Although this is similar to the situation encountered in section 3 it is more cumbersome to prove this. A brief derivation proceeds as follows. Start with (5.7) and replace $q \to 1/q$ using (2.3). This gives the polynomial identity

$$\sum_{j \in \mathbb{Z}} (-1)^j q^{\frac{1}{2}j((2ab-1)j + 4a\bar{b}-1) + \bar{a}\bar{b}} \mathcal{B}(L, M, aj + \bar{a}, bj + \bar{b}) = q^{2LM} H_{a,b}(L, M; 1/q).$$

Letting M tend to infinity establishes the positivity of $G(L - \bar{a}, L + \bar{a}; b - 1/a + 2\bar{b}, b - 2\bar{b}, a)$. All of this is of course for $a < 2b$ and $\mathrm{cf}(a, b)$ of even order, or $a > 2b$ and $\mathrm{cf}(a, b)$ of odd order. Next replace $b \to a - b$ which has the effect of changing $\bar{b} \to \bar{a} - \bar{b}$. Hence $G(L - \bar{a}, L + \bar{a}; a - b - 1/a + 2(\bar{a} - \bar{b}), a - b - 2(\bar{a} - \bar{b}), a)$ is positive for $a > 2b$ and $\mathrm{cf}(a, b)$ of even order, or $a < 2b$ and $\mathrm{cf}(a, b)$ of odd order. Letting $q \to 1/q$ in item (ii) of the Corollary 5.1 using (1.6) yields the same result. Starting with (5.7) instead of (5.6) reproduces the $q \to 1/q$ analogue of item (i) of the corollary in much the same way.

Before we come to the proof of Theorem 5.1 let us consider the example $(a, b) = (3, 1)$. Then $(\bar{a}, \bar{b}) = (1, 0)$ and $\mathrm{cf}(3, 1) = [2]$ which is of order 0. Since also $a > 2b$ we are to use (5.7) leading to

$$(5.8) \quad \sum_{j \in \mathbb{Z}} (-1)^j q^{\frac{1}{2}j(7j+1)} \mathcal{B}(L, M, 3j + 1, j)$$
$$= \sum_{m_1, m_2 \geq 0} q^{(L-m_1)^2 + (m_1 - m_2 - 1)^2} \begin{bmatrix} L + M + m_1 \\ 2L \end{bmatrix} \begin{bmatrix} L + m_2 \\ 2m_1 - 1 \end{bmatrix} \begin{bmatrix} m_1 - 1 \\ m_2 \end{bmatrix}.$$

The reason for giving this rather atypical example with trivial $(\bar{a}, \bar{b})$ is its $q \to 1/q$ counterpart, which after the replacements $m_1 \to L - i - n$ and $m_2 \to L - 2i - n - 1$ becomes

$$\sum_{j \in \mathbb{Z}} (-1)^j q^{\frac{1}{2}j(5j+3)} \mathcal{B}(L, M, 3j + 1, j)$$
$$= \sum_{n, i \geq 0} q^{n(n+1) + i(L+n+1)} \begin{bmatrix} 2L + M - i - n \\ 2L \end{bmatrix} \begin{bmatrix} 2L - 2i - n - 1 \\ n \end{bmatrix} \begin{bmatrix} L - i - n - 1 \\ i \end{bmatrix}.$$

This doubly-bounded analogue of the second Rogers–Ramanujan yields (1.12) of the introduction in the large M limit.

6. Proof of Theorem 5.1

Though considerably more involved, the proof proceeds along the same lines as the proof of Theorem 3.1 given in section 4. Again we carry out induction on $d(a, b)$, but the first difference is that the case $(a, b) = (2, 1)$ which has $d(2, 1) = 1$, is special and is not included in either (5.4) or (5.5). We therefore have to first prove the cases $(a, b) = (3, 1)$ and $(3, 2)$ which are the only solutions to $d(a, b) = 2$. These can then serve as starting point for our induction. Applying (2.6) to (5.1) gives

$$
(6.1) \quad \sum_{j\in\mathbb{Z}}(-1)^j q^{\frac{1}{2}j(13j+9)+1}\mathcal{B}(L,M,3j+1,2j+1)
$$
$$
= \sum_{i,n\geq 0} q^{i^2+n^2}\begin{bmatrix}2L+M-i\\2L\end{bmatrix}\begin{bmatrix}L+i-n-1\\2i-1\end{bmatrix}\begin{bmatrix}i-1\\n\end{bmatrix}.
$$

For $(a,b)=(3,2)$ one finds $\mathrm{cf}(a,b)=[2]$ of even order, $a<2b$ and $(\bar{a},\bar{b})=(1,1)$. Hence the left side of (5.6) for $(a,b)=(3,2)$ agrees with the left side of (6.1). To see that also the right sides agree we rewrite the right side of (6.1) in terms of the summation variables $m_1=i$ and $m_2=i-n-1$. This gives

$$
\sum_{m_1,m_2\geq 0} q^{m_1^2+(m_1-m_2-1)^2}\begin{bmatrix}2L+M-m_1\\2L\end{bmatrix}\begin{bmatrix}L+m_2\\2m_1-1\end{bmatrix}\begin{bmatrix}m_1-1\\m_2\end{bmatrix}
$$

in accordance with the $(a,b)=(3,2)$ case of (5.4). Similarly, applying (2.5) to (5.1) gives

$$
\sum_{j\in\mathbb{Z}}(-1)^j q^{\frac{1}{2}j(7j+1)}\mathcal{B}(L,M,3j+1,j) = \sum_{i,n\geq 0} q^{i^2+n^2}\begin{bmatrix}2L+M-i\\2L\end{bmatrix}\begin{bmatrix}2L-i-n-1\\2L-2i-1\end{bmatrix}\begin{bmatrix}L-i-1\\n\end{bmatrix}.
$$

Introducing $m_1=L-i$ and $m_2=L-i-n-1$ on the right transforms this into (5.8) which is the $(a,b)=(3,1)$ case of Theorem 5.1.

Now we are prepared for the induction step, and we assume the theorem to be correct for all (a',b') such that $d(a',b')=d$ in order to prove its validity for all (a,b) such that $d(a,b)=d+1$. There are eight cases to be considered depending on the relative values of a and b and on the order of $\mathrm{cf}(a,b)$. For convenience let us introduce the notation $B_{a,b}^{\mathrm{e},<}(L,M)$ for the left side of (5.6) when $a<2b$ and $\mathrm{cf}(a,b)$ has even order. In the same way we define $B_{a,b}^{\mathrm{o},>}(L,M)$, $B_{a,b}^{\mathrm{e},>}(L,M)$ and $B_{a,b}^{\mathrm{o},<}(L,M)$.

6.1. Proof for $\frac{3}{2}b<a<2b$ with $\mathrm{cf}(a,b)$ of even order. Let (a,b) be a pair such that $3b/2<a<2b$, $\mathrm{cf}(a,b)$ of even order n and $d(a,b)=d+1$. For such (a,b) we will show that

$$
(6.2) \quad B_{a,b}^{\mathrm{e},<}(L,M) = \sum_{i\geq 0} q^{i^2}\begin{bmatrix}2L+M-i\\2L\end{bmatrix}B_{b,a-b}^{\mathrm{o},<}(i,L-i),
$$

with $d(b,a-b)=d$. First write $a'=b$ and $b'=a-b$. Then $3b<2a$ gives $a'<2b'$ (in accordance with (6.2)) and hence $\mathrm{cf}(a',b')=[a_0',\ldots,a_{n'}']=b'/(a'-b')$. Since $\mathrm{cf}(a,b)=b/(a-b)=1+1/\mathrm{cf}(a',b')$ one finds $\mathrm{cf}(a,b)=[1,a_0',\ldots,a_{n'}']$. This yields $n'=n-1$ is odd and $d(a',b')=d(a,b)-1=d$. We may now conclude that at least the labels in (6.2) are correct. Next we use the definition of $B_{a',b'}^{\mathrm{o},<}$ and the Burge transform (2.6) to compute the right-hand side of (6.2) as

$$
\sum_{j\in\mathbb{Z}}(-1)^j q^{\frac{1}{2}j((2a'(a'+b')+1)j+4a'(\bar{a}'+\bar{b}')+1)+\bar{a}'(\bar{a}'+\bar{b}')}
$$
$$
\times \mathcal{B}(L,M,(a'+b')j+\bar{a}'+\bar{b}',a'j+\bar{a}')
$$

Since $\bar{a}/\bar{b}=[1,a_0,\ldots,a_{n-1}]=[1,1,a_0',\ldots,a_{n'-1}']=1+1/[1,a_0,\ldots,a_{n'-1}]=(\bar{a}'+\bar{b}')/\bar{a}'$ we find that $\bar{a}'=\bar{b}$ and $\bar{b}'=\bar{a}-\bar{b}$. Also using $a'=b$ and $b'=a-b$ we can simplify the above expression to

$$\sum_{j\in\mathbb{Z}}(-1)^j q^{\frac{1}{2}j((2ab+1)j+4\bar{a}b+1)+\bar{a}\bar{b}}\mathcal{B}(L,M,aj+\bar{a},bj+\bar{b}).$$

This is precisely the left side of (6.2) as we set out to prove.

The remaining part of the proof proceeds exactly as the proof given in section 4.1.1. In a few words, we can use the induction hypothesis to replace the right side of (6.2) by $H_{a',b'}(i, L-i)$ given in (5.4). Making the necessary variable changes gives the expression claimed by the theorem.

6.2. Proof for $\frac{3}{2}b < a < 2b$ with cf(a,b) of odd order. Let (a,b) be a pair such that $3b/2 < a < 2b$, cf(a,b) of odd order n and $d(a,b) = d+1$. Then (6.2) is replaced by

(6.3) $$B_{a,b}^{\text{o},<}(L,M) = \sum_{i\geq 0} q^{i^2} \begin{bmatrix} 2L+M-i \\ 2L \end{bmatrix} B_{b,a-b}^{\text{e},<}(i, L-i),$$

with $d(b, a-b) = d$. Again we define $a' = b$ and $b' = a - b$. Copying the paragraph below (6.2) replacing the one occurrence of 'odd' by 'even', shows that the labels are again correct. Computing the right-hand side of (6.3) using $\bar{a}' = \bar{b}$ and $\bar{b}' = \bar{a} - \bar{b}$ results in

$$\sum_{j\in\mathbb{Z}}(-1)^j q^{\frac{1}{2}j((2ab+1)j+4a\bar{b}+1)+\bar{a}\bar{b}}\mathcal{B}(L,M,aj+\bar{a},bj+\bar{b})$$

in agreement with the left-hand side of (6.3). The rest of the proof follows that of section 4.1.1.

6.3. Proof of the remaining six cases. The proofs of the remaining cases are simple modifications of the previous two and are therefore omitted. For completeness we just state the six key-identities

$$B_{a,b}^{\text{p},<}(L,M) = \sum_{i\geq 0} q^{i^2} \begin{bmatrix} 2L+M-i \\ 2L \end{bmatrix} B_{b,a-b}^{\text{p},>}(i, L-i) \quad \text{for } 2a < 3b$$

and

$$B_{a,b}^{\text{p},>}(L,M) = \sum_{i\geq 0} q^{i^2} \begin{bmatrix} 2L+M-i \\ 2L \end{bmatrix} B_{a-b,b}^{\bar{\text{p}},<}(L-i, i) \quad \text{for } 2b < a < 3b$$

$$B_{a,b}^{\text{p},>}(L,M) = \sum_{i\geq 0} q^{i^2} \begin{bmatrix} 2L+M-i \\ 2L \end{bmatrix} B_{a-b,b}^{\text{p},>}(L-i, i) \quad \text{for } a > 3b,$$

where p = e,o, $\bar{\text{e}}$ = o and $\bar{\text{o}}$ = e.

7. Rogers–Ramanujan-type identities for even moduli

Closely related to the Andrews–Gordon identities (3.19) are Bressoud's identities for even moduli [8]

(7.1) $$\sum_{n_1,\ldots,n_{k-1}\geq 0} \frac{q^{N_1^2+\cdots+N_{k-1}^2}}{(q)_{n_1}\cdots(q)_{n_{k-2}}(q^2;q^2)_{n_{k-1}}} = \frac{(q^k, q^k, q^{2k}; q^{2k})_\infty}{(q;q)_\infty}$$

for $k \geq 2$, $N_j = n_j + \cdots + n_{k-1}$ and $|q| < 1$. An obvious question is whether also these identities can be embedded in an infinite tree of Rogers–Ramanujan-type

identities. The answer to this is 'yes' and is already implicit in Burge's original paper on his transform. In [**11**, page 217] Burge states the following identity for $L, M \geq 0$:
$$\sum_{j \in \mathbb{Z}} (-1)^j q^{j^2} \mathcal{B}(L, M, j, j) = \begin{bmatrix} L + M \\ M \end{bmatrix}_{q^2}.$$
This identity is very similar to (3.1) and if we define $I_{a,b}(L, M)$ by (3.8) and (3.9) but with each q-binomial coefficient dressed with a subscript $q^{\bar{\tau}_j}$ where $\bar{\tau}_j = 3 - \tau_j$, then the following theorem is immediate.

THEOREM 7.1. *For L, M nonnegative integers and (a, b) a pair of coprime integers such that $1 \leq b < a$ there holds*
$$\sum_{j \in \mathbb{Z}} (-1)^j q^{abj^2} \mathcal{B}(L, M, aj, bj) = I_{a,b}(L, M).$$

As far as the generalized Borwein conjecture goes this is not interesting, but letting both L and M tend to infinity yields an even modulus version of Theorem 3.2.

THEOREM 7.2. *For $|q| < 1$ and (a, b) a pair of coprime integers such that $1 \leq b < a$ there holds*
$$\sum_{m \in \mathbb{Z}_+^{d(a,b)}} \frac{q^{mC(a,b)m}}{(q)_{2m_1}} \prod_{j=2}^{d(a,b)} \begin{bmatrix} \tau_j m_j + n_j \\ \tau_j m_j \end{bmatrix}_{q^{\bar{\tau}_j}} = \frac{(q^{ab}, q^{ab}, q^{2ab}; q^{2ab})_\infty}{(q; q)_\infty}$$

for $a < 2b$, and

$$\sum_{\substack{n_1, \ldots, n_{a_0} \geq 0 \\ m_{a_0+1}, \ldots, m_{d(a,b)} \geq 0}} \frac{q^{(N_1 + m_{a_0+1})^2 + \cdots + (N_{a_0} + m_{a_0+1})^2}}{(q)_{n_1} \cdots (q)_{n_{a_0-1}} (q^{\bar{\tau}_{a_0}}; q^{\bar{\tau}_{a_0}})_{n_{a_0}} (q^{\bar{\tau}_{a_0}+1}; q^{\bar{\tau}_{a_0}+1})_{\tau_{a_0+1} m_{a_0+1}}}$$
$$\times q^{\sum_{j,k=a_0+1}^{d(a,b)} m_j C(a,b)_{j,k} m_k} \prod_{j=a_0+2}^{d(a,b)} \begin{bmatrix} \tau_j m_j + n_j \\ \tau_j m_j \end{bmatrix}_{q^{\bar{\tau}_j}} = \frac{(q^{ab}, q^{ab}, q^{2ab}; q^{2ab})_\infty}{(q; q)_\infty}$$

for $a \geq 2b$.

For $(a, b) = (k, 1)$ this is Bressoud's (7.1). The most interesting other examples again feature the Fibonacci numbers, and for $(a, b) = (F_k, F_{k-1})$ and $(a, b) = (F_k, F_{k-2})$ one finds

$$\sum_{m_1, \ldots, m_{k-2} \geq 0} \frac{q^{m_1^2 + \cdots + m_{k-2}^2}}{(q)_{2m_1}} \left(\prod_{j=2}^{k-3} \begin{bmatrix} m_{j-1} + m_j - m_{j+1} \\ 2m_j \end{bmatrix} \right) \begin{bmatrix} m_{k-3} \\ m_{k-2} \end{bmatrix}_{q^2}$$
$$= \frac{(q^{F_k F_{k-1}}, q^{F_k F_{k-1}}, q^{2F_k F_{k-1}}; q^{2F_k F_{k-1}})_\infty}{(q; q)_\infty}$$

for $k \geq 4$, and

$$\sum_{m_1, \ldots, m_{k-2} \geq 0} \frac{q^{(m_1+m_2)^2 + m_2^2 + \cdots + m_{k-2}^2}}{(q)_{m_1}(q)_{2m_2}} \left(\prod_{j=3}^{k-3} \begin{bmatrix} m_{j-1} + m_j - m_{j+1} \\ 2m_j \end{bmatrix} \right) \begin{bmatrix} m_{k-3} \\ m_{k-2} \end{bmatrix}_{q^2}$$
$$= \frac{(q^{F_k F_{k-2}}, q^{F_k F_{k-2}}, q^{2F_k F_{k-2}}; q^{2F_k F_{k-2}})_\infty}{(q; q)_\infty}$$

for $k \geq 5$, where we have replaced $n_1 \to m_1$ in comparison with Theorem 7.2. These two series can again be extended to all $k \geq 3$ by taking $\sum_{m \geq 0} q^{m^2}/(q^2;q^2)_m$ as left-hand sides when $k = 3$ and $\sum_{m_1,m_2 \geq 0} q^{(m_1+m_2)^2+m_2^2}/(q)_{m_1}(q^2;q^2)_{m_2}$ as left-hand side of the second series when $k = 4$. We also note that if $k = 4$ in the first series one can perform the sum over m_2 by the q-binomial theorem to yield $(-q;q^2)_{m_1}$. The resulting identity is [20, Eq. (29)] in Slater's list of Rogers–Ramanujan-type identities.

8. Outlook: how tractable is the Borwein conjecture?

Using the Burge transform we have established two types of positivity results for $G(N, M; \alpha, \beta, K)$. The first type has $N = M$ with either α or β being an integer, and the second type has $N \neq M$ where both α and β can be noninteger. This might lead one to suspect that proving the positivity of coefficients of $G(n, n; \alpha, \beta, K)$ for noninteger α and β is perhaps a much more difficult problem. The sceptical reader might even doubt that nice identities for such G exist in the first place, casting doubt on the claim made in the introduction that our failure to prove the positivity of $A_n(q) = G(n, n; 4/3, 5/3, 3)$ is possibly just a practical and not a fundamental problem.

Indeed we believe that the original Borwein conjecture is quite a bit deeper than the positivity results proven in this paper. However, in our subsequent papers on this topic we will introduce new types of transformations that settle more complicated cases of Bressoud's conjecture. Some appealing examples are the final entries of the following two sequences of identities:

$$G(n,n;1/2,2/2,2) = \sum_{m_1 \geq 0} q^{m_1 n} \begin{bmatrix} n \\ m_1 \end{bmatrix}$$

$$G(n,n;3/3,4/3,3) = \sum_{m_1,m_2 \geq 0} q^{(n-m_1)^2+(m_1-m_2)^2} \begin{bmatrix} n+m_2 \\ 2m_1 \end{bmatrix} \begin{bmatrix} m_1 \\ m_2 \end{bmatrix}$$

$$G(n,n;5/4,6/4,4) = \sum_{m_1,m_2,m_3 \geq 0} q^{n(n-m_1)+m_2(m_2+m_3)} \begin{bmatrix} n \\ m_1 \end{bmatrix} \begin{bmatrix} m_1 \\ 2m_2 \end{bmatrix} \begin{bmatrix} m_2 \\ m_3 \end{bmatrix}$$

and

$$G(n,n;2/2,3/2,2) = \sum_{m_1 \geq 0} q^{m_1^2} \begin{bmatrix} n \\ m_1 \end{bmatrix}$$

$$G(n,n;4/3,5/3,3) = A_n(q) = \text{???}$$

$$G(n,n;6/4,7/4,4) = \sum_{m_1,m_2,m_3 \geq 0} q^{n(n-m_1)+m_2^2+m_3^2} \begin{bmatrix} n \\ m_1 \end{bmatrix} \begin{bmatrix} m_1 \\ 2m_2 \end{bmatrix} \begin{bmatrix} m_2 \\ m_3 \end{bmatrix}.$$

The pattern is of course clear and we leave it to the reader to fill in the missing item.

Note. Alexander Berkovich kindly informed me that he has independently obtained several of the results established in this paper. In particular he has obtained the corollaries 3.1 and 3.2 and all results implied by these.

References

1. G. E. Andrews, *An analytic generalization of the Rogers–Ramanujan identities for odd moduli*, Prod. Nat. Acad. Sci. USA **71** (1974), 4082–4085.
2. G. E. Andrews, *The Theory of Partitions*, Encyclopedia of Mathematics and its Applications, Vol. 2, (Addison-Wesley, Reading, Massachusetts, 1976).
3. G. E. Andrews, *Multiple series Rogers–Ramanujan type identities*, Pacific J. Math. **114** (1984), 267–283.
4. G. E. Andrews, *On a conjecture of Peter Borwein*, J. Symbolic Comput. **20** (1995), 487–501.
5. G. E. Andrews, R. J. Baxter, D. M. Bressoud, W. H. Burge, P. J. Forrester and G. Viennot, *Partitions with prescribed hook differences*, Europ. J. Combinatorics **8** (1987), 341–350.
6. A. Berkovich and B. M. McCoy, *Continued fractions and fermionic representations for characters of $M(p, p')$ minimal models*, Lett. Math. Phys. **37** (1996), 49–66.
7. A. Berkovich, B. M. McCoy and A. Schilling, *Rogers-Schur-Ramanujan type identities for the $M(p, p')$ minimal models of conformal field theory*, Commun. Math. Phys. **191** (1998), 325–395.
8. D. M. Bressoud, *An analytic generalization of the Rogers–Ramanujan identities with interpretation*, Quart. J. Maths. Oxford (2) **31** (1980), 385–399.
9. D. M. Bressoud, *Some identities for terminating q-series*, Math. Proc. Camb. Phil. Soc. **89** (1981), 211–223.
10. D. M. Bressoud, *The Borwein conjecture and partitions with prescribed hook differences*, Electron. J. Combin. **3** (1996), #4.
11. W. H. Burge, *Restricted partition pairs*, J. Combin. Theory Ser. A **63** (1993), 210–222.
12. O. Foda, K. S. Lee and T. A. Welsh, *A Burge tree of Virasoro-type polynomial identities*, Int. J. Mod. Phys. A **13** (1998), 4967–5012.
13. G. Gasper and M. Rahman, *Basic Hypergeometric Series*, Encyclopedia of Mathematics and its Applications, Vol. 35, (Cambridge University Press, Cambridge, 1990).
14. M. E. H. Ismail, D. Kim and D. Stanton, *Lattice paths and positive trigonometric sums*, Constr. Approx. **15** (1999), 69–81.
15. P. Paule, *On identities of the Rogers–Ramanujan type*, J. Math. Anal. Appl. **107** (1985), 255–284.
16. L. J. Rogers, *Second memoir on the expansion of certain infinite products*, Proc. London Math. Soc. **25** (1894), 318–343.
17. L. J. Rogers, *On two theorems of combinatory analysis and some allied identities*, Proc. London Math. Soc. (2) **16** (1917), 315–336.
18. A. Schilling and S. O. Warnaar, *Supernomial coefficients, polynomial identities and q-series*, The Ramanujan Journal **2** (1998), 459–494.
19. A. Schilling and S. O. Warnaar, *A generalization of the q-Saalschütz sum and the Burge transform*, in *Physical Combinatorics*, pp. 163–183, M. Kashiwara and T. Miwa eds., Progr. Math. **191**, (Birkhäuser, Boston, 2000).
20. L. J. Slater, *Further identities of the Rogers–Ramanujan type*, Proc. London Math. Soc. (2) **54** (1952), 147–167.

DEPARTMENT OF MATHEMATICS AND STATISTICS, THE UNIVERSITY OF MELBOURNE, VIC 3010, AUSTRALIA

E-mail address: warnaar@ms.unimelb.edu.au

Mock ϑ-Functions and Real Analytic Modular Forms

S.P. Zwegers

ABSTRACT. In this paper we examine three examples of Ramanujan's third order mock ϑ-functions and relate them to Rogers' false ϑ-series and to a real-analytic modular form of weight 1/2.

1. Introduction

Mock ϑ-functions were introduced by S. Ramanujan in the last letter he wrote to G.H. Hardy, dated January, 1920. For a photocopy of the mathematical part of this letter see [**Ra**, pp. 127–131] (also reproduced in [**A2**]). In this letter he provided a list of 17 mock ϑ-functions (4 of "order three", 10 of "order five" and 3 of "order seven"), together with identities they satisfy.

In [**AH**] we find a definition of the concept of a mock ϑ-function. Slightly rephrased it reads: a mock ϑ-function is a function f of the complex variable q, defined by a q-series of a particular type (Ramanujan calls this the Eulerian form), which converges for $|q| < 1$ and satisfies the following conditions:

(1) infinitely many roots of unity are exponential singularities,
(2) for every root of unity ξ there is a ϑ-function $\vartheta_\xi(q)$ such that the difference $f(q) - \vartheta_\xi(q)$ is bounded as $q \to \xi$ radially,
(3) there is no ϑ-function that works for all ξ, i.e. f is not the sum of two functions, one of which is a ϑ-function and the other a function which is bounded in all roots of unity.

(When Ramanujan refers to ϑ-functions, he means sums, products, and quotients of series of the form $\sum_{n \in \mathbf{Z}} \epsilon^n q^{an^2+bn}$ with $a, b \in \mathbf{Q}$ and $\epsilon = -1, 1$).

The 17 functions given by Ramanujan indeed satisfy condition (1) and (2) (see [**W1**], [**W2**] and [**S**]). However no proof has ever been given that they also satisfy condition (3). Watson (see [**W1**]) proved a very weak form of condition (3) for the "third order" mock ϑ-functions, namely, that they are not equal to ϑ-functions.

In section 3 we will see that condition (3) is not satisfied if we weaken it slightly. Indeed, we shall discuss a vector-valued third order mock ϑ-function F for which there is a real analytic modular form H such that $F - H$ is bounded in all roots of unity.

2000 *Mathematics Subject Classification.* Primary 11F37; Secondary 11F27.
Key words and phrases. q-series, mock ϑ-functions, modular forms.

©2001 American Mathematical Society

Before that, we discuss in the next section a connection between mock ϑ-functions and Rogers' false ϑ-series. Again we look at the behaviour of a mock theta function when q approaches a root of unity radially. But now we extend the function across the unit circle.

2. False ϑ-series

We will consider the mock ϑ-function ν, which is not mentioned in Ramanujan's letter, but which was found by Watson in [**W1**], and can also be found in Ramanujan's "lost" notebook [**Ra**]:

(2.1)
$$\nu(q) = \sum_{n=0}^{\infty} \frac{q^{n^2+n}}{(-q;q^2)_{n+1}}$$
$$= \frac{1}{1+q} + \frac{q^2}{(1+q)(1+q^3)} + \frac{q^6}{(1+q)(1+q^3)(1+q^5)} + \cdots$$

We can easily see that the defining sum for ν converges not only for $|q| < 1$, but also for $|q| > 1$. We will now study the function that is defined by the sum outside the unit disk. In order to do so, we replace q by q^{-1} in the sum, take $|q| < 1$ and call this new function ν_-. We get

(2.2)
$$\nu_-(q) = \sum_{n=0}^{\infty} \frac{q^{n+1}}{(-q;q^2)_{n+1}}$$
$$= \frac{q}{1+q} + \frac{q^2}{(1+q)(1+q^3)} + \frac{q^3}{(1+q)(1+q^3)(1+q^5)} + \cdots$$

In Ramanujan's "lost" notebook [**Ra**] we find the following identity for $|q| < 1$ (which was proved by Andrews in [**A1**]):

(2.3) $$\nu_-(q) = \sum_{n=0}^{\infty} (-1)^n q^{6n^2+4n+1}(1 + q^{4n+2})$$

(2.4) $$= \left(\sum_{n=0}^{\infty} - \sum_{n=-\infty}^{-1} \right) (-1)^n q^{6n^2+4n+1} = q^{\frac{1}{3}} \sum_{n=0}^{\infty} (-1)^{n+1} \left(\frac{-3}{n} \right) q^{\frac{2}{3}n^2}$$

From these identities we see that ν_- has a very simple power series expansion. This expansion looks very much like a ϑ-function, only the signs are somewhat different. Rogers uses the term *false ϑ-series* for this type of functions (see [**Ro**, pp. 328]).

The following proposition (see [**LZ**]) shows that for every root of unity ξ the function ν_- is bounded as $q \to \xi$ radially. We can even compute the complete asymptotic expansion.

PROPOSITION 2.1. *Let $C : \mathbf{Z} \to \mathbf{C}$ be a periodic function with mean value 0. Then the associated L-series $L(s, C) = \sum_{n=1}^{\infty} C(n)n^{-s}$ (Re$(s) > 1$) extends holomorphically to $\mathbf{C}$. The two functions $\sum_{n=1}^{\infty} C(n)e^{-nt}$ and $\sum_{n=1}^{\infty} C(n)e^{-n^2 t}$ ($t > 0$) have the asymptotic expansions*

(2.5)
$$\sum_{n=1}^{\infty} C(n)e^{-nt} \sim \sum_{r=0}^{\infty} L(-r, C) \frac{(-t)^r}{r!}$$
$$\sum_{n=1}^{\infty} C(n)e^{-n^2 t} \sim \sum_{r=0}^{\infty} L(-2r, C) \frac{(-t)^r}{r!}$$

as $t \searrow 0$. The numbers $L(-r, C)$ are given explicitly by

$$(2.6) \qquad L(-r, C) = -\frac{M^r}{r+1} \sum_{n=1}^{M} C(n) B_{r+1}\left(\frac{n}{M}\right) \qquad (r = 0, 1, \ldots)$$

where $B_k(x)$ denotes the kth Bernoulli polynomial and M is any period of the function C.

In order to get the asymptotic expansion of ν_- as $q \to \xi$ radially, with ξ a root of unity, we write $q = \xi e^{-t}$. Thus we have to find the asymptotic expansion of $\sum_{n=0}^{\infty}(-1)^{n+1}\left(\frac{-3}{n}\right)\xi^{\frac{2}{3}n^2} e^{-\frac{2}{3}tn^2}$ as $t \searrow 0$. We can now use the proposition provided we check that $C(n) := (-1)^{n+1}\left(\frac{-3}{n}\right)\xi^{\frac{2}{3}n^2}$ is a periodic function with mean value 0. Indeed, if K is the order of ξ then $6K$ is a period for C, while $C(6K-n) = -C(n)$. Hence the mean value of C is zero.

The behaviour of ν outside the unit circle is thus completely known. A question that now arises is whether the behaviour of ν outside the unit circle is related to the behaviour of ν inside the unit circle. Numerical computations in this and related examples led me to the following:

CONJECTURE 2.2. *If ξ is a root of unity where ν is bounded (as $q \to \xi$ radially inside the unit circle), for example $\xi = 1$, then ν is C^∞ over the line radially through ξ.*

If ξ is a root of unity where ν is not bounded, for example $\xi = -1$, then the asymptotic expansion of the bounded term in condition (2) in the introduction is the same as the asymptotic expansion of ν as $q \to \xi$ radially outside the unit circle.

Let us proceed a bit, assuming this conjecture. Let $\tilde{\nu}$ be a function which is defined in- and outside the unit circle and also at all roots of unity, such that (a) $\tilde{\nu}$ is holomorphic in- and outside the unit circle, (b) $\tilde{\nu}$ is C^∞ over all radial lines through roots of unity and (c) $\tilde{\nu} = \nu$ outside the unit circle. If we can find such a function $\tilde{\nu}$, then $\nu - \tilde{\nu}$ is zero outside the unit circle, it has asymptotic expansion zero for $q \to \xi$ if ξ is a root of unity where ν is bounded, and the bounded term in condition (2) for mock ϑ- functions also has asymptotic expansion zero for $q \to \xi$. Because of this one might expect $\nu - \tilde{\nu}$ to be modular. If indeed this is the case we have written ν as the sum of two functions $\nu - \tilde{\nu}$ and $\tilde{\nu}$, one of which is a ϑ-function and the other a function which is bounded in all roots of unity. This contradicts condition (3) in the definition of a mock ϑ-function.

Ramanujan probably had this idea in mind when he wrote in his letter to Hardy: "... I have constructed a number of examples in which it is inconceivable to construct a ϑ-function to cut out the singularties of the original function. Also I have shown that if *it is necessarily so* then it leads to the following assertion—viz. it is possible to construct two power series in x, namely $\sum a_n x^n$ and $\sum b_n x^n$, both of which have *essential singularities* on the unit circle, are convergent when $|x| < 1$, and tend to *finite limits at every point* $x = e^{2i\pi r/s}$, and that at the same time the limit of $\sum a_n x^n$ at the point $x = e^{2i\pi r/s}$ is equal to the limit of $\sum b_n x^n$ at the point $x = e^{-2i\pi r/s}$."

Although it's possible to construct two such power series (see [**A2**, pp. 284]), it might not be possible to construct a function $\tilde{\nu}$ that satisfies the conditions (a), (b) and (c).

3. Mock ϑ-Functions and Real Analytic Modular Forms

In this section we will consider the following third order mock ϑ-functions:

(3.1)
$$f(q) = \sum_{n=0}^{\infty} \frac{q^{n^2}}{(-q;q)_n^2}$$
$$= 1 + \frac{q}{(1+q)^2} + \frac{q^4}{(1+q)^2(1+q^2)^2} + \cdots$$
$$\omega(q) = \sum_{n=0}^{\infty} \frac{q^{2n^2+2n}}{(q;q^2)_{n+1}^2}$$
$$= \frac{1}{(1-q)^2} + \frac{q^4}{(1-q)^2(1-q^3)^2} + \frac{q^{12}}{(1-q)^2(1-q^3)^2(1-q^5)^2} + \cdots$$

Ramanujan mentioned f in his letter, and ω can be found in [**W1**] and [**Ra**].

DEFINITION 3.1. Define $F = (f_0, f_1, f_2)^T$ by:

(3.2)
$$f_0(\tau) = q^{-\frac{1}{24}} f(q)$$
$$f_1(\tau) = 2q^{\frac{1}{3}} \omega(q^{\frac{1}{2}})$$
$$f_2(\tau) = 2q^{\frac{1}{3}} \omega(-q^{\frac{1}{2}}),$$

with $q = e^{2\pi i \tau}$, $\tau \in \mathcal{H}$.

In [**W1**] Watson gave the modular transformation properties of f and ω. If we rewrite them in terms of F we get

LEMMA 3.2. For $\tau \in \mathcal{H}$ we have

(3.3)
$$F(\tau+1) = \begin{pmatrix} \zeta_{24}^{-1} & 0 & 0 \\ 0 & 0 & \zeta_3 \\ 0 & \zeta_3 & 0 \end{pmatrix} F(\tau)$$

and

(3.4)
$$\frac{1}{\sqrt{-i\tau}} F(-1/\tau) = \begin{pmatrix} 0 & 1 & 0 \\ 1 & 0 & 0 \\ 0 & 0 & -1 \end{pmatrix} F(\tau) + R(\tau),$$

with $\zeta_n = e^{2\pi i/n}$, $R(\tau) = 4\sqrt{3}\sqrt{-i\tau}(j_2(\tau), -j_1(\tau), j_3(\tau))^T$, where

(3.5)
$$j_1(\tau) = \int_0^{\infty} e^{3\pi i \tau x^2} \frac{\sin 2\pi \tau x}{\sin 3\pi \tau x} dx$$
$$j_2(\tau) = \int_0^{\infty} e^{3\pi i \tau x^2} \frac{\cos \pi \tau x}{\cos 3\pi \tau x} dx$$
$$j_3(\tau) = \int_0^{\infty} e^{3\pi i \tau x^2} \frac{\sin \pi \tau x}{\sin 3\pi \tau x} dx.$$

PROOF. The transformation formula for $\tau \to \tau + 1$ is trivial.

If we take the first formula from the set of transformation formulae on p. 78 in [**W1**], with $\alpha = -2\pi i \tau$, and multiply both sides by -1, we get

(3.6)
$$\frac{1}{\sqrt{-i\tau}} f_1(-1/\tau) - f_0(\tau) = -\frac{4\sqrt{3}}{\sqrt{-i\tau}} J_1(-2\pi i \tau) = -\frac{4\sqrt{3}}{\sqrt{-i\tau}} j_1(\tau),$$

which is the second component of equation (3.4).

If we take the last formula from the set of transformation formulae on p. 78 in
[**W1**], with $\alpha = -\pi i\tau$, and multiply both sides by -2, we get

$$(3.7) \qquad \frac{1}{\sqrt{-i\tau}}f_0(-1/\tau) - f_1(\tau) = \frac{2\sqrt{3}}{\sqrt{-i\tau}}J_2(-\frac{\pi i\tau}{2}) = \frac{4\sqrt{3}}{\sqrt{-i\tau}}j_2(\tau),$$

where we have replaced x by $2x$ in the integral. This equation is the first component of equation (3.4).

If we take the formula on the middle of p. 79 in [**W1**], with $\alpha = -\pi i\tau$, and multiply both sides by 2, we get

$$(3.8) \qquad \frac{1}{\sqrt{-i\tau}}f_2(-1/\tau) + f_2(\tau) = \frac{4\sqrt{3}}{\sqrt{-i\tau}}J_3(-\pi i\tau) = \frac{4\sqrt{3}}{\sqrt{-i\tau}}j_3(\tau),$$

which is the third component of equation (3.4). $\square$

In a moment we will define a (nonholomorphic) function G that satisfies the same modular transformation properties as F. Before that, we rewrite R in terms of period integrals of the following theta functions of weight $3/2$:

$$(3.9) \qquad \begin{aligned} g_0(z) &= \sum_{n\in\mathbf{Z}}(-1)^n(n+1/3)e^{3\pi i(n+\frac{1}{3})^2 z} \\ g_1(z) &= -\sum_{n\in\mathbf{Z}}(n+1/6)e^{3\pi i(n+\frac{1}{6})^2 z} \\ g_2(z) &= \sum_{n\in\mathbf{Z}}(n+1/3)e^{3\pi i(n+\frac{1}{3})^2 z}. \end{aligned}$$

These theta functions have the following modular transformation properties, which can be verified using standard methods:

$$(3.10) \qquad \begin{pmatrix} g_0(z+1) \\ g_1(z+1) \\ g_2(z+1) \end{pmatrix} = \begin{pmatrix} 0 & 0 & \zeta_6 \\ 0 & \zeta_{24} & 0 \\ \zeta_6 & 0 & 0 \end{pmatrix} \begin{pmatrix} g_0(z) \\ g_1(z) \\ g_2(z) \end{pmatrix}$$

and

$$(3.11) \qquad \begin{pmatrix} g_0(-1/z) \\ g_1(-1/z) \\ g_2(-1/z) \end{pmatrix} = -(-iz)^{3/2}\begin{pmatrix} 0 & 1 & 0 \\ 1 & 0 & 0 \\ 0 & 0 & -1 \end{pmatrix}\begin{pmatrix} g_0(z) \\ g_1(z) \\ g_2(z) \end{pmatrix}.$$

From these transformation properties and the Fourier expansions, we see that the g_j's are cusp forms.

LEMMA 3.3. *For $\tau \in \mathcal{H}$ we have*

$$(3.12) \qquad R(\tau) = -2i\sqrt{3}\int_0^{i\infty}\frac{g(z)}{\sqrt{-i(z+\tau)}}dz,$$

where g is the vector $(g_0, g_1, g_2)^T$, and we have to integrate each component of the vector.

PROOF. (sketch)
If we replace τ by $-1/\tau$ in equation (3.4), multiply both sides by $\frac{1}{\sqrt{-i\tau}}\begin{pmatrix} 0 & 1 & 0 \\ 1 & 0 & 0 \\ 0 & 0 & -1 \end{pmatrix}$ and subtract equation (3.4), then we see that

$$(3.13) \qquad R(\tau) = \frac{-1}{\sqrt{-i\tau}}\begin{pmatrix} 0 & 1 & 0 \\ 1 & 0 & 0 \\ 0 & 0 & -1 \end{pmatrix}R(-1/\tau).$$

If we now take $\tau = it$ with $t \in \mathbf{R}$, $t > 0$, we have

$$R(it) = \frac{-1}{\sqrt{t}} \begin{pmatrix} 0 & 1 & 0 \\ 1 & 0 & 0 \\ 0 & 0 & -1 \end{pmatrix} R(i/t) = \frac{4\sqrt{3}}{t} \begin{pmatrix} j_1(i/t) \\ -j_2(i/t) \\ j_3(i/t) \end{pmatrix}. \tag{3.14}$$

We now consider the first component:
$$\frac{4\sqrt{3}}{t} j_1(i/t) = \frac{4\sqrt{3}}{t} \int_0^\infty e^{-3\pi x^2/t} \frac{\sinh 2\pi x/t}{\sinh 3\pi x/t} dx = 4\sqrt{3} \int_0^\infty e^{-3\pi t y^2} \frac{\sinh 2\pi y}{\sinh 3\pi y} dy, \tag{3.15}$$

where we have substituted $x = ty$ in the integral.

From the theory of partial fraction decompositions (see [**WW**, pp. 134–136]) we get

$$\frac{\sinh 2\pi y}{\sinh 3\pi y} = -\frac{i\sqrt{3}}{6\pi} \sum_{n \in \mathbf{Z}} \frac{(-1)^n}{y - i(n + \frac{1}{3})} - \frac{i\sqrt{3}}{6\pi} \sum_{n \in \mathbf{Z}} \frac{(-1)^n}{-y - i(n + \frac{1}{3})}. \tag{3.16}$$

Using this we see

$$\frac{4\sqrt{3}}{t} j_1(i/t)$$
$$= -\frac{2i}{\pi} \int_0^\infty e^{-3\pi t y^2} \left(\sum_{n \in \mathbf{Z}} \frac{(-1)^n}{y - i(n + \frac{1}{3})} + \sum_{n \in \mathbf{Z}} \frac{(-1)^n}{-y - i(n + \frac{1}{3})} \right) dy$$
$$= -\frac{2i}{\pi} \int_{-\infty}^\infty e^{-3\pi t y^2} \left(\sum_{n \in \mathbf{Z}} \frac{(-1)^n}{y - i(n + \frac{1}{3})} \right) dy \tag{3.17}$$
$$= -\frac{2i}{\pi} \sum_{n \in \mathbf{Z}} (-1)^n \int_{-\infty}^\infty \frac{e^{-3\pi t y^2}}{y - i(n + \frac{1}{3})} dy.$$

It's not immediately clear that interchanging the order of integration and summation in the last equation is justified. However, it can be proven rigorously if we consider $\int_{-\infty}^\infty e^{-3\pi t y^2} \left(\sum_{n \in \mathbf{Z}} (-1)^n \left(\frac{1}{y - i(n + \frac{1}{3})} + \frac{1}{i(n + \frac{1}{3})} \right) \right) dy$ (here we can interchange the order of integration and summation because of absolute convergence).

We have for $r \in \mathbf{R}$, $r \neq 0$

$$\int_{-\infty}^\infty \frac{e^{-\pi t y^2}}{y - ir} dy = \pi i r \int_0^\infty \frac{e^{-\pi r^2 u}}{\sqrt{u + t}} du, \tag{3.18}$$

(both sides are solutions of $(-\frac{\partial}{\partial t} + \pi r^2) f(t) = \frac{\pi i r}{\sqrt{t}}$ and have the same limit 0 if $t \to \infty$, and hence are equal). If we use this with $r = (n + 1/3)$ and t replaced by $3t$, we obtain

$$\frac{4\sqrt{3}}{t} j_1(i/t) = 2 \sum_{n \in \mathbf{Z}} (-1)^n (n + 1/3) \int_0^\infty \frac{e^{-\pi (n+1/3)^2 u}}{\sqrt{u + 3t}} du$$
$$= 2 \int_0^\infty \frac{\sum_{n \in \mathbf{Z}} (-1)^n (n + 1/3) e^{-\pi (n + \frac{1}{3})^2 u}}{\sqrt{u + 3t}} du. \tag{3.19}$$

Again it's not immediately clear that interchanging the order of integration and summation in the last step is justified. It can be proven rigorously by first using

partial integration on the integral
(3.20)
$$\int_0^\infty \frac{e^{-\pi(n+1/3)^2 u}}{\sqrt{u+3t}} du = \frac{1}{\pi(n+1/3)^2} \frac{1}{\sqrt{3t}} - \frac{1}{2\pi(n+1/3)^2} \int_0^\infty \frac{e^{-\pi(n+1/3)^2 u}}{(u+3t)^{3/2}} du,$$
then interchanging the order of integration and summation, which is justified by absolute convergence, and finally using partial integration again. By partial integration we introduce some "boundary terms". To get rid of them we have to use Abel's theorem on continuity up to the circle of convergence, see [**WW**, pp. 57–58].

If we now substitute $u = -3iz$ in the integral we get

(3.21) $$\frac{4\sqrt{3}}{t} j_1(i/t) = -2i\sqrt{3} \int_0^{i\infty} \frac{g_0(z)}{\sqrt{-i(z+it)}} dz,$$

so we have proven the first component of equation (3.12) for $\tau = it$. Since both sides are analytic on $\mathcal{H}$, the identity holds for all $\tau \in \mathcal{H}$.

The second and third component of equation (3.12) can be proven along the same lines. Here we have to use

(3.22)
$$\frac{\cosh \pi y}{\cosh 3\pi y} = -\frac{i\sqrt{3}}{6\pi} \sum_{n\in\mathbf{Z}}^* \frac{1}{y - i(n+\frac{1}{6})} - \frac{i\sqrt{3}}{6\pi} \sum_{n\in\mathbf{Z}}^* \frac{1}{-y - i(n+\frac{1}{6})}$$
$$\frac{\sinh \pi y}{\sinh 3\pi y} = -\frac{i\sqrt{3}}{6\pi} \sum_{n\in\mathbf{Z}}^* \frac{1}{y - i(n+\frac{1}{3})} - \frac{i\sqrt{3}}{6\pi} \sum_{n\in\mathbf{Z}}^* \frac{1}{-y - i(n+\frac{1}{3})},$$

where $\sum_{n\in\mathbf{Z}}^*$ means $\lim_{m\to\infty} \sum_{n=-m}^m$. □

DEFINITION 3.4. For $\tau \in \mathcal{H} \cup \mathbf{Q}$ we define

(3.23) $$G(\tau) := 2i\sqrt{3} \int_{-\bar{\tau}}^{i\infty} \frac{(g_1(z), g_0(z), -g_2(z))^T}{\sqrt{-i(z+\tau)}} dz.$$

The integrals converge, even if $\tau \in \mathbf{Q}$, because the g_j's are cusp forms.
The function G satisfies the same modular transformation properties as F:

LEMMA 3.5. *For $\tau \in \mathcal{H}$ we have*

(3.24) $$G(\tau+1) = \begin{pmatrix} \zeta_{24}^{-1} & 0 & 0 \\ 0 & 0 & \zeta_3 \\ 0 & \zeta_3 & 0 \end{pmatrix} G(\tau)$$

and

(3.25) $$\frac{1}{\sqrt{-i\tau}} G(-1/\tau) = \begin{pmatrix} 0 & 1 & 0 \\ 1 & 0 & 0 \\ 0 & 0 & -1 \end{pmatrix} G(\tau) + R(\tau).$$

PROOF. The first equation follows from (3.10) by replacing z by $z - 1$ in the integral. We have

(3.26)
$$\frac{1}{\sqrt{-i\tau}} G(-1/\tau) = \frac{2i\sqrt{3}}{\sqrt{-i\tau}} \int_{1/\bar{\tau}}^{i\infty} \frac{(g_1(z), g_0(z), -g_2(z))^T}{\sqrt{-i(z-1/\tau)}} dz$$
$$= 2i\sqrt{3} \int_0^{-\bar{\tau}} \frac{(g_1(-1/z), g_0(-1/z), -g_2(-1/z))^T}{\sqrt{1+\tau/z}} \frac{dz}{(-iz)^2},$$

where we have replaced z by $-1/z$ in the integral. Using equation (3.11), we get

$$\frac{1}{\sqrt{-i\tau}} G(-1/\tau) = -2i\sqrt{3} \int_0^{-\overline{\tau}} \frac{g(z)}{\sqrt{-i(z+\tau)}} \, dz; \tag{3.27}$$

hence, by Lemma 3.3 we get

$$\frac{1}{\sqrt{-i\tau}} G(-1/\tau) - \begin{pmatrix} 0 & 1 & 0 \\ 1 & 0 & 0 \\ 0 & 0 & -1 \end{pmatrix} G(\tau)$$

$$= -2i\sqrt{3} \int_0^{-\overline{\tau}} \frac{g(z)}{\sqrt{-i(z+\tau)}} \, dz - 2i\sqrt{3} \int_{-\overline{\tau}}^{i\infty} \frac{g(z)}{\sqrt{-i(z+\tau)}} \, dz \tag{3.28}$$

$$= -2i\sqrt{3} \int_0^{i\infty} \frac{g(z)}{\sqrt{-i(z+\tau)}} \, dz = R(\tau).$$

$\square$

THEOREM 3.6. *The function H defined by*

$$H(\tau) = F(\tau) - G(\tau), \tag{3.29}$$

is a (vector-valued) real-analytic modular form of weight $1/2$, satisfying

$$H(\tau+1) = \begin{pmatrix} \zeta_{24}^{-1} & 0 & 0 \\ 0 & 0 & \zeta_3 \\ 0 & \zeta_3 & 0 \end{pmatrix} H(\tau),$$

$$\frac{1}{\sqrt{-i\tau}} H(-1/\tau) = \begin{pmatrix} 0 & 1 & 0 \\ 1 & 0 & 0 \\ 0 & 0 & -1 \end{pmatrix} H(\tau), \tag{3.30}$$

and H is an eigenfunction of the Casimir operator $\Omega_{1/2} = -4y^2 \frac{\partial^2}{\partial \tau \partial \overline{\tau}} + iy \frac{\partial}{\partial \overline{\tau}} + \frac{3}{16}$ with eigenvalue $\frac{3}{16}$, where $\tau = x + iy$, $\frac{\partial}{\partial \tau} = \frac{1}{2}\left(\frac{\partial}{\partial x} - i \frac{\partial}{\partial y}\right)$ and $\frac{\partial}{\partial \overline{\tau}} = \frac{1}{2}\left(\frac{\partial}{\partial x} + i \frac{\partial}{\partial y}\right)$.

PROOF. The modular transformation properties of H are a direct consequence of the transformation properties of F and G given in Lemma 3.2 and Lemma 3.5.

Since F is a holomorphic function of τ, we have $\frac{\partial}{\partial \overline{\tau}} F(\tau) = 0$; hence

$$\frac{\partial}{\partial \overline{\tau}} H(\tau) = -\frac{\partial}{\partial \overline{\tau}} G(\tau) = -2i\sqrt{3} \frac{(g_1(-\overline{\tau}), g_0(-\overline{\tau}), -g_2(-\overline{\tau}))^T}{\sqrt{-i(\tau - \overline{\tau})}} \tag{3.31}$$

$$= -\frac{i\sqrt{6}}{\sqrt{y}} (g_1(-\overline{\tau}), g_0(-\overline{\tau}), -g_2(-\overline{\tau}))^T.$$

We see that $\sqrt{y} \frac{\partial}{\partial \overline{\tau}} H(\tau)$ is anti-holomorphic, so

$$\frac{\partial}{\partial \tau} \sqrt{y} \frac{\partial}{\partial \overline{\tau}} H(\tau) = 0. \tag{3.32}$$

We can write the operator $\Omega_{1/2} = -4y^2 \frac{\partial^2}{\partial \tau \partial \overline{\tau}} + iy \frac{\partial}{\partial \overline{\tau}} + \frac{3}{16}$ as

$$\Omega_{1/2} = \frac{3}{16} - 4y^{3/2} \frac{\partial}{\partial \tau} \sqrt{y} \frac{\partial}{\partial \overline{\tau}}. \tag{3.33}$$

Hence we get $\Omega_{1/2} H = \frac{3}{16} H$. $\square$

Writing F as $H + G$, we get the following:

COROLLARY 3.7. *The vector-valued third order mock ϑ-function F can be written as the sum of a real analytic modular form H and a function G that is bounded in all rational points.*

4. Other mock ϑ-functions

In the previous section we have only dealt with the third order mock ϑ-functions f and ω. However, I have found similar results for most other mock ϑ-functions, and I expect that it can be done for all known ones. I hope to present these results, and the details omitted in the previous section, in my Ph.D.-thesis, which should appear somewhere near the end of 2002.

References

[A1] G.E. Andrews, *An introduction to Ramanujan's "lost" notebook*, Amer. Math. Monthly **86** (1979), 89–108.
[A2] _____, *Mock theta functions*, Theta functions—Bowdoin 1987, Part 2 (Brunswick, 1987), Proc. Symp. Pure Math., vol. 49, Amer. Math. Soc., Providence, RI, 1989, pp. 283–298.
[AH] G.E. Andrews and D. Hickerson, *Ramanujan's "lost" notebook: the sixth order mock theta functions*, Adv. Math. **89** (1991), 60–105.
[LZ] R. Lawrence and D.B. Zagier, *Modular forms and quantum invariants of 3-manifolds*, Asian J. Math. **3** (1999), no. 1, 93–107.
[Ra] S. Ramanujan, *The lost notebook and other unpublished papers*, Narosa Publishing House, New Delhi, 1987.
[Ro] L.J. Rogers, *On two theorems of combinatory analysis and some allied identities*, Proc. London Math. Soc. (2) **16** (1917), 316–336.
[S] A. Selberg, *Über die Mock-Thetafunktionen siebenter Ordnung*, Arch. Math. og Naturvidenskab **41** (1938), 3–15.
[W1] G.N. Watson, *The final problem: an account of the mock theta functions*, J. London Math. Soc. **11** (1936), 55–80.
[W2] _____, *The mock theta functions (2)*, Proc. London Math. Soc. (2) **42** (1937), 274–304.
[WW] E.T. Whittaker and G.N. Watson, *Modern Analysis*, Fourth Edition, Cambridge at the University Press, 1927.

MATHEMATISCH INSTITUUT, UNIVERSITEIT UTRECHT, POSTBUS 80010, 3508 TA UTRECHT, THE NETHERLANDS
E-mail address: zwegers@math.uu.nl